# Interdisciplinary Applied Mathematics

Volumes published are listed at the end of the book.

**Springer**

*New York*
*Berlin*
*Heidelberg*
*Hong Kong*
*London*
*Milan*
*Paris*
*Tokyo*

# Interdisciplinary Applied Mathematics

## Volume 25

Problems in engineering, computational science, and the physical and biological sciences are using increasingly sophisticated mathematical techniques. Thus, the bridge between the mathematical sciences and other disciplines is heavily traveled. The correspondingly increased dialog between the disciplines has led to the establishment of the series: *Interdisciplinary Applied Mathematics.*

The purpose of this series is to meet the current and future needs for the interaction between various science and technology areas on the one hand and mathematics on the other. This is done, firstly, by encouraging the ways that that mathematics may be applied in traditional areas, as well as point towards new and innovative areas of applications; and, secondly, by encouraging other scientific disciplines to engage in a dialog with mathematicians outlining their problems to both access new methods and suggest innovative developments within mathematics itself.

The series will consist of monographs and high-level texts from researchers working on the interplay between mathematics and other fields of science and technology.

Anne Beuter    Leon Glass
Michael C. Mackey    Michèle S. Titcombe
Editors

# Nonlinear Dynamics in Physiology and Medicine

With 162 Illustrations

 Springer

Anne Beuter
Institut de Biologie
Universite de Montpellier 1
4, Boulevard Henri IV
Montpellier Cedex 1, 34060
France
anne.beuter@wanadoo.fr

Leon Glass
Department of Physiology
McGill University
Montreal, Quebec H3G 1Y6
Canada
glass@cnd.mcgill.ca

Michael C. Mackey
Department of Physiology
McGill University
Montreal, Quebec H3G 1Y6
Canada
mackey@cnd.mcgill.ca

Michèle S. Titcombe
Department of Physiology
McGill University
Montreal, Quebec H3G 1Y6
Canada
titcombe@cnd.mcgill.ca

*Editors*
S.S. Antman
Department of Mathematics
and
Institute for Physical Science and Technology
University of Maryland
College Park, MD 20742
USA
ssa@math.umd.edu

J.E. Marsden
Control and Dynamical Systems
Mail Code 107-81
California Institute of Technology
Pasadena, CA 91125
USA
marsden@cds.caltech.edu

L. Sirovich
Division of Applied Mathematics
Brown University
Providence, RI 02912
USA
chico@camelot.mssm.edu

S. Wiggins
School of Mathematics
University of Bristol
Bristol BS8 1TW
UK
s.wiggins@bris.ac.uk

Mathematics Subject Classification (2000): 92Cxx, 37-01

Library of Congress Cataloging-in-Publication Data
Nonlinear dynamics in physiology and medicine / editors, Anne Beuter ... [et al.].
      p.   cm.—(Interdisciplinary applied mathematics ; v.25)
   Includes bibliographical references and index.

    1. Physiology—Mathematical models.   2. Nonlinear systems.   3. Dynamics.   I. Beuter,
Anne.   II. Series.
QP33.6.M36N663      2003
612'.001'5118—dc21                                                           2003044936

ISBN  978-1-4419-1821-5          e-ISBN 978-0-387-21640-9

www.springer-ny.com

Springer-Verlag   New York  Berlin  Heidelberg
*A member of BertelsmannSpringer Science+Business Media GmbH*

*We dedicate this book to the memory of Arthur Winfree, a man before his time in life and death.*

# Contributors

Jacques Bélair
Département de Mathématiques
  et de Statistique
Université de Montréal

Anne Beuter
Laboratoire de Physiologie
Institut de Biologie
Université de Montpellier 1

Marc Courtemanche
Centre de Recherche
Institut de Cardiologie de Montréal

Roderick Edwards
Department of Mathematics
  and Statistics
University of Victoria

Leon Glass
Department of Physiology
McGill University

Michael R. Guevara
Department of Physiology
McGill University

Caroline Haurie
Department of Physiology
McGill University

André Longtin
Department of Physics
University of Ottawa

Michael C. Mackey
Department of Physiology
McGill University

John Milton
Department of Neurology
The University of
  Chicago Hospitals

Michèle S. Titcombe
Department of Physiology
McGill University

Alain Vinet
Département de Physiologie
Université de Montréal

# Preface

*Je tiens impossible de connaître les parties sans connaître
le tout, non plus que de connaître le tout sans connaître
particulièrement les parties*
                                                    —Pascal

*The eternal mystery of the world is its comprehensibility*
                                                    —Einstein

This book deals with the application of mathematical tools to the study
of physiological systems. It is directed toward an audience of physiologists,
physicians, physicists, kinesiologists, psychologists, engineers, mathematicians, and others interested in finding out more about the complexities
and subtleties of rhythmic physiological processes from a theoretical perspective. We have attempted to give a broad view of the underlying notions
behind the dynamics of physiological rhythms, sometimes from a theoretical
perspective and sometimes from the perspective of the experimentalist.

This book can be used in a variety of ways, ranging from a more traditional approach such as a textbook in a biomathematics course (at
either the advanced undergraduate or graduate level) to a research resource in which someone interested in a particular problem might look at
the corresponding discussion here to guide their own thinking. We hope
that researchers at all levels will find inspiration from the way we have
dealt with particular research problems to tackle completely new areas of
investigation, or even approach these in totally new ways.

Mathematically, we expect that readers will have a solid background in
differential and integral calculus as well as a first course in ordinary differential equations. From the physiological side, they should have a minimal
training in general physiology at the organismal level. We have endeav-
ored in every area to make the book as self-contained as possible given
this mathematical and physiological preparation. Furthermore, many of
the later chapters stand on their own and can be read independently of the
others. When necessary we have added references to other chapters or to
appendices.

This book is divided roughly into two sections. Following an introductory Chapter 1, the first section (sort of theory) mainly introduces concepts
from nonlinear dynamics using an almost exclusively biological setting for

motivation. This is contained in Chapters 2 through 5. Chapter 2 introduces concepts from nonlinear dynamics using the properties of difference and differential equations as a basis. Chapter 3 extends and draws on this material in considering bifurcations from fixed points and limit cycles, illustrating a number of bifurcation patterns with examples drawn from data. This is continued in Chapter 4, which focuses on excitable cell electrophysiology. The first section concludes with a consideration in Chapter 5 of the properties of biological oscillators when perturbed by single stimuli or periodic inputs.

The second part (applications, more or less) consists of five in-depth examples of how the authors have used the concepts of the first part of this book in their research investigations of biological and physiological systems. Thus, Chapter 6 examines the influence of noise in nonlinear dynamical systems using examples drawn from neurobiology. Chapter 7 looks at the properties of spatially extended nonlinear systems as illustrated by the properties of excitable cardiac tissue. In Chapter 8 attention is directed to the dynamics of cellular replication, illustrating how periodic diseases can illuminate underlying dynamics. Chapter 9 returns to the neurobiology arena in considering the properties of a simple neural feedback system, the pupil light reflex. The book continues with this neurobiological emphasis to conclude in Chapter 10 with an examination of the dynamics of tremor.

We have not considered data analysis techniques as a separate subject, but rather have included an Appendix C, explaining those techniques we have found valuable (e.g., Fourier analysis, Lomb periodogram analysis). No doubt, other investigators will have their favorites that are not mentioned.

Since a combination of both analytical and computational techniques are essential to maximize understanding of complex physiological dynamics, we have included both analytic and computer exercises throughout the book using either the freeware XPP or the generally available commercial package Matlab. Introductions to both XPP and Matlab are to be found in Appendices A and B, respectively. The source code and data files (and in some cases, help files) for the computer exercises are available on the book's web site, available via www.springer-ny.com.

## About the Authors and the Audience

The Centre for Nonlinear Dynamics in Physiology and Medicine was officially created in 1988 by the University Senate of McGill University, and from its inception was conceived of as a multiuniversity grouping of researchers and their graduate students and postdoctoral fellows that encompassed people from McGill, the Université de Montréal, the Université du Québec à Montréal, the University of Ottawa, and the University of Chicago. Since then, as young investigators have left to establish their own groups, it has grown to include Concordia University, the University of Victoria, and the University of Waterloo. Thus, in a very real sense, the

Centre for Nonlinear Dynamics in Physiology and Medicine has become a virtual center.

Back in 1995, members of the Centre for Nonlinear Dynamics in Physiology and Medicine hatched the idea of running a novel type of summer school, one that not only tried to convey the art and craft of mathematical modeling of biological systems through lectures, but also employed the use of daily afternoon computer simulation laboratories that would directly engage the students in exploring what they had heard about in morning lectures. The first of these summer schools was held in May 1996 (and dubbed "Montreal96"). Its objective was to teach what our group had learned about modeling physiological systems to interested individuals. Members of the Centre for Nonlinear Dynamics in Physiology and Medicine had as their goal to deliver lectures and laboratories whose material was coordinated between topics, and which were not merely recitals of disparate research topics as is so often the case in summer schools.

This first summer school was an unqualified success. More than 60 students attended "Montreal96" from 16 countries, ranging in subject specialization from biology, medicine, psychology, physiology, and theoretical physics through applied mathematics. At that time a minimal mathematical background of differential and integral calculus resulted in a wide diversity of expertise in the students, though a majority of the students had more advanced levels of preparation. Career levels varied from advanced undergraduates through graduate students, postdoctoral fellows, university faculty members, and physicians. The summer school was repeated in 1997 and 2000, with some variations in the makeup of students and the teaching team.

This book has evolved out of the notes written from these three summer schools. Thus the materials that we used for this book were gathered in a rough form during the 1996 summer school of the Centre for Nonlinear Dynamics in Physiology and Medicine. Therefore, this book is not a collection of disjointed chapters. Rather, a real effort has been made to organize the chapters in an integrated coherent whole even if this was not an easy thing to do, since there is a relatively wide spectrum of contents, some chapters being more mathematical, others more physiological, and still others lying in between. This is one of the few books in the "biomath" field that deals specifically with physiological systems. The description of the makeup of the attendees of these summer schools tells exactly for whom this book is written: a diverse collection of individuals with backgrounds as varied as those who have contributed to the history of this field over the past three hundred years. The one thing that will unite the intended readers of this book will be an absolute driving passion to understand the workings of physiological systems and the willingness to use any available technique (even mathematics!) to achieve that understanding.

We thank the authors and publishers for permission to reproduce many of the figures we have used in the book, and which are acknowledged in the figure captions. In particular, we thank Tom Inoué of Studio i-Design for his tireless efforts in the figure preparation, Dr. Wei Zong of the Research Resource for Complex Physiologic Signals for designing and implementing a collaborative authoring system that greatly facilitated the completion of this project, and Caroline Haurie, who was instrumental in drafts of the book. We thank all the people who attended or taught in the summer schools that were held in 1996, 1997, and 2000 in Montreal, because their feedback and efforts were essential for whatever success this book may achieve. We also thank Springer Verlag, and especially Ms. Achi Dosanjh, for their support during the production of this book. Finally, we thank all the people, colleagues, friends, and students who tested the computer exercises in the summer of 2002 at the Centre for Nonlinear Dynamics in Physiology and Medicine Computer Exercise Pizza Testing Party! All of the remaining errors are, of course, the responsibility of the authors. Our research and the preparation of this book have benefited from funds from many sources, notably the Natural Sciences Engineering and Research Council of Canada, the Canadian Institutes of Health Research, the Mathematics of Information Technology and Complex Systems National Center of Excellence, the Québec Fonds pour la formation de Chercheurs et l'aide à la recherche, and the Research Resource for Complex Physiologic Signals funded by the National Center for Research Resources of the National Institutes of Health (USA).

Montréal and Montpellier, November 2002

# Sources and Credits

The sources of previously published figures are given in the figure captions. Additional information is provided below. In most cases, the lettering in the figures was modified slightly. We are grateful to the authors and publishers for granting permission to reproduce these figures.

**Figure:**

**2.6, 2.7** Guevara, M.R., Ward, G., Shrier, A., and Glass, L. 1984. Electric alternans and period doubling bifurcations. *IEEE Computers in Cardiology*, pages 167–170 (Copyright ©1984 IEEE).

**2.11** Edelstein-Keshet, L. 1988. *Mathematical Models in Biology.* Copyright ©1988 by McGraw-Hill. Reprinted with permission of The McGraw-Hill Companies.

**2.12** Breda, E., Cavaghan, M.K., Toffolo, G., Polonsky, K.S., and Cobelli, C. 2001. Oral glucose tolerance test minimal model indexes of $\beta$-cell function and insulin sensitivity. *Diabetes*, Volume 50, pages 150–158. Reprinted with permission from the American Diabetes Association.

**2.14** Glass, L., and Mackey, M.C.: *From Clocks to Chaos.* Copyright ©1988 by Princeton University Press. Reprinted by permission of Princeton University Press.

**3.1** Adapted from Tasaki, I. 1959. Demonstration of two stable states of the nerve membrane in potassium-rich media. *Journal of Physiology (London)*, Volume 148, pages 306–331. Adapted with permission of the Physiological Society and I. Tasaki.

**3.7** Adapted from Aihara, K., and Matsumoto, G. 1983. Two stable steady states in the Hodgkin–Huxley axons. *Biophysical Journal*, Volume 41, pages 87–89. Adapted with permission of the Biophysical Society and K. Aihara.

**3.8** Chialvo, D.R., and Apkarian, V. 1993. Modulated noisy biological dynamics: Three examples. *Journal of Statistical Physics*, Volume 70, pages 375–391. Reprinted with permission of Kluwer Academic/Plenum Publishers and V. Apkarian.

**3.13** From *Nonlinear Dynamics and Chaos* by Steven H. Strogatz. Copyright ©1994. Reprinted by permission of Perseus Book Publishers, a member of Perseus Books, L.L.C.

**3.17** Reprinted with permission from Jalife, J., and Antelevitch, C. 1979. Phase resetting and annihilation of pacemaker activity in cardiac tissue. *Science*, Volume 206, pages 695–697. Copyright ©1979 American Association for the Advancement of Science.

**3.18** Guevara, M.R., and Jongsma, H.J. 1992. Phase resetting in a model of sinoatrial nodal membrane: Ionic and topological aspects. *American Journal of Physiology*, Volume 258, pages H734–H747. Reprinted by permission of the American Physiological Society.

**3.19, 3.20** Guttman, R., Lewis, S., and Rinzel, J. 1980. Control of repetitive firing in squid axon membrane as a model for a neuron oscillator. *Journal of Physiology (London)*, Volume 305, pages 377–395. Reprinted by permission of the Physiological Society and J. Rinzel.

**3.21** Guevara, M.R., and Jongsma, H.J. 1992. Phase resetting in a model of sinoatrial nodal membrane: Ionic and topological aspects. *American Journal of Physiology*, Volume 258, pages H734–H747. Reprinted by permission of the American Physiological Society.

**3.22** Noma, A., and Irisawa, H. 1975. Effects of $Na^+$ and $K^+$ on the resting membrane potential of the rabbit sinoatrial node cell. *Japanese Journal of Physiology*, Volume 25, pages 287–302. Reprinted with permission of the Japanese Journal of Physiology.

**3.25, 3.27** Guttman, R., Lewis, S., and Rinzel, J. 1980. Control of repetitive firing in squid axon membrane as a model for a neuroneoscillator. *Journal of Physiology (London)*, Volume 305, pages 377–395. Reprinted by permission of the Physiological Society and J. Rinzel.

**3.29** Abraham, R.H., and Shaw, C.D. 1982. *Dynamics: The Geometry of Behavior*. Reprinted by permission of R.H. Abraham, Aerial Press.

**3.30** Yehia, A.R., Jeandupeux, D., Alonso, F., and Guevara, M.R. 1998. Hysteresis and bistability in the direct transition from 1:1 to 2:1 rhythm in periodically driven single ventricular cells. *Chaos*, Volume 9, pages 916–931. Reprinted by permission of the American Institute of Physics.

**3.31** Abraham, R.H., and Shaw, C.D. 1982. *Dynamics: The Geometry of Behavior*. Reprinted by permission of R.H. Abraham, Aerial Press.

**3.32, 3.33** Guevara, M.R., Alonso, F., Jeandupeux, D. and van Ginneken, A.G.G. 1989. Alternans in periodically stimulated isolated ventricular myocytes: Experiment and model. In *Cell to Cell Signalling: From*

*Experiments to Theoretical Models*, pages 551–563. Reprinted with permission of M.R. Guevara.

**3.34** Kaplan, D.T., Clay, J.R., Manning, T., Glass, L., Guevara, M.R., and Shrier, A. 1996. Subthreshold dynamics in periodically stimulated squid giant axons. *Physical Review Letters*, Volume 76, pages 4074–4076.

**3.35** Guevara, M.R., and Jongsma, H.J. 1992. Phase resetting in a model of sinoatrial nodal membrane: Ionic and topological aspects. *American Journal of Physiology*, Volume 258, pages H734–H747. Reprinted by permission of the American Physiological Society.

**3.37** Panels A and B from Seydel, R. 1994. *From Equilibrium to Chaos: Practical Bifurcation and Stability Analysis*. Reprinted with permission of Springer-Verlag.

**3.37** Panels C and D from Abraham, R.H., and Shaw, C.D. 1982. *Dynamics: The Geometry of Behavior*. Reprinted by permission of R.H. Abraham, Aerial Press.

**3.38** Abraham, R.H., and Shaw, C.D. 1982. *Dynamics: The Geometry of Behavior*. Reprinted by permission of R.H. Abraham, Aerial Press.

**3.39** Panels A and B from Crutchfield, J., Farmer, D., Packard, N., Shaw, R., Jones, G., and Donnelly, R.J. 1980. Power spectral analysis of a dynamical system. *Physical Letters A*, Volume 76, pages 1–4, Copyright ©1980, with permission of Elsevier.

**3.39** Panel C from Olsen, L.F., and Degn, H. 1985. Chaos in biological systems. *Quarterly Review of Biophysics*, Volume 18, pages 165–225. Reprinted with the permission of Cambridge University Press.

**3.40** Panels A (left) and B (left) from Abraham, R.H., and Shaw, C.D. 1982. *Dynamics: The Geometry of Behavior*. Reprinted by permission of R.H. Abraham, Aerial Press.

**3.40** Panels A (right), B (right) and C from Seydel, R. 1994. *From Equilibrium to Chaos: Practical Bifurcation and Stability Analysis*. Reprinted with permission of Springer-Verlag.

**3.44** Guevara, M.R., and Jongsma, H.J. 1992. Phase resetting in a model of sinoatrial nodal membrane: Ionic and topological aspects. *American Journal of Physiology*, Volume 258, pages H734–H747. Reprinted by permission of the American Physiological Society.

**3.45** Guevara, M.R., and Jongsma, H.J. 1992. Phase resetting in a model of sinoatrial nodal membrane: Ionic and topological aspects. *American Journal of Physiology*, Volume 258, pages H734–H747. Reprinted by permission of the American Physiological Society.

**4.2A** Hodgkin, A.L., and Keynes, R.D. 1956. Experiments on the injection of substances into squid giant axons by means of a microsyringe. *Journal of Physiology (London)*, Volume 131. Reprinted with permission of The Physiological Society.

**4.2B** Hille, B. 2001. *Ionic Channels of Excitable Membranes*, Sinauer Sunderland. Reprinted with the permission of Sinauer Associates, Inc.

**4.2** Panels C and D from Baker, P.F., Hodgkin, A.L., and Shaw, T.I. 1961. Replacement of the protoplasm of a giant nerve fibre with artificial solutions. *Nature*, Volume 190 (no. 4779), pages 885–887. Reprinted with permission.

**4.3** Nicholls, J.G., Martin, A.R., Wallace, B.G., and Fuchs, P.A. 2001. *From Neuron to Brain*, Sinauer Sunderland. Reprinted with the permission of Sinauer Associates, Inc.

**4.4** Alonso, A., and Klink, R. 1993. Differential responsiveness of stellate and pyramidal-like cells of medial entorhinal cortex layer II. *Journal of Neurophysiology*, Volume 70, pages 128–143. Reprinted with permission of the American Physiological Society.

**4.5** Hodgkin, A.L. 1958. The Croonian Lecture: Ionic movements and electrical activity in giant nerve fibres. *Proceedings of the Royal Society of London*, Series B, Volume 148, pages 1–37, Fig. 12, page 19, reprinted with the permission of the Royal Society of London.

**4.6** Sánchez, J.A., Dani, J.A., Siemen, D., and Hille, B. 1986. Slow permeation of organic cations in acetylcholine receptor channels. *Journal of General Physiology*, Volume 87, pages 985–1001. Reproduced by copyright permission of The Rockefeller University Press.

**4.8A** Hille, B. 2001. *Ionic Channels of Excitable Membranes*, Sinauer Sunderland. Reprinted with the permission of Sinauer Associates, Inc.

**4.10** Nicholls, J.G., Martin, A.R., Wallace, B.G., and Fuchs, P.A. 2001. *From Neuron to Brain*, Sinauer Sunderland. Reprinted with the permission of Sinauer Associates, Inc.

**4.11** Hodgkin, A.L. 1958. The Croonian Lecture: Ionic movements and electrical activity in giant nerve fibres. *Proceedings of the Royal Society of London*, Series B, Volume 148, pages 1–37, Fig. 9, page 15, reprinted with the permission of the Royal Society of London.

**4.12** Permission to reproduce this figure granted by F. Bezanilla. See Hille (2001), Figure 3.17, page 91.

**4.13** Panel A from Hodgkin, A.L. 1958. The Croonian Lecture: Ionic movements and electrical activity in giant nerve fibres. *Proceedings of the*

*Royal Society of London*, Series B, Volume 148, pages 1–37, Fig. 11, right panel, page 16, reprinted with the permission of the Royal Society of London.

**4.15**  Permission to reproduce this figure granted by J.B. Patlak. See Hille (2001), Figure 3.16, page 90.

**4.16**  Panel A from Hodgkin, A.L. 1958. The Croonian Lecture: Ionic movements and electrical activity in giant nerve fibres. *Proceedings of the Royal Society of London*, Series B, Volume 148, pages 1–37, Fig. 11, left panel, page 16, reprinted with the permission of the Royal Society of London.

**5.1**  Josephson, M.E., Callans, D., Almendral, J.M., Hook, B.G., and Kleiman, R.B. 1993. Resetting and entrainment of ventricular tachycardia associated with infarction: Clinical and experimental studies. In *Tachycardia: Mechanisms and Management*, pages 505–536. Reprinted by permission of Blackwell Publishing.

**5.2, 5.3**  Zeng, W., Glass, L., and Shrier, A. 1992. The topology of phase response curves induced by single and paired stimuli. *Journal of Biological Rhythms*, Volume 7, pages 89–104, copyright ©1992 by Sage Publications. Reprinted by permission of Sage Publications, Inc.

**5.4, 5.5**  Glass, L., and Mackey, M.C.: *From Clocks to Chaos*. Copyright ©1988 by Princeton University Press. Reprinted by permission of Princeton University Press.

**5.6**  Glass, L., and Winfree, A.T. 1984. Discontinuities in phase-resetting experiments. *American Journal of Physiology*, Volume 246, pages R251–R258. Reprinted with permission of the American Physiological Society.

**5.7, 5.8**  Glass, L., and Sun, J. 1994. Periodic forcing of a limit cycle oscillator: Fixed points, Arnold tongues, and the global organization of bifurcations. *Physical Review E*, Volume 50, pages 5077–5084.

**5.9, 5.10**  Glass, L., Nagai, Y., Hall, K., Talajic, M., and Nattel, S. 2002. Predicting the entrainment of reentrant cardiac waves using phase resetting curves. *Physical Review E*, Volume 65, 021908.

**5.11**  Glass, L., Guevara, M.G., and Shrier, A. 1987. Universal bifurcations and the classification of cardiac arrhythmias. *Annals of the New York Academy of Sciences*, Volume 504, pages 168–178. Reprinted with permission of the New York Academy of Sciences.

**5.12**  Glass, L., Guevara, M.R., Bélair, J., and Shrier, A. 1984. Global bifurcations of a periodically forced biological oscillator. *Physical Review A*, Volume 29, pages 1348–1357.

**5.13, 5.14** Glass, L., Nagai, Y., Hall, K., Talajic, M., and Nattel, S. 2002. Predicting the entrainment of reentrant cardiac waves using phase resetting curves. *Physical Review E*, Volume 65, 021908.

**6.2, 6.4** Longtin, A. 1991. Nonlinear dynamics of neural delays feedback. In *1990 Lectures in Complex Systems, Santa Fe Institute Studies in the Sciences of Complexity*, Lectures Volume III, L. Nadel and D. Stein, Eds. Copyright ©1991. Reprinted by permission of Perseus Book Publishers, a member of Perseus Books, L.L.C.

**6.12** Braun, H.A., Schäfer, K., and Wissing, H. 1990. Theories and models of temperature transduction. In *Thermoreception and Temperature Regulation*. Reprinted with permission of Springer-Verlag.

**7.1** Chialvo, D.R., Michaels, D.C., and Jalife, J. 1990. Supernormal excitability as a mechanism of chaotic activation in cardiac Purkinje fibers. *Circulation Research*, Volume 66, pages 525–545. Copyright ©1990, reprinted with permission of Lippincott Williams & Wilkins.

**7.2** Top panel from Chialvo, D.R., Michaels, D.C., and Jalife, J. 1990. Supernormal excitability as a mechanism of chaotic activation in cardiac Purkinje fibers. *Circulation Research*, Volume 66, pages 525–545. Copyright ©1990, reprinted with permission of Lippincott Williams & Wilkins.

**7.2** Bottom panel reprinted from Franz, M.R. 1991. Method and theory of monophasic action potential recording. *Progress in Cardiovascular Diseases*, Volume 33, Number 6, pages 347–368, Copyright ©1991, with permission from Elsevier.

**7.3** Left panel reprinted from Vinet, A., and Roberge, F.A. 1994. Excitability and repolarization in an ionic model of the cardiac cell membrane. *Journal of theoretical Biology*, Volume 170, pages 568–591, Copyright ©1994, with permission from Elsevier.

**7.3** Right panel from Vinet, A., Chialvo, D.R., Michaels, D.C., and Jalife, J. 1990. Nonlinear dynamics of rate-dependent activation in models of single cardiac cells. *Circulation Research*, Volume 66, pages 1510–1524. Copyright ©1990, reprinted with permission of Lippincott Williams & Wilkins.

**7.4** Frame, L.H.F., and Simpson, M.B. 1988. Oscillations of conduction, action potential duration, and refractoriness: A mechanism for spontaneous termination of reentrant tachycardias. *Circulation*, Volume 77, pages 1277–1287. Copyright ©1988, reprinted with permission of Lippincott Williams & Wilkins.

**7.5** Greenberg, J.M., Hassard, N.D., and Hastings, S.P. 1978. Pattern formation and periodic structures in systems modeled by reaction diffusion equation. *Bulletin of the American Mathematical Society*, Volume 84, pages 1296–1326. Reprinted with permission of the American Mathematical Society.

**7.6, 7.7, 7.8, 7.9, 7.10** Courtemanche, M., Glass, L., and Keener, J.P. 1996. A delay equation representation of pulse circulation on a ring of excitable media. *SIAM Journal on Applied Mathematics*, Volume 56, pages 119–142. Reprinted with permission of the Society of Industrial and Applied Mathematics.

**7.12** Perstov, A.M., Davidenko, J.M., Salomonsz, R., Baxter, W.T., and Jalife, J. 1993. Spiral waves of excitation underlie reentrant activity in isolated cardiac muscle. *Circulation Research*, Volume 72, pages 631–650. Copyright ©1993, reprinted with permission of Lippincott Williams & Wilkins.

**7.14** Ruskin, J.N., DiMarco, J.P., and Garan, H.G. 1980. Out-of-hospital cardiac arrest – Electrophysiological observations and selection of long-term antiarrhythmic therapy. *New England Journal of Medicine*, Volume 303, pages 607–613. Copyright ©1980 Massachusetts Medical Society. All rights reserved. Reprinted with permission.

**8.1** Haurie, C., and Mackey, M.C., and Dale, D.C. 1998. Cyclical neutropenia and other periodic hematological diseases: A review of mechanisms and mathematical models. *Blood*, Volume 92, pages 2629–2640. Copyright American Society of Hematology, used with permission.

**8.2** Adapted from Mackey, M.C. 1996. Mathematical models of hematopoietic cell replication. In *Case Studies in Mathematical Modeling* by Othmer/Adler/Lewis, ©1996. Reprinted by permission of Pearson Education, Inc., Upper Saddle River, NJ.

**8.3** Haurie, C., Mackey, M.C., and Dale, D.C. 1999. Occurrence of periodic oscillations in the differential blood counts of congenital, idiopathic, and cyclical neutropenic patients before and during treatment with G-CSF. *Experimental Hematology*, Volume 27, pages 401–409, Copyright ©1999, with permission from International Society for Experimental Hematology.

**8.4, 8.5** Haurie, C., Person, R., Mackey, M.C., and Dale, D.C. 1999. Hematopoietic dynamics in grey collies. *Experimental Hematology*, Volume 27, pages 1139–1148, Copyright ©1999, with permission from International Society for Experimental Hematology.

**8.6, 8.7, 8.8** Hearn, T., Haurie, C., and Mackey, M.C. 1998. Cyclical neutropenia and the peripherial control of white blood cell production. *Journal of theoretical Biology*, Volume 192, pages 167–181, Copyright ©1998, with permission from Elsevier.

**8.10** Adapted from Mackey, M.C. 1996. Mathematical models of hematopoietic cell replication. In *Case Studies in Mathematical Modeling* by Othmer/Adler/Lewis, ©1996. Reprinted by permission of Pearson Education, Inc., Upper Saddle River, NJ.

**8.11, 8.12** Mackey, M.C. 1978. A unified hypothesis for the origin of aplastic anemia and periodic haematopoiesis. *Blood*, Volume 51, pages 941–956. Copyright American Society of Hematology, used with permission.

**8.13** Adapted from Mackey, M.C. 1996. Mathematical models of hematopoietic cell replication. In *Case Studies in Mathematical Modeling* by Othmer/Adler/Lewis, ©1996. Reprinted by permission of Pearson Education, Inc., Upper Saddle River, NJ.

**9.3** Longtin, A., and Milton, J.G. 1989. Longtin, A. and J.G. Milton, Modelling autonomous oscillations in the human pupil light reflex using non-linear delay-differential equations. *Bulletin of Mathematical Biology*, Volume 51, pages 605–624, Copyright ©1989, with permission from Elsevier.

**9.8** Milton, J.G., and Longtin, A. 1991. Evaluation of pupil constriction and dilation from cycling measurements. Vision Research, Volume 30, pages 515–525, Copyright ©1990, with permission from Elsevier.

**10.1** Kandel, E.R., Schwartz, J.H., and Jessell, T.M. 1991. *Principles of Neural Science*, 3rd edition.

**10.11** Edwards, R.E., Beuter, A., and Glass, L. 1999. Parkinsonian tremor and simplification in network dynamics. *Bulletin of Mathematical Biology*, Volume 51, pages 157–177, Copyright ©1999, with permission from Elsevier.

**10.13** Beuter, A., and Edwards, R.E. 1999. Using frequency domain characteristics to discriminate physiologic and Parkinsonian tremors. *Journal of Clinical Neurophysiology*, Volume 16, pages 484–494. Copyright ©1999, reprinted with permission of Lippincott Williams & Wilkins.

**10.14** Edwards, R.E., Beuter, A., and Glass, L. 1999. Parkinsonian tremor and simplification in network dynamics. *Bulletin of Mathematical Biology*, Volume 51, pages 157–177, Copyright ©1999, with permission from Elsevier.

# Contents

# 1

# Theoretical Approaches in Physiology

## Michael C. Mackey
## Anne Beuter

## 1.1 Introduction

Introductory chapters are usually boring, and many (most) readers skip them. Please feel free to do the same with this one, but be warned that you will miss some pretty interesting historical motivation for why this book was written.

## 1.2 A Wee Bit of History to Motivate Things

For decades (maybe centuries, but our memories do not go back that far) scientists in the physical and biological sciences have collectively thought of themselves as having little in common with each other. From a historical perspective, however, this could not be further from the truth. Consider the fact that many of the most famous "physicists" of the past few centuries (such as Copernicus, Galileo, Galvani, Young, Foucault, and Helmholtz) were not really physicists at all. Rather, they were trained as physicians! This is not as surprising as one might think, however, if you realize that at the time that they were educated (with the exception of Helmholtz) the only university training available was in medicine, law, or theology. If that were the end of the story, then you might rightly protest that this is a pretty silly connection. However, there is more to the story, and tracing out pieces of it is the whole point of the next few paragraphs.

### 1.2.1 Excitable Cells

Let us start with thinking about early investigations into the phenomena of "animal electricity" that were initiated by Luigi Galvani in the 1700s.

Galvani* had heard about the investigations of Benjamin Franklin into the nature of lightning and its relation to static electricity (many American school children have heard the apocryphal stories of Franklin flying kites with strings soaked in salt water during thunderstorms), and wondered whether there was any relation between the electricity that Franklin was looking at and the "animal electricity" that he had observed in his dissections of frogs. Animal electricity was manifested by muscular contraction when a nerve or muscle was touched with dissecting instruments made of different metals. Galvani's investigations intrigued his friend and colleague Vito Volta, who eventually came up with a totally different (and correct) explanation for the phenomena that Galvani was trying to explain. In the process, Galvani and Volta maintained their friendship (in spite of their differences of scientific opinion), and Volta developed the ideas that eventually led to the invention of the Voltaic pile (forerunner of the modern battery). Incidentally, Volta only turned in the equivalent of his doctoral dissertation when he was 50 years old.

The next stop in our list of famous "physicists" is Hermann von Helmholtz, revered by physicists everywhere for his broadly based expository treatment of physics in his eight-volume treatise published in 1898 (von Helmholtz 1898). What most physicists do not realize is that Helmholtz the physicist was also very much Helmholtz the physiologist, publishing a study of *The Mechanism of the Ossicles and the Membrana Tympani* (von Helmholtz 1874) related to audition, and a *Handbook of Physiological Optics* (von Helmholtz 1867). The latter has been reprinted many times over the past 150 years because of the valuable information and observations it contains.

In the early part of the twentieth century, the physiologist A.V. Hill made seminal contributions to our understanding of the mechanics and thermodynamics of muscle contraction, and one wonders how Hill's beautifully crafted papers (Hill 1965) combining experiment and theory were written until it is realized that he initially trained in physics and mathematics at Cambridge University before being seduced (figuratively, not literally) by a physiologist.

A few years later, Bernard Katz, working at University College London made, with his student Paul Fatt, a major advance in our understanding of synaptic transmission when they published their paper "Spontaneous subthreshold activity at motor nerve endings" (Fatt and Katz 1952), which was a marvel of experimental investigation combined with mathematical modeling of stochastic processes. This paper contains one of the first suggestions that deterministic processes might lead to data with a stochastic appearance.

---

*See http://www.english.upenn.edu/~jlynch/Frank/People/galvani.html

Jumping back a few decades, the German physical chemist Walter Nernst was interested in the transport of electrical charge in electrolyte solutions (Nernst 1888; Nernst 1889). His work intrigued another physicist, Max Planck, one of the fathers of modern quantum theory, who extended Nernst's experimental and theoretical work, eventually writing down a transport equation describing the current flow in an electrolyte under the combined action of an electric field and a concentration gradient (Planck 1890a; Planck 1890b). This work lay forgotten until the 1930s, when it was picked up by the physicist Kenneth C. Cole at Columbia University and his graduate student David Goldman (originally trained in physics). They realized that the work of Nernst and Planck (in the form of the Nernst–Planck equation) could be used to describe ion transport through biological membranes and did so with great effect. Their work resulted in the development of the "Goldman equation", which describes the membrane equilibrium potential in terms of intra- and extracellular ionic concentrations and ionic permeabilities (Goldman 1943). This background theoretical work of Nernst and Planck was also instrumental (see the excellent book *Membranes, Ions and Impulses*, Cole 1968) in helping Cole first experimentally demonstrate that there was a massive increase in membrane conductance during an action potential (Cole and Curtis 1939).

The Second World War interrupted these investigations, and Cole, like hundreds of other physicists, went off to work at the Massachusetts Institute of Technology (MIT) radiation labs. When the war was over, one of the few positive outcomes was the existence of highly sophisticated high-input impedance vacuum tubes that had been developed for the amplifiers in radar receivers. Cole (at the National Institutes of Health (NIH) in Bethesda by this time and at Woods Hole Biological Laboratory in the summer; Cole was no fool, and Bethesda is beastly hot in the summer), working with Marmont, used these new electronic advances to build an electronic feedback circuit that allowed him to control the action potential and actually measure the membrane changes occurring during the action potential (Marmont 1949). Allan Hodgkin (Cambridge) was visiting the United States about this time, and realized the power of the "voltage clamp" apparatus. On his return to England he teamed up with his old buddy A.F. Huxley, mathematician turned physiologist, to really measure what was going on during the generation of an action potential in the squid giant axon. This work was published in a brilliant series of five papers in the *Journal of Physiology* in 1952. The final one (Hodgkin and Huxley 1952) is an intellectual *tour de force* combining both experimental data analysis and mathematical modeling (the Hodgkin–Huxley equations; cf. Chapter 4) that eventually won Hodgkin and Huxley the Nobel Prize in 1963. Huxley the mathematician/physiologist was not content to stop there, however, and went on to publish his celebrated review (Huxley 1957) of muscle contraction data and its synthesis into the mathematically formulated cross bridge theory in 1957, a theory that still stands in its essential ingredients today.

The Hodgkin–Huxley model for excitability in the membrane of the squid giant axon is complicated and consists of one nonlinear partial differential equation coupled to three ordinary differential equations. In the early 1960s Richard FitzHugh, a young physiologist (more of that later) who had originally trained as a mathematician at the University of Colorado, applied some of the techniques that he had learned from the Russian applied mathematics literature to an analysis of the Hodgkin–Huxley equations (FitzHugh 1960; FitzHugh 1961; FitzHugh 1969). That reduction of the Hodgkin–Huxley equations (cf. Chapter 3) later became known as the FitzHugh–Nagumo model (cf. Section 4.5.6) and has given us great insight into the mathematical and physiological complexities of the excitability process. Another consequence of the Hodgkin–Huxley model, if taken to its interpretational extreme, was the implication that there were microscopic "channels" in the membrane through which ions would flow and which were controlled by membrane potential. Although there were hints in the 1960s that such channels might really exist, it was left to the German physicist Erwin Neher, in conjunction with the physiologist Bert Sakmann, to develop the technology and techniques that eventually allowed them to demonstrate the existence of these ion channels (Neher and Sakmann 1976). They were awarded the Nobel Prize in 1991 for this work. Modifications of the Hodgkin–Huxley equations were soon proposed for cardiac tissue as well as a myriad of other excitable cells. Extensions of the work of Hodgkin and Huxley by the physicist Walt Woodbury and Wayne Crill in cardiac tissue, for example, led them to the conclusions that a syncytium of excitable cells were coupled by so-called gap junctions or tight junctions, and it is now known that such junctions operate in many if not all tissues (Woodbury 1962). Woodbury also offered one of the first analyses of the molecular energy profiles in an ion channel, long before ion channels had been demonstrated to exist, using theoretical physical chemistry techniques (Erying, Lumry, and Woodbury 1949) learned from his famous "foster" *Doktorvater* Henry Erying.

## 1.2.2   Little Nervous Systems

While all of this activity was going on in terms of trying to understand how the electrical behavior of single cells could be understood, others were interested in explaining how collections of these cells might behave, especially in simple nervous systems. One of the early attempts to study how assemblies of cells could be involved in "computation" was by the mathematicians Warren McCulloch and Walter Pitts when they wrote their paper "A logical calculus of the ideas immanent in nervous activity" (McCulloch and Pitts 1943) and demonstrated how collections of neurons could perform logical calculations using binary logic. This paper helped shape much of the later research in computational neurobiology and forms the basis for much of the subsequent work in "neural networks" (which are really not neural networks at all) and the work of John Hopfield (Hopfield 1984) (cf. Section 10.5.2). At about the same time the polymath Norbert Wiener was making his mark in the mathematical world with his work in stochastic systems, in quantum mechanics, and information theory (Masani 1976). What most of Wiener's mathematical colleagues at MIT and elsewhere did not realize was that he had also formed a close personal friendship with the Mexican cardiologist Arturo Rosenblueth and that they were busily mathematically examining the behavior of sheets of excitable cardiac cells (Wiener and Rosenblueth 1946) (cf. Section 7.3.1). Simultaneously, Wiener was developing his ideas of living organisms as merely complex collections of control systems that led to his introduction of the concepts of *cybernetics* and the publication of several books (Wiener 1948; Wiener 1950) detailing his ideas bridging mathematics and physiology.

One of the most remarkable individuals interested in the dynamic behavior of simple nervous systems was H.K. Hartline of the John Hopkins University. Hartline was trained as a physiologist, and following receipt of his Ph.D. spent an additional two years at Hopkins taking mathematics and physics courses. For some unaccountable reason he was still not satisfied with his training and obtained funding to study for a further year in Leipzig with the famous physicist Werner Heisenberg and a second year in Munich with Arthur Sommerfeld. Armed with this rather formidable training in the biological, mathematical, and physical sciences he then devoted the majority of his professional life at Hopkins to the experimental study of the physiology of the retina of the horseshoe crab *Limulus*. His papers are a marvel of beautiful experimental work combined with mathematical modeling designed to explain and codify his findings, and his life work (Hartline 1974) justly earned him the Nobel Prize in 1967. As an aside we should point out that FitzHugh (of the FitzHugh–Nagumo reduction of the Hodgkin–Huxley model, Section 4.5.6) received his Ph.D. in physiology under Hartline after completing his mathematical studies at the University of Colorado.

One can hardly underestimate the impact that this work in excitable cell physiology has had on the biological sciences; the impact is so broad and pervasive. The *Notices of the American Mathematical Society* (December 1999) has a very nice article by Nancy Kopell (Kopell 1999) with some of the mathematical side of the story, and *Nature Neuroscience* (November 2000) featured some of this from a biological perspective in an interesting and lively survey.

## 1.2.3  Some Other Examples

By now, you, dear and gentle reader, may be convinced (if you were not before) that the biological, mathematical, and physical sciences have a lot of overlap and interaction. So, just a couple of more examples of these ties in other fields.

There is currently much excitement about the field of "molecular biology" (the excitement, in our opinion, is mostly about the wrong things, but that is another story). Molecular biology as a field would probably not exist in its present form were it not for the impact of the beautiful little book *What is Life?* (Schrödinger 1944) written by Erwin Schrödinger (yes, one of the other fathers of quantum mechanics along with Max Planck, whom we met earlier). *What is Life?* was the written account of a series of lectures that Schrödinger gave at the Dublin Institute for Advanced Studies in 1943. *What is Life?* was absolutely instrumental in recruiting a whole generation of *theoretical physicists* like Max Delbruck, Walter Gilbert, Leo Szilard, Seymor Benzer, Sidney Brenner, Jim Watson, and Francis Crick away from pure physics and into the exciting new area of "molecular biology" (Olby 1974). The results are history, but it has not ended there. For example, the current director of the Whitehead Center for Genomic Research at MIT is Eric Lander, whose D. Phil. from Oxford was not in molecular biology but rather in mathematics (algebraic coding theory)! The mathematical and physical sciences continue to have a great role in the whole area of molecular biology, and some of the potential areas of impact are detailed in the March 2001 issue of *Chaos* (see Bibliography under Chaos Focus Issue).

Finally, we cannot resist mentioning the ultimate tale of how mathematics can often have unexpected applications elsewhere. The number theorist G.H. Hardy said in his autobiographical writings *A Mathematicians Apology* (Hardy 1940) that "I have never done anything 'useful'. No discovery of mine has made, or is likely to make, directly or indirectly, for good or ill, the least difference to the amenity of the world." How wrong he was, as witness the consequence of typing in "Hardy-Weinberg Law" into any currently available Internet search engine! You will probably receive about 15,000 hits related to this fundamental relationship used daily by geneticists in counseling, and developed by Hardy and Weinberg in the 1930s to look at the stability of gene and genotype frequencies in populations.

## 1.2.4  Impact & Lessons

The impact that these (and others we have not mentioned) examples have had in terms of the development of modern biology and medicine are virtually incalculable. In spite of the fact that we find it almost impossible to assess their complete impact, we can draw some lessons for biologists and for physicists/mathematicians from this history spanning several years:

- Mathematics and physics have played a major role in defining and shaping the life sciences as we know them today.

- This will continue at an ever accelerated pace in our opinion.

- Biologists can ignore physics and mathematics if they want, but they do so at their own peril!

- Though much of mathematics has been driven by problems from physics for the last century, increasingly we are seeing mathematics driven by problems from biology.

- Thus the converse lesson for applied mathematicians: Ignore biology and medicine if you want, but do so at your own peril!

We collectively feel that in all interdisciplinary research of this type it is essential to constantly remember that one of the goals is to understand how living organisms work. This implies the obvious, but too often forgotten, corollary that *integrative* aspects are essential and must always be kept in mind. In all research one must constantly guard against learning more and more about less and less until everything is known about nothing.

## 1.2.5 Successful Collaborations

The history of biology and medicine is a history of successful collaboration between these sciences and physics and mathematics, either between individuals or within a given individual using tools from both areas. In the past this interdisciplinary background could be resident in one individual, but in the future we may expect to see interdisciplinary teams of individuals each bringing their expertise to bear on given problems. In excitable cell physiology another essential key to this success was the availability of high-resolution temporal data plus a certain obvious connection between the biology and physics. The same has been true in neurophysiology. To the extent that there have been, or will be in the future, highly successful collaborations in other areas of biology, these will ultimately depend on the availability of equally good data.

An example of a new and potentially important collaboration between theoreticians and neurophysiologists is found in the exploration of mechanisms underlying chronic electrical brain stimulation. This technique is used in patients with movement disorders but also in patients with epilepsy or pain. It has been known for many years now that high-frequency electrical stimulation of neural structures located deep in the brain such as ventral intermediate nucleus of the thalamus, internal globus pallidus, or subthalamic nucleus produced spectacular improvements of a variety of symptoms without impeding the function of these neural structures. No acceptable explanation for these improvements exists yet. But new collaborations are developing between physicists and mathematicians on one side and neurophysiologists and clinicians on the other side (Titcombe, Glass, Guehl, and Beuter 2001; Montgomery Jr and Baker 2000; Tass 2000). These collaborations are important for obvious reasons now and will probably in a not-too-distant future completely change these fields.

# 2

# Introduction to Dynamics in Nonlinear Difference and Differential Equations

Jacques Bélair
Leon Glass

Nonlinear dynamics is a branch of mathematics that deals with rhythms in nonlinear equations. There is an emphasis on qualitative aspects of dynamics. For example, does a particular set of equations display a steady state, a cycle, or a more complex rhythm? For what range of parameters are the rhythms stable? When the rhythms become unstable, what new rhythms are observed?

From a purely mathematical perspective, nonlinear dynamics has a rich history with important contributions from leading mathematicians going back to Henri Poincaré at the end of the nineteenth century. The recognition of the potential of nonlinear dynamics to help understand dynamics in the natural sciences, particularly in the biological sciences, has been slower to develop. Most strikingly in the biomedical sciences, there are still whole disciplines in which mathematics has a negligible impact in the training of students.

This chapter focuses on introducing the main concepts of nonlinear dynamics in a simplified format, along with some examples to illustrate the applications to medicine. More complete introductions to the field of nonlinear dynamics can be found in a number of forums. We especially recommend, objectively, Kaplan and Glass (1995) for an elementary introduction and Strogatz (1994), Devaney (1989), and Alligood, Sauer, and Yorke (1997) for a more mathematically rigorous introduction. Since the current volume is directed toward a broad audience including individuals from the biological sciences, we introduce the main mathematical concepts needed in an accurate yet digestible style.

## 2.1   Main Concepts in Nonlinear Dynamics

A number of concepts characterize the field of nonlinear dynamics. Most of these concepts can be associated with everyday experience, and so can often be appreciated with little reference to formal mathematics (although there is inherent danger in taking semantic associations too seriously). The most important concepts are those of **fixed point**, **oscillation**, **stability**, **bifurcation**, and **chaos**. These concepts can be associated with an equation that describes the dynamics in a system, or they can refer to the properties of the system itself. Thus, biological and physical systems display oscillations, as do the equations that describe the naturally occurring systems.

We now give a brief description of each concept in turn. Each concept really describes the possible behaviors in a system.

- **Fixed point**. A fixed point is a behavior that is constant in time. The terms *steady state* and *equilibrium point* express a similar concept and are sometimes used interchangeably.

- **Oscillation**. An oscillation is a behavior that repeats periodically in time. The term *cycle* is synonymous and is used interchangeably.

- **Chaos**. Chaos is a behavior that fluctuates irregularly in time; it is a behavior that is not a fixed point and not a cycle. However, not all behavior that fluctuates irregularly in time represents chaotic behavior. Chaotic dynamics, as used in nonlinear dynamics, represents behavior that is observed in deterministic dynamical equations. Chaotic dynamics has the additional property that small differences in the initial value will grow over time, but that the dynamics will be finite, and not grow indefinitely. This property is often called sensitive dependence on initial conditions. Because of sensitive dependence on initial conditions, it is impossible to make accurate long-term predictions about the state of the system without knowing *exactly* its initial state. Any minor uncertainty will be amplified so as to render actual knowledge of the state of the system impossible. In a real system (as opposed to an equation) it is always impossible to know the state of a system exactly, since there is always some measure of experimental uncertainty.

- **Stability**. Stability is a term with many different meanings, which must therefore be qualified. *Local* stability refers to a behavior that is reestablished following a small perturbation. For example, if a fixed point or cycle is reestablished following a small perturbation, then the point or cycle are called (locally) stable. *Global* stability refers to the return to a cycle or fixed point following arbitrary (i.e., not necessarily small) perturbations. *Structural* stability refers to the stability of the dynamic features of a system, under changes to the system itself.

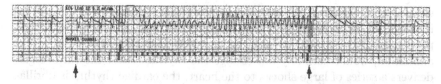

Figure 2.1. Traces showing the testing of an implantable defibrillator. The trace at the left shows the normal sinus rhythm. A series of shocks delivered to the heart at the first arrow induces ventricular fibrillation. The implantable defibrillator senses the abnormal rhythm and then automatically delivers a large shock, second arrow, that restores the normal sinus rhythm. This record illustrates the concept that the human heart can sustain two stable dynamically different states: the normal sinus rhythm and ventricular fibrillation. Tracing provided by the Harvard–Thorndike Cardiac Electrophysiology Laboratory, Beth Israel Deaconess Medical Center, Boston, MA.

- **Bifurcation**. Bifurcation describes the qualitative change in dynamics that can be observed as the parameters in a system vary. For example, under variation in a parameter value, a fixed point may change from being stable to unstable, or a cycle might suddenly appear.

These terms can all be mathematically rigorously defined, and we will give some illustrative examples, with mathematics, shortly. However, the application of the strict mathematical definitions can be difficult in the context of the natural sciences. In most natural systems, for example, there are intrinsic random fluctuations that may arise from a variety of factors such as random thermal fluctuations in the physical and biological sciences and random opening of channels in membranes in biology. Thus, even though it is often convenient to refer to fixed points and cycles in biology, invariably there is an intrinsic fluctuation in the behaviors that are not present in a deterministic mathematical model of the biological system.

We illustrate these concepts with a dramatic example taken from cardiology. Some patients, who are at risk of dying suddenly from an abnormal cardiac rhythm called ventricular fibrillation, have implanted cardiac defibrillators. These devices detect ventricular fibrillation if it is present, and deliver a shock directly to the heart to reverse this dangerous rhythm and restore the normal rhythm called sinus rhythm. However, in order to test whether these devices are working, the cardiologist delivers a series of large electrical shocks directly to the patient's heart that converts the normal sinus rhythm to fibrillation, and then the device delivers a shock to convert the rhythm back to the sinus rhythm. Figure 2.1 shows an illustration.

At the beginning of the record, the patient's heart is in the normal sinus rhythm. The sinus rhythm is a locally stable cycle, since a small perturbation, such as a small electric shock to the heart, would leave the heart still in the normal sinus rhythm. It is also a structurally stable rhythm, since a small change in the structure of the patient's heart, which might be induced, for example, by giving a drug that changes some of the electrical properties of the heart, or removing a small piece of heart tissue for examination, would normally not change the rhythm. After the cardiologist delivers a series of large shocks to the heart, the cardiac rhythm is fibrillation. Since this rhythm is irregular, there has been debate about whether this irregular rhythm is chaotic. If the fibrillation rhythm persists for a long time, (fortunately, this is not the case in the illustration!) then the patient would die, and the dynamics would not change in time. There would thus be a fixed point. In this example, both the normal sinus rhythm and the fibrillation are locally stable behaviors, but they are not reestablished after large perturbations. Thus a given dynamical system can have two or more locally stable stable behaviors, i.e., there can be bistability or multistability.

From a dynamical point of view, we must assume that patients who are at risk for sudden death have some abnormality. This anomaly in their heart anatomy or physiology makes it possible that some parameter change in the person leads to a bifurcation in the dynamics. The normal sinus rhythm may then no longer be locally stable and may be replaced by the abnormal and potentially fatal rhythm of ventricular fibrillation. One final subtlety: Although the sinus rhythm appears to be periodic in time and represent a cycle, there are usually fluctuations in the times between the normal heart beats, and some have claimed that these fluctuations are chaotic.

These concepts appear in a variety of different types of equations that have been used to model biological systems. In this chapter we discuss two different types of equations, which model phenomena that are described using either discrete or continuous times. In the former case, the model takes the form of difference equations, in which the behavior of a system at one time is a function of the behavior at a previous time; in the latter case, differential equations are used to represent a system evolving continuously in time, under a governing rule specifying rates of change (derivatives).

## 2.2   Difference Equations in One Dimension

A difference equation can be written as

$$x_{n+1} = f(x_n), \tag{2.1}$$

where $x_n$ is the value of a variable $x$ at some time $n$, and $f$ is a function that describes the relationship between the value of $x$ at time $n$ and $n+1$. This equation can be analyzed using a simple geometric interpretation, called the cobweb construction, which is illustrated in Figure 2.2.

To implement this "cobweb" construction, proceed as follows. Given an initial condition $x_0$, draw a vertical line until it intersects the graph of $f$: that height is given by $x_1$. At this stage we may return to the horizontal axis and repeat the procedure to get $x_2$ from $x_1$, but it is more convenient to

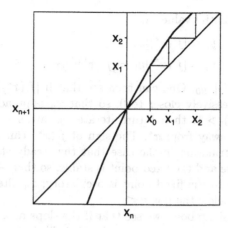

Figure 2.2. Cobweb method to iterate the quadratic map in equation (2.3).

simply trace a horizontal line until it intersects the diagonal line $x_{n+1} = x_n$, and then move vertically to the curve again. The process is then iterated to find the subsequent points in the series. Cobweb diagrams may be useful in providing global geometrical information when more analytical methods fail.

We are interested in the long-term behavior of the sequence $x_1, x_2, \ldots$ generated from an initial point $x_0$ in equation (2.1). The simplest possibility is that the iterates converge to a fixed point satisfying the equation $x^* = f(x^*)$. Fixed points are located by finding the intersection points of the graph of the function $f$ with the line $x_{n+1} = x_n$.

One important property of a fixed point is its local stability. In order to be locally stable, any initial condition nearby to the fixed point must lead to a sequence of iterates that converges to the fixed point. The local stability can be determined from the graph of $f$ near the fixed point $x^*$. If we replace the actual graph of $f$ by a linear approximation, the latter will have as slope the derivative of the function $f$ evaluated at $x^*$, $f'(x^*)$. If we now let $y_n$ denote the deviation of $x_n$ from the steady-state value $x^*$, $y_n = x_n - x^*$, we must understand the behavior of the iterates of the linear finite-difference equation

$$y_{n+1} = [f'(x^*)]y_n, \qquad (2.2)$$

to see whether this difference is increasing, in which case points are being repelled from the steady state, or decreasing, in which case nearby points approach the fixed point.

## 2.2.1  Stability and Bifurcations

The behavior in the linear approximation can be determined explicitly, since an algebraic solution is possible. Consider equation (2.2), and compute, from an initial value $y_0$,

$$y_1 = [f'(x^*)]y_0,$$
$$y_2 = [f'(x^*)]y_1 = [f'(x^*)]^2 y_0, \ldots,$$

so that $y_n = [f'(x^*)]^n y_0$. One can then see that if $|f'(x^*)| < 1$, future iterates $y_n$ are successively closer to 0, so that $x_n$ approaches $x^*$. On the contrary, if $|f'(x^*)| > 1$, then future iterates $y_n$ are successively bigger, and $x_n$ is further away from $x^*$. The sign of $f'(x^*)$ can also be used to obtain further information in the case that the steady state is stable: If this sign is negative and the fixed point is stable, so that $-1 < f'(x^*) < 0$, then the approach to the fixed point is oscillatory; $x_n$ alternates between values above and below the point $x^*$.

From the discussion above, we see that if the slope at a fixed point of a difference equation is either $+1$ or $-1$, then it is likely that a small change in a parameter of the difference equation will change the stability of the fixed point and lead to a bifurcation of the dynamics. Several different bifurcations can occur.

To illustrate bifurcations in finite-difference equations, we will consider the equation with $f$ given by a quadratic function so that

$$x_{n+1} = \mu x_n (1 - x_n), \tag{2.3}$$

where $\mu$ is a constant. This example is sometimes called the quadratic, or logistic, map. As $\mu$ changes, there can be qualitative changes, or bifurcations in the dynamics.

## Transcritical Bifurcation

When $\mu = 1$, the finite-difference equation (2.3) has the unique fixed point $x^* = 0$, and the linearized function there has slope 1. In general, for equation (2.3), the slope of the map at the fixed point $x^*$ is $\mu(1 - 2x^*)$, and the fixed points are 0 and $(\mu - 1)/\mu$. For the value $\mu = 1$, we thus see that both fixed points are the same, and that $f'(0)$ is then 1. This is a bifurcation point. This *transcritical* bifurcation, illustrated in Figure 2.3, results in an exchange of stability between the fixed points 0 and $(\mu - 1)/\mu$. For values of $\mu < 1$, the fixed point at 0 is stable, and the fixed point at $(\mu - 1)/\mu$ is unstable. When $\mu > 1$, the fixed point at 0 is now unstable, and the fixed point at $(\mu - 1)/\mu$ is stable.

This bifurcation occurs when the slope of the function at the fixed point is 1, and some property of the finite-difference equation prevents this fixed point from disappearing.

## Period-Doubling Bifurcations and Chaos

The most famous bifurcation in one-dimensional difference equations is the *period-doubling* bifurcation, which occurs when the slope of the map has the value $-1$ at a fixed point.

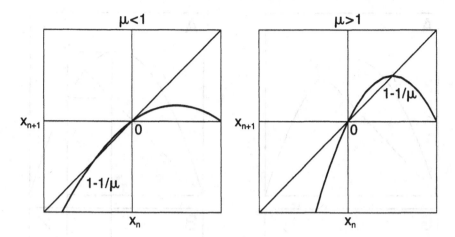

Figure 2.3. The transcritical bifurcation.

The transition from a single steady-state to a cycle of period-two also occurs in the quadratic map in equation (2.3), at the value $\mu = 3$. This phenomenon is best illustrated by considering, in addition to the graph of $f(x_n)$, that of the second iterate, $f(f(x_n)) = f^2(x_n)$, as shown in Figure 2.4. A fixed point of the second iterate satisfies $x_{n+2} = x_n$, and can thus be either a fixed point or a cycle of period-two, in which case $x_{n+1} \neq x_n$ but $x_{n+2} = x_n$.

A cobweb representation of the first and second iterates indicates that as the steady state becomes unstable, a pair of fixed points for the second iterate $f^2$ is born, and thus a period-doubling bifurcation occurs. The stability of this new cycle can be determined as for the original map, by looking at the slope at the fixed point, which corresponds, in the original map, to the product of the slopes at the two points constituting the periodic orbit.

To summarize, we have seen for the quadratic map equation (2.3) that: (i) when $0 < \mu < 1$, the steady state 0 is stable; (ii) there is a transcritical bifurcation at $\mu = 1$, and the steady state $(\mu - 1)/\mu$ is stable for $1 < \mu < 3$; (iii) there is a period-doubling bifurcation at $\mu = 3$, and a stable cycle of period two for $\mu > 3$. We do not really expect the cycle of period two to be stable for all values of $\mu$, and indeed it is not. It can be shown, by computing the slope of $f^2$ at one point of the period-two cycle, that another period-doubling bifurcation takes place when $\mu = 1 + \sqrt{6}$. The period-two cycle becomes unstable and there is creation of a stable cycle of period 4.

There is an infinite number of period-doubling bifurcations in equation (2.3) as $\mu$ goes from 3 to about 3.57, as shown in the **bifurcation diagram** of Figure 2.5.

This diagram is obtained by iterating numerically, for each value of $\mu$, equation (2.3) and displaying a number of points from the sequence of iterates at this value of $\mu$. The "clouds" visible for values of $\mu$ above 3.57

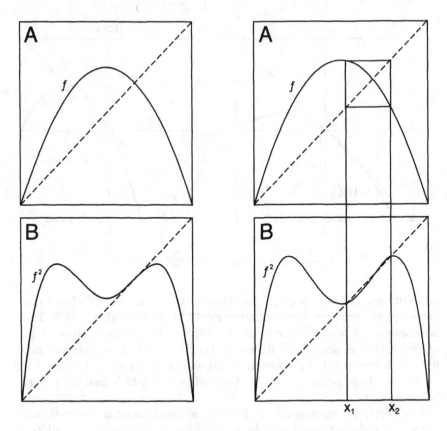

Figure 2.4. Convergence to a stable fixed point (left) before the period-doubling bifurcation, and to a period-two cycle (right) after the period-doubling bifurcation. In each illustration, the function $f$ is displayed in panel A, and its second iterate is displayed in panel B.

correspond to the absence of a stable cycle. There exist values of $\mu$ for which all the iterates of some initial points remain in the interval $(0, 1)$, and the iterates do not converge to any finite cycle. The equation displays chaos (Feigenbaum 1980).

The quantitative measure of this sensitive dependence on initial conditions is the **Lyapunov exponent** $\lambda$ of an orbit starting at the initial point $x_0$, which is defined as

$$\lambda = \lim_{n \to \infty} \left[ \frac{1}{n} \sum_{i=0}^{n-1} \ln |f'(x_i)| \right], \qquad (2.4)$$

and essentially corresponds to the average slope of the cycle starting at the point $x_0$. When $\lambda$ is negative, that average slope is less than 1, so the dynamics are stable, whereas for positive Lyapunov exponent, the distance between neighboring points grows exponentially.

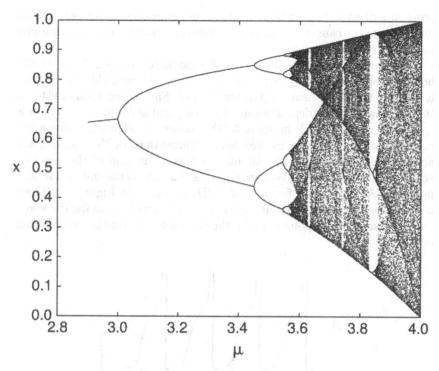

Figure 2.5. Bifurcation diagram for the quadratic map.

A word of caution is in order in interpreting the bifurcation diagram of the quadratic map. For values of $\mu$ slightly above 3.83, one can see a stable cycle of period three that attracts almost all initial conditions. An initial point $x_0$ taken at random in the interval $(0, 1)$ will almost certainly converge to this cycle. So why did Li and Yorke (1975) claim "Period three implies chaos?" (Li and Yorke 1975).

The answer lies in the definition of chaos. For Li and Yorke, chaos meant an infinite number of cycles, and also an uncountable number of points that do not converge to any of these cycles. That definition does not involve the stability of the cycles. As is seen in this periodic window, it is possible for the only *observable* cycle to be a stable cycle, even though some initial conditions may be associated with unstable fixed points or cycles.

This example shows that one must pay attention to definitions. The most common and relevant definition of chaos for applications involves the sensitive dependence on initial conditions, i.e., positivity of Lyapunov exponents. However, this definition is difficult to implement in the investigation of experimental systems, since it is applicable only to deterministic systems. As mentioned at the beginning of this chapter, biological systems in particular are in most cases not described entirely accurately by deterministic systems of equations (the modeling process involving a simplification procedure of some sort). Consequently, a determination of a positive Lyapunov exponent cannot be taken as definite evidence for chaotic dynamics in an experimental system.

As an example, consider the effects of periodic stimulation on a spatially homogeneous excitable aggregate of cells from embryonic chick heart that is not spontaneously beating (Guevara, Ward, Shrier, and Glass 1984). In this preparation, electrical stimulations are applied at regular time intervals $t_s$, and an excitation of duration APD is observed. Figure 2.6 shows the response at two different frequencies of stimulation. In this system, the action-potential duration can be obtained as a function of the previous action-potential duration. We can graph the duration of the $(n+1)$st action-potential, $APD_{n+1}$, as a function of $APD_n$ as shown in Figure 2.7. There is alternation between two different values of the action-potential duration, and an unstable fixed point between the two values, with slope greater than one.

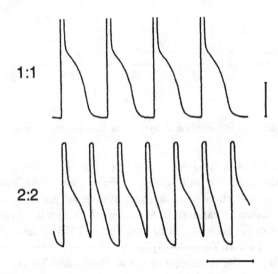

Figure 2.6. Intracellular recording of transmembrane potential from a periodically stimulated quiescent heart-cell aggregate. In the upper trace, the period of the stimulation is 300 msec, and in the lower trace it is 180 msec. From Guevara, Ward, Shrier, and Glass (1984).

## Saddle-Node Bifurcation

Another bifurcation in finite-difference equations is the saddle-node bifurcation. This bifurcation occurs when two fixed points, one stable and one unstable, coalesce and annihilate one another in a *saddle-node* bifurcation, sometimes called the *tangent* bifurcation.

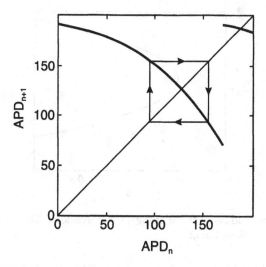

Figure 2.7. Cobweb diagram for the iteration of $APD_n$ when heart cell aggregates are stimulated with a period 170 msec. There is a stable cycle with APD alternating between 94 msec and 156 msec, and a stable steady state with $APD = 187$ msec. From Guevara, Ward, Shrier, and Glass (1984).

This bifurcation can be illustrated in the following modification of the quadratic map,

$$x_{n+1} = c + \mu x_n(1 - x_n), \quad \mu > 0. \tag{2.5}$$

When $c = 0$, this is the quadratic map.

To illustrate the saddle-node bifurcation, we consider what happens as $c$ decreases from 0, for $\mu = 3$. The fixed points are given by

$$x^* = \frac{1 \pm \sqrt{1 + c}}{3}. \tag{2.6}$$

Figure 2.8 shows the values of the fixed points as a function of $c$. As $c$ passes through $-1$, the two bifurcation points coalesce and annihilate each other: for $c < -1$, there are no fixed points. In this case, the dynamics would approach $-\infty$ under iterations.

## 2.3    Ordinary Differential Equations

Many biological systems are best suited to a representation in continuous time, so that they can be described at any instant and not just at isolated times as in difference equations. In addition, realistic systems involve the interaction of several variables. Depending on the setting, these variables may represent, for example, animal species, ionic currents, different types

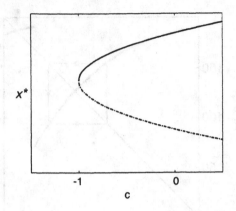

Figure 2.8. The saddle-node or tangent bifurcation. The fixed points in equation (2.5) as a function of $c$.

of circulating blood cells, or drug concentrations in different body compartments. The dynamics in these systems are often well approximated at first by ordinary differential equations that define the rates of change of the variables in terms of the current state. In the physical sciences, there is strong emphasis on analytic computation of the dynamics as a function of time. Our approach is more geometric. In biology, we start with the knowledge that any equation is an approximation of the real system. Further, the qualitative features of the dynamics, whether there are steady states, cycles, or chaos, are the most compelling aspects of the experimental setting. Therefore, in the analysis of biological systems, there is often a strong emphasis on adopting a geometric approach to investigate qualitative aspects of the dynamics.

### 2.3.1  One-Dimensional Nonlinear Differential Equations

First consider a nonlinear differential equation with one variable,

$$\frac{dx}{dt} = f(x). \tag{2.7}$$

This equation contains a derivative, and we assume that $f(x)$ is a single-valued nonlinear function.

The global flow of equation (2.7) can be determined from a graphical representation of the function $f$ on its right-hand side. We consider, nevertheless, a construction that will become essential in higher-dimensional systems, namely, the approximation of the full equation in (2.7) by a linear equation.

We replace the function $f(x)$ on the right-hand side of equation (2.7) by a linear function. Consider values of $x$ close to a fixed point $x^*$. If we let $X(t) = x(t) - x^*$, the slope at the fixed point is given by the derivative

$f'(x^*)$, and the linear approximation becomes

$$\frac{dX}{dt} = [f'(x^*)]\, X = a\, X,\qquad (2.8)$$

where $a = f'(x^*)$.

Equation (2.8) is an example of a linear differential equation. In most cases, such equations have exponential functions as their solution. Consequently, the general form of a solution of equation (2.8) is $Ke^{at}$. This last function decreases to 0, in absolute value, when $a < 0$, and becomes unbounded when $a > 0$. The fixed point $x^*$ is thus locally stable when $f'(x^*) < 0$, and locally unstable when $f'(x^*) > 0$.

We illustrate a one-dimensional differential equation with a simple model for drug administration (Glantz 1979).

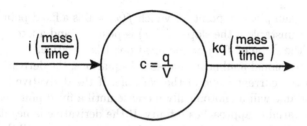

Figure 2.9. One-compartment description for drug distribution and metabolism.

Consider the one-compartment description for drug distribution and metabolism illustrated in Figure 2.9. In this representation, we assume that a drug is infused at a constant rate $i$ into a compartment (organ or system) of constant volume $V$, and that it is removed at a constant rate $kq$, where $q$ is the quantity of drug in the compartment. Under homogeneity of distribution of material in the compartment, it is possible to derive, from the conservation principle, the governing equation for the concentration $c = q/V$ of drug in the compartment as

$$\frac{dc}{dt} = \frac{i}{V} - kc .\qquad (2.9)$$

This equation states that the rate of change of the concentration $c$ as a function of time is the difference between the intake $i/V$ and the removal $kc$. A fixed-point solution $c^*$ of equation (2.9) satisfies $c^* = i/kV$. The right-hand side of equation (2.9) has a derivative (with respect to $c$) that is the negative constant $-k$. The steady state $c^*$ is stable, and in fact, the monotonicity of the decreasing function $(i/V) - kc$ implies that this stationary solution attracts all solutions of equation (2.9). The time evolution of the concentration $c(t) = (i/kV)(1 - e^{-kt})$ is illustrated in Figure 2.10.

For an arbitrary nonlinear function $f(x)$, it might be difficult or impossible to determine a solution analytically. However, if we can draw a graph of $f(x)$, then we can determine important qualitative features of the dy-

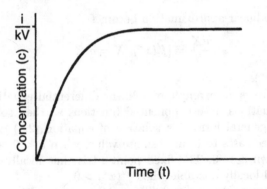

Figure 2.10. Drug distribution evolution in the one-compartment model.

namics. For example, any point for which $f(x) = 0$ is a fixed point. A fixed point will be unstable if the slope of $f(x)$ is positive, and a fixed point will be stable if the slope of $f(x)$ at the fixed point is negative.

In one-dimensional nonlinear differential equations, the derivative always depends on the current value of the variable. If the derivative is positive, then the variable will monotonically increase until a fixed point is reached, or until the variable approaches infinity. If the derivative is negative, then the variable will decrease until a fixed point is reached, or until the variable approaches negative infinity. Neither oscillations nor chaos are possible: more variables or time delays are needed to observe these types of dynamics in nonlinear differential equations.

## 2.3.2   Two-Dimensional Differential Equations

Two-dimensional nonlinear ordinary differential equations are written in the form

$$\frac{dx}{dt} = f(x, y),$$

$$\frac{dy}{dt} = g(x, y), \tag{2.10}$$

where $f(x, y)$ and $g(x, y)$ are functions of $x$ and $y$ which may be nonlinear.

In a fashion similar to the analysis of stability in one-dimensional ordinary differential equations, we analyze the dynamics of equation (2.10) in a neighborhood of a fixed point at $(x^*, y^*)$. By definition, a fixed point $(x^*, y^*)$ satisfies $f(x^*, y^*) = g(x^*, y^*) = 0$ in equation (2.10). We carry out an expansion of the nonlinear functions $f(x, y)$ and $g(x, y)$ in a neighborhood of $(x^*, y^*)$. The Taylor expansion of a function $f(x, y)$ is

$$f(x, y) = f(x^*, y^*) + \left.\frac{\partial f}{\partial x}\right|_{x^*, y^*} (x - x^*) + \left.\frac{\partial f}{\partial y}\right|_{x^*, y^*} (y - y^*) + \cdots, \tag{2.11}$$

where the dots represent terms with higher-order derivatives. If we now make the change of variables

$$X = x - x^*, \quad Y = y - y^*,$$
(2.12)

so that $X$ and $Y$ denote the deviations of the solutions from the steady state, we can expand equation (2.10) to obtain

$$\frac{dX}{dt} = AX + BY + \cdots ,$$

$$\frac{dY}{dt} = CX + DY + \cdots ,$$
(2.13)

where

$$A = \left. \frac{\partial f}{\partial x} \right|_{x^*,y^*}, \quad B = \left. \frac{\partial f}{\partial y} \right|_{x^*,y^*},$$

$$C = \left. \frac{\partial g}{\partial x} \right|_{x^*,y^*}, \quad D = \left. \frac{\partial g}{\partial y} \right|_{x^*,y^*}.$$
(2.14)

Equation (2.13) can be written in matrix form as

$$\begin{pmatrix} \frac{dX}{dt} \\ \frac{dY}{dt} \end{pmatrix} = \begin{pmatrix} A & B \\ C & D \end{pmatrix} \begin{pmatrix} X \\ Y \end{pmatrix}.$$
(2.15)

Suppose we could define two variables, $\xi$ and $\eta$, such that for some constants $\alpha$, $\beta$, $\gamma$ and $\delta$,

$$X = \alpha \xi + \beta \eta,$$

$$Y = \gamma \xi + \delta \eta,$$
(2.16)

and such that

$$\begin{pmatrix} \frac{d\xi}{dt} \\ \frac{d\eta}{dt} \end{pmatrix} = \begin{pmatrix} \lambda_1 & 0 \\ 0 & \lambda_2 \end{pmatrix} \begin{pmatrix} \xi \\ \eta \end{pmatrix}.$$
(2.17)

This equation is much easier to solve because it is two uncoupled equations

$$\frac{d\xi}{dt} = \lambda_1 \xi \text{ and } \frac{d\eta}{dt} = \lambda_2 \eta,$$

each of which is one-dimensional. From the previous section, we know that these two equations have the solution $\xi(t) = K_\xi e^{\lambda_1 t}$ and $\eta(t) = K_\eta e^{\lambda_2 t}$. Now it would be easy to find $X(t)$ and $Y(t)$ simply by applying (2.16). The problem of finding $\xi$ and $\eta$ that satisfy equations (2.16) and (2.17) is well known in linear algebra as the eigenvalue problem. It involves solving the equation

$$\det \begin{vmatrix} A - \lambda & B \\ C & D - \lambda \end{vmatrix} = 0,$$
(2.18)

where det denotes the determinant of the matrix. The determinant of a matrix with 2 rows and 2 columns is defined by

$$\det \begin{vmatrix} a & b \\ c & d \end{vmatrix} = ad - bc.$$

Equation (2.18) is called the characteristic equation. Computing the determinant in (2.18) yields

$$\lambda^2 - (A+D)\lambda + (AD - BC) = 0, \tag{2.19}$$

which is equivalent to equation (2.18). This equation will have two solutions given by

$$\lambda_1 = \frac{A+D}{2} + \frac{\sqrt{(A-D)^2 + 4BC}}{2},$$

$$\lambda_2 = \frac{A+D}{2} - \frac{\sqrt{(A-D)^2 + 4BC}}{2}.$$

The roots of the characteristic equation are called the *characteristic roots*, or *eigenvalues*.

The solutions of equation (2.15) are given by

$$X(t) = K_1 e^{\lambda_1 t} + K_2 e^{\lambda_2 t},$$
$$Y(t) = K_3 e^{\lambda_1 t} + K_4 e^{\lambda_2 t}, \tag{2.20}$$

where $K_1$, $K_2$, $K_3$, and $K_4$ are constants that are typically set by the initial conditions.

In equation (2.10) there are two variables. One way to display the solutions is to plot the values of $x$ and $y$ as a function of time. A different way to display the solution is to consider time an implicit variable and to plot the evolution of $x$ and $y$ in the $(x, y)$ plane. The $(x, y)$ plane is called the state space or phase plane. Starting at some initial condition, $x(0), y(0)$, the equations determine a path or trajectory. The trajectories defined by equation (2.10) have distinctive geometries in a neighborhood of fixed points. These geometric features are classified by the linearized equations.

The classification is illustrated in Figure 2.11, where the vertical axis is the determinant $(AD - BC)$, while the horizontal axis is the trace $(A+D)$ of the matrix in equation (2.15). The insets in the figure are schematic illustrations of the phase plane portraits of the flows in a neighborhood of the fixed points.

There are three types of flows:

- **Focus.** $(A - D)^2 + 4BC < 0$. The eigenvalues are complex numbers. The flow winds around the fixed point in a spiral. The size of the imaginary part tells how fast the winding occurs. The real part is $\frac{A+D}{2}$. If $\frac{A+D}{2} < 0$, the focus is stable, and if $\frac{A+D}{2} > 0$, the focus is unstable. The special case where $\frac{A+D}{2} = 0$ is called a **center**.

- **Node.** $(A - D)^2 + 4BC > 0$ and $|A + D| > \left|\sqrt{(A-D)^2 + 4BC}\right|$. The eigenvalues are both real and of the same sign. If $\frac{A+D}{2} < 0$, the node is stable, and if $\frac{A+D}{2} > 0$, the node is unstable.

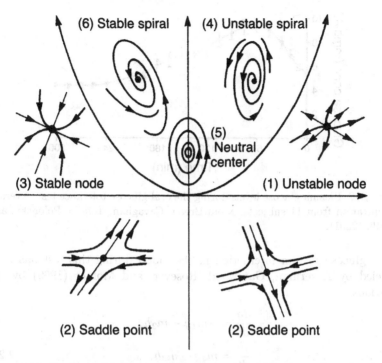

Figure 2.11. Qualitative behavior of two-dimensional linear systems of ordinary differential equations. The vertical axis is the determinant, while the horizontal axis is the trace of the matrix in equation (2.15). From Edelstein-Keshet (1988).

- **Saddle point.** $(A-D)^2+4BC > 0$ and $|A + D| < \left|\sqrt{(A - D)^2 + 4BC}\right|$.

  In this case, the eigenvalues are both real, but with different signs.

### Glucose tolerance test

We illustrate the dynamics in two-dimensional ordinary differential equations by considering the glucose tolerance test. Diabetes mellitus is a disease of metabolism that is characterized by too much sugar (glucose) in the blood and urine. A diabetic's body is unable to burn off its carbohydrates because of an insufficient, or ineffective, supply of insulin. Diabetes is quite complicated, since it involves a number of interacting hormones and systems of the organism.

A common diagnostic technique to determine whether an individual has a problem controlling the blood sugar level is the oral glucose tolerance test (American Diabetes Association 1995). This test consists in having a patient who has fasted for 10 hours ingest 75 g of glucose. Blood samples are collected before the oral administration, and then at 30-minute intervals for two hours (Figure 2.12). The plasma glucose value taken before the test, the fasting value, is considered to be the basal value for the patient.

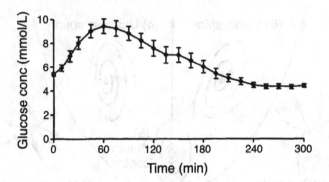

Figure 2.12. Plasma glucose levels during the oral glucose tolerance test. Average concentration from 11 subjects. From Breda, Cavaghan, Toffolo, Polonsky, and Cobelli. (2001).

The glucose–insulin interaction in the blood during this test has been modeled by Ackerman, Gatewood, Rosevar, and Molnar (1969) by the equations

$$\frac{dg}{dt} = -m_1 g - m_2 h,$$
$$\frac{dh}{dt} = m_4 g - m_3 h, \tag{2.21}$$

where $g$ is the displacement of the plasma glucose from its basal value, $h$ is the displacement of insulin from its basal value, and $m_1$, $m_2$, $m_3$, and $m_4$ are positive constants. The glucose displacement $g(t)$ satisfies, under this assumption, a second-order linear equation, whose characteristic equation is given by

$$\lambda^2 + (m_1 + m_3)\lambda + (m_1 m_3 + m_2 m_4) = 0, \tag{2.22}$$

with roots

$$\lambda_{1,2} = \frac{-(m_1 + m_3) \pm \sqrt{(m_3 - m_1)^2 + 4m_2 m_4}}{2}. \tag{2.23}$$

Since $(m_3 - m_1)^2 + 4m_2 m_4 > 0$, both characteristic roots are real. Furthermore, the null solution is stable, since $m_1 + m_3 > 0$. This is consistent with a return to basal value after the initial ingestion. We display in Figure 2.13 the solution $g(t) = c_1 e^{\lambda_1 t} + c_2 e^{\lambda_2 t}$ for one particular combination of the constants $c_1$ and $c_2$.

### 2.3.3  Three-Dimensional Ordinary Differential Equations

The techniques of the previous sections can be generalized to systems of three or more dimensions: an $n$-dimensional system of first-order ordinary differential equations can be derived from a Taylor series expansion in a

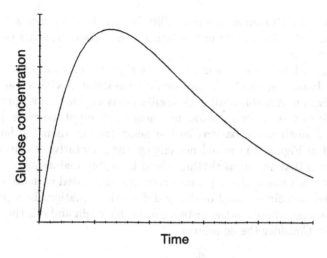

Figure 2.13. A solution $g(t)$ to equation (2.21). The initial conditions are $g(0) = 0$ and $g'(0) > 0$.

neighborhood of a fixed point, in which case the associated linear system will have a Jacobian matrix as matrix of coefficients. The classification of the type and stability of fixed points is not as clean and elegant as in the case of two-dimensional systems, and, of course, phase spaces of high dimension are difficult to represent graphically! The matrix approach is necessary in these cases to carry out a systematic investigation.

For the systems of dimension three,

$$\frac{dx}{dt} = f(x, y, z),$$
$$\frac{dy}{dt} = g(x, y, z), \qquad (2.24)$$
$$\frac{dy}{dt} = h(x, y, z) ,$$

there will be three eigenvalues to characterize the stability of a fixed point, corresponding to three roots of a cubic polynomial, which is given as the determinant of a matrix. Depending on these roots, and whether they are complex or not, we obtain phase space representations for linear systems close to the steady state. The latter is stable when the real parts of all three roots $\lambda_1$, $\lambda_2$, and $\lambda_3$ are negative.

## 2.4   Limit Cycles and the Hopf Bifurcation

Although it is extremely useful to understand the stability of fixed points and the geometric structure of the flows in a neighborhood of fixed points, this still leaves open the question of the nature of the dynamics away from the fixed points. Of course, one possibility is that the dynamics will always approach a stable fixed point in the limit $t \to \infty$. However, other behaviors are also possible.

In Figure 2.1 we showed a normal sinus rhythm. From a perspective of nonlinear dynamics, such a rhythm would be associated with a stable limit cycle oscillation. A stable limit cycle oscillation is a cycle that is approached in the limit at $t \to \infty$ for all points in a neighborhood of the cycle. Luckily, the normal small perturbations to the heart (rather than the large one illustrated in Figure 2.1) would not change the qualitative features of the dynamics, so that the sinus rhythm would be stably maintained.

To illustrate some of the important concepts associated with limit cycles, we consider two-dimensional ordinary differential equations in a polar coordinate system where $r$ is the distance from the origin and $\phi$ is the angular coordinate. Consider the equations

$$\frac{dr}{dt} = f(r, \mu),$$
$$\frac{d\phi}{dt} = 2\pi, \tag{2.25}$$

where $f(r, \mu)$ is a nonlinear function that depends on a parameter $\mu$. Since $d\phi/dt$ is constant, any fixed point of the function $f(r, \mu)$ would correspond to a limit cycle. It follows from the properties of one-dimensional differential equations that in equation (2.25) a stable fixed point of $f(r, \mu)$ corresponds to a stable limit cycle, and an unstable fixed point of $f(r, \mu)$ corresponds to an unstable limit cycle. Consequently, changes in the number or stability of the fixed points of $f(r, \mu)$ would lead to changes in the number and stability of limit cycle oscillations in equation (2.25).

We illustrate these concepts with a prototypical example. First assume that

$$f(r, \mu) = r(\mu - r^2). \tag{2.26}$$

If $\mu < 0$, there is one stable fixed point of $f(r, \mu)$ at $r = 0$, whereas for $\mu > 0$, there is an unstable fixed point of $f(r, \mu)$ at $r = 0$, and a stable fixed point of $f(r, \mu)$ at $r = \sqrt{\mu}$ (remember that in a polar coordinate system the radial coordinate $r$ corresponds to the distance of a point from the origin and is always taken to be positive). The phase plane portrait of the dynamics is shown in Figure 2.14(a).

If $\mu$ increases from negative to positive, a stable limit cycle will be born as $\mu$ crosses 0. The amplitude of the limit cycle is shown in Figure 2.14(b). This is called a supercritical Hopf bifurcation, and it reflects one of the classic ways in which limit cycles can gain their stability.

A different scenario for generating a limit cycle, albeit an unstable limit cycle, occurs when

$$f(r, \mu) = r(c + r^2). \tag{2.27}$$

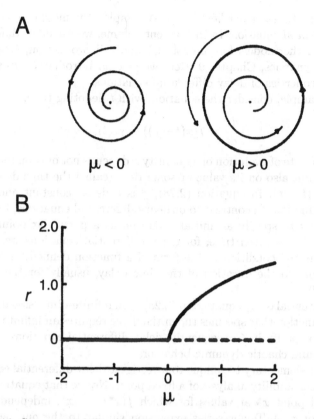

Figure 2.14. Supercritical Hopf bifurcation. (a) Phase plane representation of the dynamics in equation (2.25) with $f(r)$ defined in equation (2.26). (b) The amplitude of the limit cycles, where the solid curve corresponds to the stable solutions and the dashed curve refers to the unstable solutions. From Glass and Mackey (1988).

Application of the methods of this chapter shows that for $\mu > 0$ there is a single stable fixed point at $r = 0$, and for $\mu < 0$ there is a single unstable fixed point at $r = 0$ and an unstable limit cycle with an amplitude of $r = \sqrt{-\mu}$. This scenario is called the subcritical Hopf bifurcation. This bifurcation is often associated with the appearance of a large-amplitude stable limit cycle oscillation (see Section 3.5.3).

## 2.5   Time-Delay Differential Equations

Another mathematical formulation is sometimes useful to describe physiological systems in continuous time, especially when the regulated process involves a number of well-defined stages, and the modeled variable is the end of these stages. The system thus depends on its history, and such systems are sometimes called *hereditary*. We employ the more common term *delay differential equation*. In the current volume, we use delay differential equations in the models for control of blood cell production, Chapter 8, and pupil dynamics, Chapter 9. Here we give an introduction to some of the basic properties of delay differential equations.

As an example, consider the equation, with $t$ denoting time,

$$\frac{dx}{dt} = f(x(t - \tau)) - \gamma x(t), \qquad (2.28)$$

in which the rate of evolution of a quantity $x$ depends not only on its current value $x(t)$, but also on its value at some time (called the time delay) $\tau$ in the past, $x(t - \tau)$. In equation (2.28), $\gamma$ is a decay constant and $f$ is a nonlinear function. In contrast to ordinary differential equations, for which we need only to specify an initial condition as a particular point in the phase space for a given time, for delay differential equations we have to specify an initial condition in the form of a function defined for a period of time equal to the duration of the time delay, usually for the interval $-\tau \leq t' \leq 0$.

The differential delay equation in (2.28) is an infinite-dimensional system, since the function that specifies the initial value requires an infinite number of points to specify it. Consequently, delay differential equations can have oscillatory and chaotic dynamic behaviors.

The most elementary technique to investigate delay differential equations involves local stability analysis of a fixed point. Notice that equation (2.28) has a fixed point $x^*$ at values for which $f(x^*) = \gamma x^*$, independently of the value of $\tau$. A Taylor series expansion similar to the one performed in the previous sections for finite-difference and differential equations can also be computed close to this stationary solution (see Exercise 9.7-1). One technical difference worth mentioning is that since the system is infinite-dimensional, there will be an infinite number of eigenvalues in the linearized equation. Conditions can also be given for the emergence of a periodic solution when a stationary solution becomes unstable, so that a Hopf bifurcation can be established in the context of delay differential equations. Also in some situations, the right-hand side of the equation can be given as a piecewise linear function, and this can allow determination of an analytic solution (Glass and Mackey 1979).

In order to illustrate the properties of time-delay equations, we consider two different forms for the nonlinear control function (Mackey and Glass 1977). Assume that the nonlinear function is a monotonically decreasing function,

$$\frac{dx}{dt} = \frac{\beta \theta^m}{x(t - \tau)^m + \theta^m} - \gamma x, \qquad (2.29)$$

where $m$ and $\theta$ are constants. Then the dynamics will either approach a fixed point, or show a stable limit cycle oscillation (Mackey and Glass

1977). For example, in Figure 2.15a, we show a stable limit cycle for the parameters $\beta = 2$, $\theta = 1$, $\gamma = 1$, $m = 10$, $\tau = 2$. In contrast, if the nonlinear function is a single-humped function,

$$\frac{dx}{dt} = \frac{\beta \theta^m x(t - \tau)}{x(t - \tau)^m + \theta^m} - \gamma x, \qquad (2.30)$$

then for the same parameters, the dynamics are chaotic; Figure 2.16a (Mackey and Glass 1977).

In the initial analyses of this delay differential equation carried out in 1977, Michael Mackey and one of us (Leon Glass) plotted out the time series using a time-delayed embedding by plotting $x(t - \tau)$ as a function of $x(t)$ for different values of the parameters. Figure 2.15b shows a time-delay embedding for the limit cycle oscillation, and 2.16b for the chaotic time series. When looking at these time-delay embedding plots, the presence of loops in the time-delayed representation of a limit cycle with one set of parameter values often heralded the appearance of chaotic dynamics for some not too different set of parameter values.

There is an interesting sidelight to this use of time-delayed embeddings in nonlinear dynamics. In a New York Academy of Sciences meeting in November 1977, one of us (Leon Glass) presented a sequence of time-delayed embeddings for equation (2.30), showing a period doubling bifurcation sequence. At the same meeting, David Ruelle made some interesting suggestions to Glass concerning the possibility of examining the Poincaré map (see next section) along a line transverse to the trajectories. The proceedings of the meeting eventually appeared in 1979, showing the same sequence of time-delayed embeddings that had been presented at the meeting. In 1980, Norman Packard and colleagues used an embedding in which the derivative of a variable is plotted as a function of the variable (Packard, Crutchfield, Farmer, and Shaw 1980). They also mention the possibility of using time-delayed embeddings to extract geometry from a time series. In 1982, Doyne Farmer developed in great detail the use of time-delay embeddings, using as an example the current delay equation (Farmer 1982). In this work, Farmer called this equation the Mackey–Glass equation, and introduced its interesting properties to a broad audience.

## 2.6   The Poincaré Map

So far we have discussed the dynamics in difference equations and differential equations as distinct topics. However, the analysis of the dynamics in differential equations can often be facilitated by analyzing a difference equation that captures the qualitative features of the differential equation.

One way to do this would be to consider the dynamics in a neighborhood of a (stable or unstable) cycle in a differential equation. To make visualization easier, we assume that there is a cycle in a three-dimensional

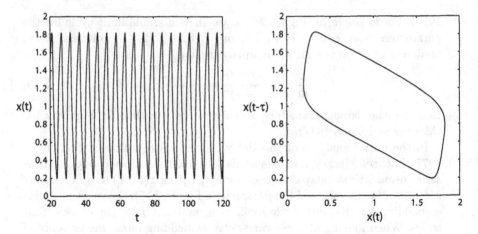

Figure 2.15. Limit cycle oscillation in a time delayed negative feedback system, equation (2.29) with $\beta = 2$, $\theta = 1$, $\gamma = 1$, $m = 10$, and $\tau = 2$. Time-delayed embedding of the time series in the left-hand panel. Thanks to S. Bernard for generating this figure.

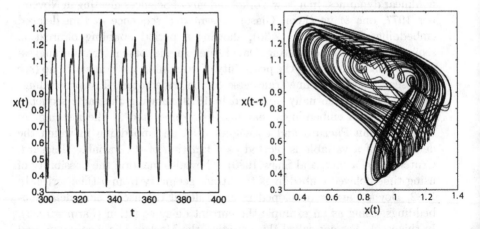

Figure 2.16. Chaotic dynamics with single-humped feedback equation (2.30) with $\beta = 2$, $\theta = 1$, $\gamma = 1$, $m = 10$, and $\tau = 2$. Time-delayed embedding of the time series in the left-hand panel. Thanks to S. Bernard for generating this figure.

differential equation. Imagine a two-dimensional plane that cuts across the cycle. Consider the point where the cycle cuts through the plane. After integrating the differential equation for a length of time that is equal to the period of the cycle, the point will once again return to the intersection of the cycle and the plane. Consider, however, other points in the plane in a neighborhood of the cycle. Assume that when the equations of motion are integrated, each of these points returns to the plane and that all trajectories transversely cross the plane.

Take the intersection of the cycle and the plane as the origin and assume that the coordinates of a point in a neighborhood of the cycle are $(x_0, y_0)$. Then when the equation is integrated, the point $(x_0, y_0)$ will return to the cross-sectional plane at coordinates $(x_1, y_1)$; Figure 2.17. More generally, we have the two-dimensional difference equation

$$x_{n+1} = f(x_n, y_n),$$
$$y_{n+1} = g(x_n, y_n), \tag{2.31}$$

where $f(x_n, y_n)$ and $g(x_n, y_n)$ are nonlinear functions. In general, a function that maps a hyperplane that is transverse to a flow back to itself under the action of the flow is called a Poincaré map. From the construction, it is clear that fixed points of the Poincaré map correspond to limit cycles in the associated differential equation, and consequently, an analysis of the stability of the fixed points of the Poincaré map might be an important step in the analysis of the dynamics of a differential equation.

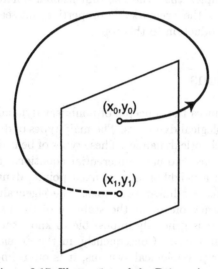

Figure 2.17. Illustration of the Poincaré map.

The stability of fixed points in equation (2.31) follows along the same procedures that have been carried out earlier in this chapter, and so we will not repeat them in detail here. Briefly, to carry out the stability analysis, there are three steps: (i) Solve for the fixed points; (ii) in a neighborhood of each fixed point, determine the linearized difference equation; and (iii) determine the eigenvalues of the linearized difference equation. In the case of the Poincaré map, the eigenvalues are also sometimes called by the unlikely term Floquet multipliers. Since this is a map, the fixed point will be stable if all the eigenvalues are within the unit circle, and it will be unstable otherwise. If both eigenvalues are real and one is inside the unit circle and one is outside the unit circle, then the fixed point is called a saddle point. If both eigenvalues are real and both are either inside or outside the unit circle, then the fixed point is called a node. On the other hand, if the eigenvalues are complex, then the fixed point is called a focus if eigenvalues lie outside the unit circle, and a center if the eigenvalues lie on the unit circle.

One point of caution. The terms node, focus, saddle point, and center are also used to describe the geometry of fixed points in differential equations. Indeed, an integration of equation (2.10) using the Euler method, starting from an initial condition $(x_0, y_0)$ for a time interval $\Delta t$, gives $(x_1, y_1) = (x_0 + f(x, y)\Delta t, y_0 + g(x, y)\Delta t)$. Thus, a nonlinear two-dimensional map can emerge naturally from a two-dimensional nonlinear differential equation. The computer exercise in Section 2.9 should help you to understand the geometric features of fixed points in maps and differential equations in two dimensions.

Just as one-dimensional difference equations are more complex than one-dimensional differential equations, the two-dimensional difference equations are much more complex than the two-dimensional differential equations, and we will not cover their remarkable properties, but see Devaney (1989) for an excellent introduction to this topic.

## 2.7   Conclusions

This chapter introduces the concepts of nonlinear dynamics and gives applications to physiological dynamics. The main types of dynamics are fixed points, cycles, and chaotic dynamics. These types of behaviors can be found in difference equations, ordinary differential equations, and delay differential equations. In a neighborhood of fixed points, dynamical equations can be approximated by linear equations. The eigenvalues of the linear equations give information about the stability of the fixed points in the original equations. It is generally impossible to know the exact equations of any physiological system. Consequently, in the development of mathematical models for physiological systems, it is often important that the main qualitative features of the physiological system be preserved in the mathematical model. The methods of nonlinear dynamics deal with these matters, so there is a close link between the physiological approach and the mathematical approach.

## 2.8   Computer Exercises: Iterating Finite-Difference Equations

These exercises introduce you to the iteration of finite-difference equations, or "maps." The particular map studied will be the quadratic map:

$$x_{n+1} = \mu(1 - x_n)x_n. \tag{2.32}$$

## Software

There are 4 Matlab* programs:

**fditer(function, xzero, mu, niter)** This program iterates the map specified in **function**. There are three additional arguments: **xzero** is the initial condition; **mu** is the bifurcation parameter $\mu$; and **niter** is the number of iterations. The output is a vector **y** of length **niter** containing the iterated values. For example, to iterate 100 times the quadratic map, with $\mu = 4$ starting at $x_0 = 0.1$:

```
y = fditer( 'quadmap',0.1,4,100);
```

**testper(y,epsilon,maxper)** This program determines whether there is a periodic orbit in the sequence given by the vector **y** whose period is less than or equal to **maxper**. The convergence criterion is that two iterates of **y** are closer than **epsilon**. The output is the period **per**. If no convergence is found the output is **-1**.

**bifurc(function, mubegin,muend)** This program plots the bifurcation diagram for the specified map, using 100 steps of the bifurcation parameter $\mu$ between **mubegin** and **muend**.

**cobweb(function, xzero,mu,niter)** This program iterates the specified finite-difference equation and displays the cobweb diagram. There are four arguments: **function** is the name of the map to use (enclosed in single quotes); **xzero** is the initial condition; **mu** is the parameter; **niter** is the number of iterations for which you will display the cobweb.

Although for the present we will be using only the quadratic map (implemented in the Matlab function **'quadmap'**, other maps are available (e.g., **'sinemap'**, **'tentmap'**), and additional ones can easily be written by you following the template of **'quadmap'**.

## How to Run the Programs

The following steps give an illustrative example of how to run these programs.

- To generate 100 iterations of the quadratic map with an initial condition of $x_1 = 0.5$, $\mu = 3.973$, type

  ```
  y=fditer('quadmap',0.5,3.973,100);
  ```

- To plot the time series from this iteration, type

  ```
  plot(y,'+');
  ```

---

*See Introduction to Matlab in Appendix B.

- To determine whether there is a period of length less than or equal to 20 with a convergence of 0.00001 of the time series **y**, type

  ```
  per=testper(y,.00001,20);
  ```

  In this case there is no period, and the program returns **per**= $-1$. If a value $\mu = 3.2$ had been used to generate the time series in the quadratic program, the program **testperiod** returns a value of **per**= 2.

- To plot a cobweb diagram for the quadratic map with an initial condition of **xzero=0.3** and $\mu = \mathbf{3.825}$ with 12 steps, type

  ```
  cobweb('quadmap',0.3,3.825,12);
  ```

## Exercises

**Ex. 2.8-1. Feigenbaum's number.** This exercise asks you to determine Feigenbaum's number.

Feigenbaum's number is defined as follows. Call $\Delta_n$ the range of $\mu$ values that give a period-$n$ orbit. Then Feigenbaum found that in a sequence of period-doubling bifurcations,

$$\lim_{n \to \infty} \frac{\Delta_n}{\Delta_{2n}} = 4.6692\ldots.$$

The constant 4.6692... is now called **Feigenbaum's number**.

According to Feigenbaum, he initially discovered this number by carrying out numerical iterations on a hand calculator. As the period of the cycle gets longer, the range of parameter values over which a given period is found gets smaller. Therefore, it is necessary to think carefully about what is involved in the calculation.

Try to numerically compute

$$\frac{\Delta_8}{\Delta_{16}}.$$

You will want to vary $\mu$ over a range of values and to determine the value of $\mu$ where the period changes. It may be useful to make use of some of the programs above. In order to do this exercise, you need to be aware that the actual parameter value numerically determined for the bifurcation points will depend on the number of iterations you have selected and the convergence criterion. Therefore, the numerically determined value of Feigenbaum's number will depend on the parameters in your test programs.

There are other period-doubling sequences for this map. For example, when $\mu = 3.83$ there is a period-3 orbit. Compute the ratio

$$\frac{\Delta_3}{\Delta_6}, \frac{\Delta_6}{\Delta_{12}}, \ldots$$

as $\mu$ is increased from 3.83. Does Feigenbaum's number also appear in this period-doubling sequence?

**Ex. 2.8-2. Universal sequence of the quadratic map.** The purpose of this exercise is to compute the sequence of periodic orbits encountered as $\mu$ is increased.

Try to find all ranges of $\mu$ that give periodic orbits up to period 6. As $\mu$ is increased you should be able to find windows that give periodic orbits in the sequence $1, 2, 4, 6, 5, 3, 6, 5, 6, 4, 6, 5, 6$. **Note:** The window giving a particular periodic orbit is not necessarily adjacent to the window giving the next periodic orbit in the series. You will need to increment $\mu$ in very small steps to find the different periodic orbits.

The sequence of periodic orbits is called the **universal** sequence. It is the same for all maps with a quadratic maximum and a single hump.

**Ex. 2.8-3. Single-humped sine map.** The behaviors found for the quadratic map here are also found in other simple maps, complicated equations, and a variety of experimental systems. It is this **universal** behavior that has attracted the attention of physicists and others.

You can use the programs to carry out similar computations for the single-humped sine map

$$x_{n+1} = \mu \sin(\pi x_n), \tag{2.33}$$

where $0 < \mu < 1$, which is implemented in Matlab as sinemap.

## 2.9 Computer Exercises: Geometry of Fixed Points in Two-Dimensional Maps

This set of exercises enables you to generate correlated random dot patterns. These exercises are based on observations made in Glass (1969) and Glass and Perez (1973). These papers deal with the patterns that are generated by superimposing a random set of dots on itself following a linear transformation on the original set of dots. This is equivalent to selecting a set of random points and then transforming them by a two-dimensional map.

The programs show a random pattern of 400 dots superimposed on itself following a rescaling **a** in the $x$-coordinate, **b** in the $y$-coordinate, and a rotation by the angle $\theta$. There is a fixed point at $x = y = 0$.

This transformation is given by the equations

$$x_{n+1} = ax_n \cos\theta - by_n \sin\theta, \tag{2.34}$$

$$y_{n+1} = ax_n \sin\theta + by_n \cos\theta. \tag{2.35}$$

The eigenvalues of this map are given by

$$\lambda_\pm = \frac{(a+b)\cos\theta \pm \sqrt{(a-b)^2 - (a+b)^2 \sin\theta}}{2}, \tag{2.36}$$

The eigenvalues of this map reflect the geometry of the map at the fixed point at $x = y = 0$. If the eigenvalues are complex numbers, the fixed point is a focus; if the eigenvalues are real and are both inside or outside the unit circle, the fixed point is a node; if the eigenvalues are real and one is inside the unit circle and the other is outside the unit circle, the fixed point is a saddle. If the eigenvalues are pure imaginary, the fixed point is a center.

## Software

There is one Matlab[†] program:

**dots(a,b,thetam,numb).** This program generates 400 random dots and **numb** iterates of each of these dots using the transformation above. (Use 4 iterates for better visualization, but a single iterate can also suffice. To see just the random dots, with no iterates, use 0 for **numb**.) The program **dots** plots the random dots and their iterates, and calculates the eigenvalues of the transformation, which are printed underneath the figure.

## How to Run the Program

To display a plot with $a = 0.95$, $b = 1.05$, and $\theta = 0.4/\pi$, type

```
dots(0.95,1.05,0.4/pi,4);
```

## Exercises

Ex. 2.9-1. **Parameter values to give centers, focuses, nodes, and saddles.** You may wish to see what happens for particular values of the parameters. Try to find parameters that give centers, focuses, nodes, and saddles.

---

[†]See Introduction to Matlab in Appendix B.

Ex. 2.9-2. **Visual perception.** Increase the angle of rotation until you can no longer perceive the geometry of the transformation but simply see a pattern of random dots. For example, if $a = 1$ and $b = 1$, then you will observe a circular geometry characteristic of a center for a limited range of rotation angles. Try to find the maximum rotation angle that allows you to visually perceive the circular geometry. Can you predict theoretically the critical parameters that destroy your ability to perceive the geometry of the transformation rather than a pattern of random dots? If so, this might be a good result in the field of visual perception.

Ex. 2.9-3. **Bifurcation from a saddle to a focus.** Here is a problem. In general, it should be impossible to find a bifurcation from a saddle to a focus except in exceptional cases. Consider the bifurcations observed with $a = 0.95$, $b = 1.05$ as $\theta$ varies.

Ex. 2.9-4. **Direct bifurcation from a saddle to a focus?** Is there a direct bifurcation from a saddle to a focus? Try to determine this (a) by looking at the pictures and (b) analytically. Which is simpler and which is more informative, (a) or (b)?

# 3

# Bifurcations Involving Fixed Points and Limit Cycles in Biological Systems

## Michael R. Guevara

## 3.1 Introduction

Biological systems display many sorts of dynamic behaviors including constant behavior, simple or complex oscillations, and irregular fluctuating dynamics. As parameters in systems change, the dynamics may also also change. For example, changing the ionic composition of a medium bathing nerve cells or cardiac cells can have marked effects on the behaviors of these cells and may lead to the stabilization or destabilization of fixed points or the initiation or termination of rhythmic behaviors. This chapter concerns the ways that constant behavior and oscillating behavior can be stabilized or destabilized in differential equations. We give a summary of the mathematical analysis of bifurcations and biological examples that illustrate the mathematical concepts. While the context in which these bifurcations will be illustrated is that of low-dimensional systems of ordinary differential equations, these bifurcations can also occur in more complicated systems, such as partial differential equations and time-delay differential equations.

We first describe how fixed points can be created and destroyed as a parameter in a system of differential equations is gradually changed, producing a bifurcation. We shall focus on three different bifurcations: the saddle-node bifurcation, the pitchfork bifurcation, and the transcritical bifurcation (Abraham and Shaw 1982; Thompson and Stewart 1986; Wiggins 1990; Strogatz 1994).

We then consider how oscillations are born and how they die or metamorphose. There are several bifurcations in which limit cycles are created or destroyed. These include the Hopf bifurcation (see Chapter 2), the saddle-node bifurcation, the period-doubling bifurcation, the torus bifurcation, and the homoclinic bifurcation (Abraham and Shaw 1982; Thompson and Stewart 1986; Wiggins 1990; Strogatz 1994).

# 3.2    Saddle-Node Bifurcation of Fixed Points

## 3.2.1    Bistability in a Neural System

We now consider the case of "two stable resting potentials" as an example of a biological situation in which the number of fixed points in the system is changed as a parameter is varied. Normally, the voltage difference across the membrane of a nerve cell (the transmembrane potential) has a value at rest (i.e., when there is no input to the cell) of about $-60$ mV. Injecting a brief depolarizing current pulse produces an action potential: There is an excursion of the transmembrane potential, with the transmembrane potential asymptotically returning to the resting potential. This shows that there is a stable fixed point present in the system. However, it is possible under some experimental conditions to obtain two stable resting potentials. Figure 3.1 shows the effect of injection of a brief-duration stimulus pulse in an experiment in which a nerve axon is bathed in a potassium-rich medium: The transmembrane potential does not return to its resting value in response to delivery of a depolarizing stimulus pulse (the second stimulus pulse delivered in Figure 3.1); rather, it "hangs up" at a new depolarized potential, and rests there in a stable fashion (Tasaki 1959). This implies the existence of a second stable fixed point in the phase space of the system. Injection of a brief-duration current pulse of the opposite polarity (a hyperpolarizing stimulus pulse) can then return the membrane back to its initial resting potential (Tasaki 1959). In either case, the stimulus pulse must be large enough in amplitude for the flip to the other stable fixed point to occur; e.g., in Figure 3.1, the first stimulus pulse delivered was too small in amplitude to result in a flip to the other stable resting potential.

The phase space of the system of Figure 3.1 must also contain some sort of divider that separates the basins of attraction of the two stable fixed points (the **basin of attraction** of a fixed point is the set of initial conditions that asymptotically lead to that point). Figure 3.2 shows how this can occur in the simple one-dimensional system of ordinary differential equations $dx/dt = x - x^3$. In addition to the two stable fixed points at $x = \pm 1$, there is an unstable fixed point present at the origin, which itself acts to separate the basins of attraction of the two stable fixed points: trajectories starting from initial conditions to the left of the unstable fixed point at the origin (i.e., $x(t = 0) < 0$) go to the stable fixed point at $x = -1$, while trajectories starting from initial conditions to the right of that point (i.e., $x(t = 0) > 0$) go to the stable fixed point at $x = +1$.

The simplest way that the coexistence of two stable fixed points can occur in a two-dimensional system is shown in Figure 3.3, in which there is a saddle point in addition to the two stable nodes. Remember that the stable manifold of the saddle fixed point (the set of initial conditions that lead to it) is composed of a pair of **separatrices** (dashed lines in Figure 3.3), which divide the plane into two halves, forming the basins of attraction

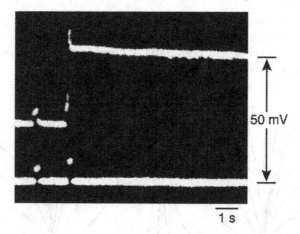

Figure 3.1. The phenomenon of two stable resting potentials in the membrane of a myelinated toad axon bathed in a potassium-rich medium. A steady hyperpolarizing bias current is injected throughout the experiment. Stimulation with a brief depolarizing current pulse that is large enough in amplitude causes the axon to go to a new level of resting potential. The top trace is the transmembrane potential; the bottom trace is the stimulus current. Adapted from Tasaki (1959).

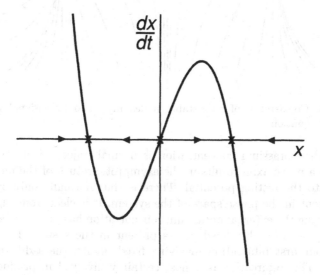

Figure 3.2. Coexistence of two stable fixed points in the one-dimensional ordinary differential equation $dx/dt = x - x^3$.

of the two stable fixed points. In Figure 3.3 the thick lines give the pair of trajectories that form the unstable manifold of the saddle point. From this phase-plane picture, can one explain why the pulse amplitude must be sufficiently large in Figure 3.1 for the transition from one resting potential to the other to occur? Can one explain why a hyperpolarizing, and not depolarizing, stimulus pulse was used to flip the voltage back to the initial resting potential once it had been initially flipped to the more depolarized resting potential in Figure 3.1 by a depolarizing pulse?

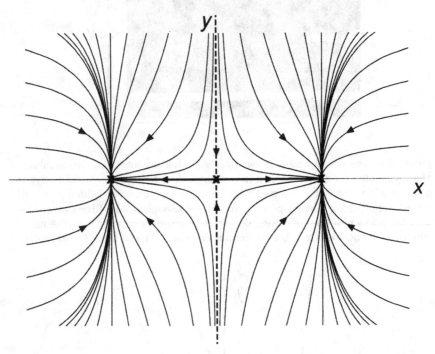

Figure 3.3. Coexistence of two stable nodes in a two-dimensional ordinary differential equation.

When the potassium concentration is normal, injection of a stimulus pulse into a nerve axon results in the asymptotic return of the membrane potential to the resting potential. There is thus normally only one fixed point present in the phase space of the system. It is clear from Figure 3.1 that elevating the external potassium concentration has produced a change in the number of stable fixed points present in the system. Let us now consider our first bifurcation involving fixed points: the saddle-node bifurcation. This bifurcation is almost certainly involved in producing the phenomenon of two stable resting potentials shown in Figure 3.1.

### 3.2.2   Saddle-Node Bifurcation of Fixed Points in a One-Dimensional System

In a one-dimensional system of ordinary differential equations, a **saddle-node bifurcation** results in the creation of two new fixed points, one stable, the other unstable. This can be seen in the simple equation

$$\frac{dx}{dt} = \mu - x^2, \tag{3.1}$$

where $x$ is the **bifurcation variable** and $\mu$ is the **bifurcation param-eter**. Figure 3.4 illustrates the situation. For $\mu < 0$, there are no fixed points present in the system ($\mu = -0.5$ in Figure 3.4A). For $\mu = 0$ (**the bifurcation value**) there is one fixed point (at the origin), which is semistable (Figure 3.4B). For $\mu > 0$ there are two fixed points, one of which ($x^* = \sqrt{\mu}$) is stable, while the other ($x^* = -\sqrt{\mu}$) is unstable ($\mu = 0.5$ in Figure 3.4C). For obvious reasons, this bifurcation is also often called a **tangent bifurcation**.

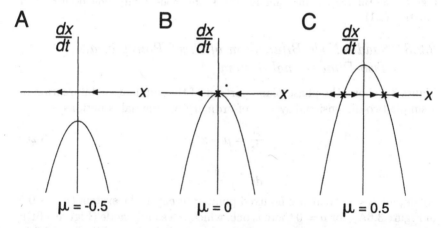

Figure 3.4. Saddle-node bifurcation in the one-dimensional ordinary differential equation of equation (3.1). (A) $\mu = -0.5$, (B) $\mu = 0$, (C) $\mu = 0.5$.

Figure 3.5 shows the corresponding bifurcation diagram, in which the equilibrium value ($x^*$) of the bifurcation variable $x$ is plotted as a function of the bifurcation parameter $\mu$. The convention used is that stable points are shown as solid lines, while unstable points are denoted by dashed lines. In such a diagram, the point on the curve at which the saddle-node bifurcation occurs is often referred to as a **knee**, **limit point**, or **turning point**. This bifurcation is also called a **fold bifurcation** and is associated with one of the elementary catastrophes, the **fold catastrophe** (Arnold 1986; Woodcock and Davis 1978).

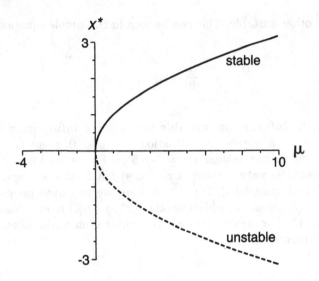

Figure 3.5. Bifurcation diagram for the saddle-node bifurcation occurring in equation (3.1).

### 3.2.3    Saddle-Node Bifurcation of Fixed Points in a Two-Dimensional System

Figure 3.6 shows the phase-plane portrait of the saddle-node bifurcation in a simple two-dimensional system of ordinary differential equations

$$\frac{dx}{dt} = \mu - x^2, \tag{3.2a}$$

$$\frac{dy}{dt} = -y. \tag{3.2b}$$

Again, for $\mu < 0$, there are no fixed points present in the system ($\mu = -0.5$ in Figure 3.6A); for $\mu = 0$ there is one, which is a saddle-node (Figure 3.6B); while for $\mu > 0$ there are two, which are a node and a saddle ($\mu = 0.5$ in Figure 3.6C), hence the name of the bifurcation. While in the particular example shown in Figure 3.6 the node is stable, the bifurcation can also be such that the node is unstable. This is in contrast to the one-dimensional case, where one cannot obtain two unstable fixed points as a result of this bifurcation. The bifurcation diagram for the two-dimensional case of Figure 3.6 is the same as that shown in Figure 3.5, which was for the one-dimensional case (Figure 3.4).

At the bifurcation point itself ($\mu = 0$), there is a special kind of fixed point, a **saddle-node**. This point has one eigenvalue at zero, the other necessarily being real (if negative, a stable node is the result of the bifurcation; if positive, an unstable node). In fact, this is the algebraic criterion for a saddle-node bifurcation: A single real eigenvalue passes through the origin in the root-locus diagram as a parameter is changed (Figures 3.4, 3.6). Note

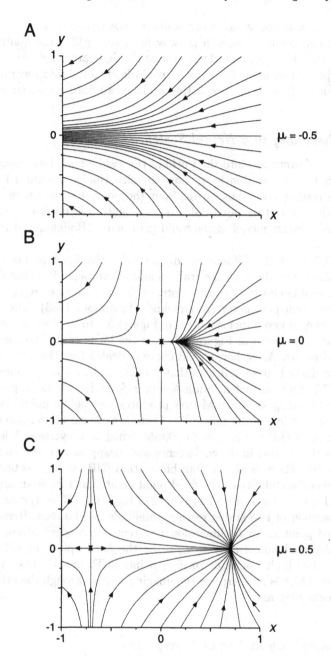

Figure 3.6. Phase-plane portrait of the saddle-node bifurcation in the two-dimensional ordinary differential equation of equation (3.2). (A) $\mu = -0.5$, (B) $\mu = 0$, (C) $\mu = 0.5$.

that the system is not structurally stable at the bifurcation value of the parameter; i.e., small changes in $\mu$ away from zero will cause qualitative changes in the phase-portrait of the system. In this particular case, such a change in parameter away from $\mu = 0$ would lead to the disappearance of the saddle-node ($\mu < 0$) or its splitting up into two fixed points ($\mu > 0$).

### 3.2.4  Bistability in a Neural System (Revisited)

After our brief excursion into the world of the saddle-node bifurcation, we now return to the question as to how the situation in Figure 3.1, with two stable resting potentials, arose from the normal situation in which there is only one resting potential. To investigate this further, we study the Hodgkin–Huxley model of the squid giant axon (Hodgkin and Huxley 1952).

Figure 3.7 gives the bifurcation diagram for the fixed points in the Hodgkin–Huxley model with the transmembrane voltage ($V$) being the bifurcation variable and the external potassium concentration ($K_{out}$) acting as the bifurcation parameter (Aihara and Matsumoto 1983). The model here is not one- or two-dimensional, as in Figures 3.2 to 3.6, but rather four-dimensional. The curve in Figure 3.7 gives the locus of the $V$-coordinate of the fixed points. As earlier, a solid curve indicates that the point is stable, while a dashed curve indicates that it is unstable. As $K_{out}$ is increased in Figure 3.7, there is first a saddle-saddle bifurcation at the upper limit point ($LP_u$) at $K_{out} \approx 51.8$ mM that produces two saddle points (which, by definition, are inherently unstable). There is a second bifurcation at the lower limit point ($LP_l$) at $K_{out} \approx 417.0$ mM, which is a reverse saddle-node bifurcation that results in the coalescence and disappearance of a node and a saddle point. There is also a Hopf bifurcation (HB) at $K_{out} \approx 66.0$ mM that converts the stability of the fixed point created at $LP_u$ from unstable to stable. There is thus quite a large range of $K_{out}$ (66–417 mM) over which the phenomenon of two stable resting potentials can be seen. Remember that a fixed point in an $N$-dimensional system has $N$ eigenvalues, which can be calculated numerically (e.g., using the Auto option in XPP*). At each of the two limit points or turning points in Figure 3.7, one of these eigenvalues, which is real, crosses the imaginary axis through the origin on the root-locus diagram.

### 3.2.5  Bistability in Visual Perception

Bistability, the coexistence in the phase space of the system of two asymptotically locally stable ("attracting") objects, is a phenomenon not limited

---

*See Appendix A for an introduction to XPP.

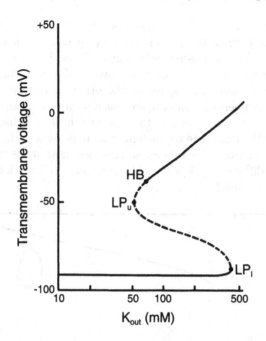

Figure 3.7. Bifurcation diagram for Hodgkin–Huxley model. The transmembrane voltage is the bifurcation variable, while the external potassium concentration ($K_{out}$) is the bifurcation parameter. As in the experimental work (Figure 3.1), a steady hyperpolarizing bias current (20 $\mu$A/cm$^2$ here) is injected throughout. Adapted from Aihara and Matsumoto (1983).

to fixed points. As we shall see later in this chapter, one can have bistability between a fixed point and a limit-cycle oscillator, which leads to the phenomena of single-pulse triggering and annihilation. In addition, one can have bistability between two stable periodic orbits (e.g., Abraham and Shaw 1982; Goldbeter and Martiel 1985; Guevara, Shrier, and Glass 1990; Yehia, Jeandupeux, Alonso, and Guevara 1999). Thus, many other phenomena in which two stable behaviors are seen in experimental work are almost certainly due to the coexistence of two stable attractors of some sort. The most appealing example of this is perhaps in the field of visual perception, where one can have bistable images. Figure 3.8 shows a nice example, in which the relative size of the basins of attraction for the perception of the two figures gradually changes as one moves from left to right.

Figure 3.8. Bistability in visual perception. Adapted from Chialvo and Apkarian (1993).

At the extreme ends of the sequence of images in Figure 3.8, only one figure is perceived, while in the middle the eye perceives one of two figures at a given time ("ambiguous figure"). Thus, Figure 3.9 shows a candidate bifurcation diagram, in which there are two saddle-node bifurcations, so that one perceives only one figure or the other at the two extremes of the diagram. The phenomenon of hysteresis also occurs: Scan the sequence of images in Figure 3.8 from left to right and note at which image the transition from the male face to the female form is perceived. Then repeat, reversing the direction of scanning, so that one now scans from right to left. Is there a difference? How does the schematic bifurcation diagram of Figure 3.9 explain this?

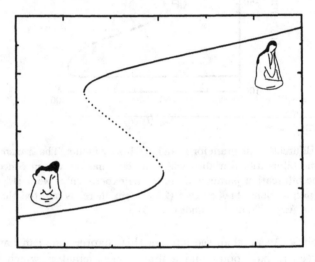

Figure 3.9. Candidate bifurcation diagram for hysteresis in visual perception.

## 3.3   Pitchfork Bifurcation of Fixed Points

### 3.3.1   Pitchfork Bifurcation of Fixed Points in a One-Dimensional System

In the **pitchfork bifurcation**, a fixed point reverses its stability, and two new fixed points are born. A pitchfork bifurcation occurs at $\mu = 0$ in the one-dimensional ordinary differential equation

$$\frac{dx}{dt} = x(\mu - x^2). \tag{3.3}$$

For $\mu < 0$, there is one fixed point at zero, which is stable ($\mu = -0.5$ in Figure 3.10A); at $\mu = 0$ there is still one fixed point at zero, which is still stable (Figure 3.10B); for $\mu > 0$, there are three fixed points, with the original fixed point at zero now being unstable, and the two new symmetrically placed points being stable ($\mu = 0.5$ in Figure 3.10C).

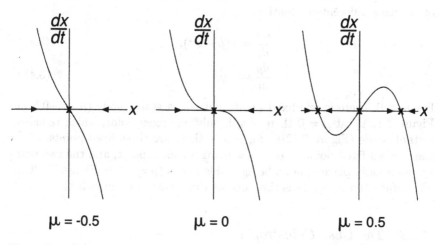

$\mu = -0.5$                $\mu = 0$                $\mu = 0.5$

Figure 3.10. Pitchfork bifurcation in the one-dimensional system of equation (3.3). (A) $\mu = -0.5$, (B) $\mu = 0$, (C) $\mu = 0.5$.

Figure 3.11 shows the bifurcation diagram for the pitchfork bifurcation. The bifurcation of Figures 3.10, 3.11 is **supercritical pitchfork bifurcation**, since there are stable fixed points to either side of the bifurcation point. Replacing the minus sign with a plus sign in equation (3.3) results in a **subcritical** pitchfork bifurcation.

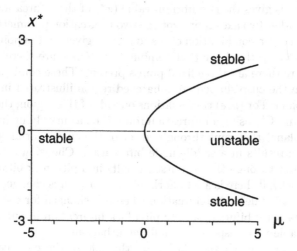

Figure 3.11. Bifurcation diagram for the pitchfork bifurcation.

### 3.3.2  Pitchfork Bifurcation of Fixed Points in a Two-Dimensional System

There is a **pitchfork bifurcation** at $\mu = 0$ for the two-dimensional system of ordinary differential equations

$$\frac{dx}{dt} = x(\mu - x^2), \tag{3.4a}$$

$$\frac{dy}{dt} = -y. \tag{3.4b}$$

For $\mu < 0$, there is one fixed point, which is a stable node ($\mu = -0.5$ in Figure 3.12A); at $\mu = 0$ there is still only one fixed point, which remains a stable node (Figure 3.12B); for $\mu > 0$, there are three fixed points, with the original fixed point at zero now being a saddle point, and the two new symmetrically placed points being stable nodes ($\mu = 0.5$ in Figure 3.12C). The bifurcation diagram is the same as that shown in Figure 3.11.

### 3.3.3  The Cusp Catastrophe

So far, we have generally considered one-parameter bifurcations, in which we have changed a single bifurcation parameter $\mu$. Let us now introduce a second bifurcation parameter $\epsilon$ into the one-dimensional equation producing the pitchfork bifurcation, equation (3.3), studying instead

$$\frac{dx}{dt} = x(\mu - x^2) + \epsilon. \tag{3.5}$$

Figure 3.13 gives the resultant two-parameter bifurcation diagram, in which the vertical axis gives the equilibrium value ($x^*$) of the bifurcation variable $x$, while the other two axes represent the two bifurcation parameters ($\mu$ and $\epsilon$). Thus, at a given combination of $\mu$ and $\epsilon$, $x^*$ is given by the point(s) lying in the surface directly above that combination. There are therefore values of ($\mu, \epsilon$) where there are three fixed points present. These combinations are found within the cusp-shaped cross-hatched region illustrated in the ($\mu, \epsilon$) parameter plane. For ($\mu, \epsilon$) combinations outside of this region, there is only one fixed point. Choosing a bifurcation route (i.e., a curve lying in the ($\mu, \epsilon$) parameter plane) that runs through either of the curves forming the cusp in that plane results in a saddle-node bifurcation. Choosing a bifurcation route that runs through the cusp itself results in a pitchfork bifurcation. It is now clear why, if there is a pitchfork bifurcation as $\mu$ is changed for $\epsilon = 0$, there will be a saddle-node bifurcation when $\mu$ is changed for $\epsilon \neq 0$. Viewed as a one-parameter bifurcation, the pitchfork bifurcation in one dimension is thus unstable with respect to small perturbations.

There has been much speculation on the role of the cusp catastrophe in phenomena encountered in many areas of life, including psychiatry, economics, sociology, and politics (see, e.g., Woodcock and Davis 1978).

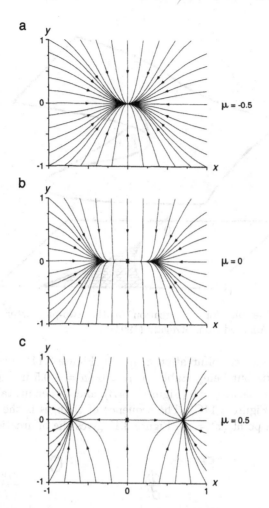

Figure 3.12. Phase-plane portrait of the pitchfork bifurcation in the two-dimensional ordinary differential equation of equation (3.4). (A) $\mu = -0.5$, (B) $\mu = 0$, (C) $\mu = 0.5$.

## 3.4    Transcritical Bifurcation of Fixed Points

### 3.4.1    Transcritical Bifurcation of Fixed Points in a One-Dimensional System

In the **transcritical bifurcation** there is an exchange of stability between two fixed points. In the one-dimensional ordinary differential equation

$$\frac{dx}{dt} = x(\mu - x), \tag{3.6}$$

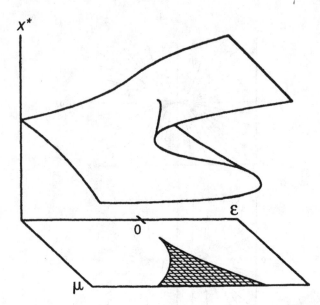

Figure 3.13. The pitchfork bifurcation in the two-parameter unfolding of equation (3.5). Adapted from Strogatz (1994).

there is a transcritical bifurcation at $\mu = 0$ (Figure 3.14). The fixed point at $x^* = 0$ starts out being stable for $\mu < 0$ ($\mu = -0.5$ in Figure 3.14A), becomes semistable at $\mu = 0$ (Figure 3.14B), and is then unstable for $\mu > 0$ ($\mu = -0.5$ in Figure 3.14C). The sequence of changes is the opposite for the other fixed point ($x^* = \mu$). Figure 3.15 gives the bifurcation diagram.

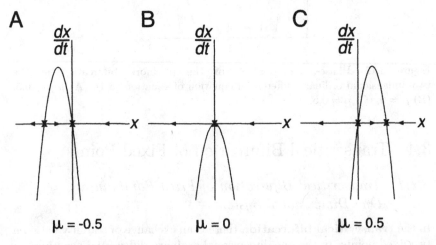

Figure 3.14. Transcritical bifurcation in the one-dimensional ordinary differential equation of equation (3.6). (A) $\mu = -0.5$, (B) $\mu = 0$, (C) $\mu = 0.5$.

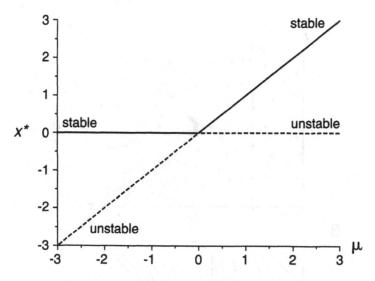

Figure 3.15. The bifurcation diagram for the transcritical bifurcation of equation (3.6).

As in the case of the pitchfork bifurcation, the transcritical bifurcation in a one-dimensional system (Figure 3.14) is not stable to small perturbations, in that should a term $\epsilon$ be added to the right-hand side of equation (3.6), the transcritical bifurcation ($\epsilon = 0$) is replaced by either no bifurcation at all ($\epsilon > 0$) or by a pair of saddle-node bifurcations ($\epsilon < 0$) (Wiggins 1990).

### 3.4.2    Transcritical Bifurcation of Fixed Points in a Two-Dimensional System

The two-dimensional ordinary differential equation

$$\frac{dx}{dt} = x(\mu - x),$$ (3.7a)

$$\frac{dy}{dt} = -y,$$ (3.7b)

also has a transcritical bifurcation at $\mu = 0$ (Figure 3.16). The fixed point at $x^* = 0$ starts out being a stable node for $\mu < 0$ ($\mu = -0.5$ in Figure 3.16A), becomes a semistable saddle-node at $\mu = 0$ (Figure 3.16B), and is then an unstable saddle point for $\mu > 0$ ($\mu = 0.5$ in Figure 3.16C). The sequence of changes is the opposite for the other fixed point ($x^* = \mu$). As with the saddle-node bifurcation (Figure 3.6), the fixed point present at the bifurcation point ($\mu = 0$) is a saddle-node.

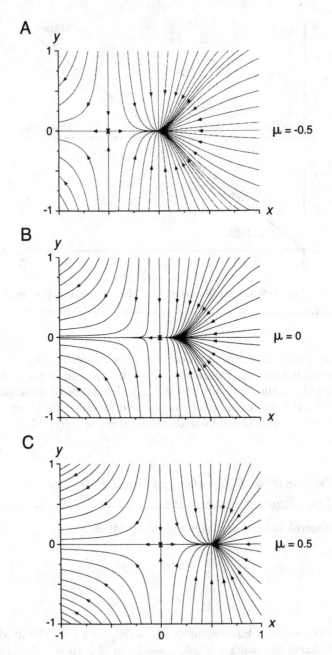

Figure 3.16. Phase-plane portrait of the transcritical bifurcation for the two-dimensional ordinary differential equation of equation (3.7). (A) $\mu = -0.5$, (B) $\mu = 0$, (C) $\mu = 0.5$.

## 3.5   Saddle-Node Bifurcation of Limit Cycles

### 3.5.1   Annihilation and Single-Pulse Triggering

The sinoatrial node is the pacemaker that normally sets the rate of the heart. Figure 3.17 shows the transmembrane potential recorded from a cell within the sinoatrial node. At the arrow, a subthreshold pulse of current is delivered to the node, and spontaneous activity ceases. This phenomenon is termed **annihilation**. Injection of a suprathreshold current pulse will then restart activity (**"single-pulse triggering"**). Both of these phenomena can be seen in an ionic model of the sinoatrial node: Figure 3.18A shows annihilation, while Figure 3.18B shows single-pulse triggering.

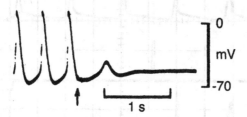

Figure 3.17. Annihilation in tissue taken from the sinoatrial node. From Jalife and Antzelevitch (1979).

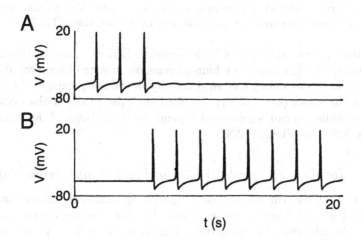

Figure 3.18. (A) Annihilation and (B) single-pulse triggering in an ionic model of the sinoatrial node. A constant hyperpolarizing bias current is injected to slow the beat rate. From Guevara and Jongsma (1992).

Annihilation has been described in several other biological oscillators, including the eclosion rhythm of fruit flies, the circadian rhythm of bioluminescence in marine algae, and biochemical oscillators (see Winfree 2000 for a synopsis). Figure 3.19 shows another example taken from electrophysiology. When a constant ("bias") current is injected into a squid axon, the axon will start to spontaneously generate action potentials (Figure 3.19A). Injection of a well-timed pulse of current of the correct amplitude annihilates this spontaneous activity (Figure 3.19B).

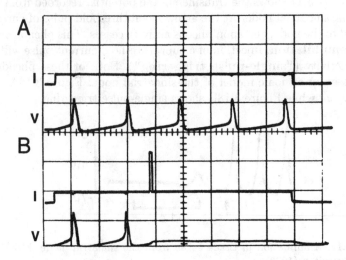

Figure 3.19. (A) Induction of periodic firing of action potentials in the giant axon of the squid by injection of a depolarizing bias current. (B) Annihilation of that bias-current-induced activity by a brief stimulus pulse. $V$ = transmembrane voltage, $I$ = injected current. From Guttman, Lewis, and Rinzel (1980).

Annihilation can also be seen in the Hodgkin–Huxley equations (Hodgkin and Huxley 1952), which are a four-dimensional system of ordinary differential equations modeling electrical activity in the membrane of the giant axon of the squid (see Chapter 4). Note that the phase of the cycle at which annihilation can be obtained depends on the polarity of the stimulus (Figure 3.20A vs. Figure 3.20B).

### 3.5.2    Topology of Annihilation and Single-Pulse Triggering

The fact that one can initiate or terminate spontaneous activity by injection of a brief stimulus pulse means that there is the coexistence of two stable attractors in the system. One is a stable fixed point, corresponding to rest or quiescence; the other is a stable limit-cycle oscillator, corresponding to spontaneous activity. The simplest topology that is consistent with this requirement is shown in Figure 3.21A. Starting from initial condition $a$ or $b$, the state point asymptotically approaches the stable limit cycle (*solid curve*), while starting from initial condition $c$, the stable fixed point is approached. The unstable limit cycle (*dashed curve*) in this two-dimensional system is thus a separatrix that divides the plane into the

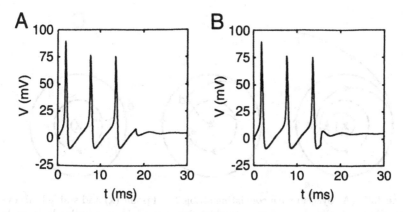

Figure 3.20. Annihilation of bias-current-induced spontaneous activity in the Hodgkin–Huxley equations by a current-pulse stimulus. (A) Hyperpolarizing, (B) depolarizing current pulse. The convention that the resting membrane potential is 0 mV is taken. From Guttman, Lewis, and Rinzel (1980).

basins of attraction of the stable fixed point and the stable limit cycle. In a higher-dimensional system, it is the stable manifold of the unstable limit cycle that can act as a separatrix, since the limit cycle itself, being a one-dimensional object, can act as a separatrix only in a two-dimensional phase space.

In Figure 3.21B, the state point is initially sitting at the stable fixed point, producing quiescence in the system. Injecting a brief stimulus of large enough amplitude will knock the state point to the point $d$, allowing it to escape from the basin of attraction of the fixed point and enter into the basin of attraction of the stable limit cycle. Periodic activity will then be seen. Figure 3.21B thus explains why the stimulus pulse must be of some minimum amplitude to trigger spontaneous activity. Figure 3.21C shows the phenomenon of annihilation. During spontaneous activity, at the point in the cycle when the state point is at $e$, a stimulus is injected that takes the state point of the system to point $f$, which is within the basin of attraction of the stable fixed point (**black hole** in the terminology of Winfree 1987). Spontaneous activity is then asymptotically extinguished. One can appreciate from this figure that for annihilation to be successful, the stimulus must be delivered within a critical window of timing, and that the location of this window will change should the polarity, amplitude, or duration of the stimulus pulse be changed. One can also see that the stimulus pulse must be of some intermediate amplitude to permit annihilation of spontaneous activity.

The phenomena of annihilation and single-pulse triggering are not seen in all biological oscillators. For example, one would think that it might not be a good idea for one's sinoatrial node to be subject to annihilation. Indeed, there are other experiments on the sinoatrial node that indicate

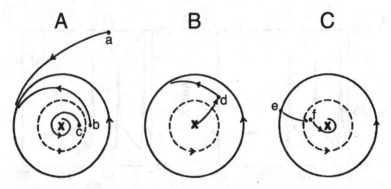

Figure 3.21. (A) System with coexisting stable fixed point (x) and stable limit cy-cle oscillation (*solid closed curve*). Dashed closed curve is an unstable limit-cycle oscillation. (B) Single-pulse triggering. (C) Annihilation. From Guevara and Jongsma (1992).

that there is only one fixed point present, and that this point is unstable (Figure 3.22). Thus, these other experiments suggest that the sinoatrial node belongs to the class of oscillators with the simplest possible topology: There is a single limit cycle, which is stable, and a single fixed point, which is unstable. This topology does not allow triggering and annihilation. The question thus naturally arises as to how the topology of Figure 3.21A can originate. There are several such ways, one of which involves a saddle-node bifurcation of periodic orbits, which we shall now discuss.

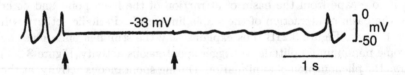

Figure 3.22. Clamping the transmembrane potential of a spontaneously beating piece of tissue taken from the sinoatrial node to its equilibrium value (*first arrow*) and then releasing the clamp (*second arrow*) results in the startup of spontaneous activity, indicating that the fixed point is unstable. From Noma and Irisawa (1975).

### 3.5.3   Saddle-Node Bifurcation of Limit Cycles

In a **saddle-node bifurcation of limit cycles**, there is the creation of a pair of limit cycles, one stable, the other unstable. Figure 3.23 illus-trates this bifurcation in the two-dimensional system of ordinary differential equations, written in polar coordinates (Strogatz 1994)

$$\frac{dr}{dt} = \mu r + r^3 - r^5, \tag{3.8a}$$

$$\frac{d\theta}{dt} = \omega + br^3. \tag{3.8b}$$

The first equation above can be rewritten as $dr/dt = r(\mu + r^2 - r^4)$, which has roots at $r^* = 0$ and at $r^* = [(1 \pm (1 + 4\mu)^{1/2})/2]^{1/2}$. The solution $r^* = 0$ corresponds to a fixed point. For $\mu < -\frac{1}{4}$, the two other roots are complex, and there are no limit cycles present (Figure 3.23A). At the bifurcation point ($\mu = -\frac{1}{4}$), there is the sudden appearance of a limit cycle of large (i.e., nonzero) amplitude with $r^* = 1/\sqrt{2}$ (Figure 3.23B). This limit cycle is semistable, since it attracts trajectories starting from initial conditions exterior to its orbit, but repels trajectories starting from initial conditions lying in the interior of its orbit. For $\mu > -\frac{1}{4}$ (Figure 3.23C), there are two limit cycles present, one stable (at $r^* = [(1 + (1 + 4\mu)^{1/2})/2]^{1/2}$) and the other unstable (at $r^* = [(1 - (1 - 4\mu)^{1/2})/2]^{1/2}$).

Figure 3.24 gives the bifurcation diagram for the saddle-node bifurcation of periodic orbits. When plotting such a diagram, one plots some characteristic of the limit cycle (such as the peak-to-peak amplitude of one variable, or the maximum and/or minimum values of that variable) as a function of the bifurcation parameter. In Figure 3.24, the diameter of the circular limit cycles of Figure 3.23 (which amounts to the peak-to-peak amplitude) is plotted as a function of the bifurcation parameter.

## 3.5.4    Saddle-Node Bifurcation in the Hodgkin–Huxley Equations

Let us now return to our example involving annihilation in the Hodgkin–Huxley equations (Figure 3.20). Figure 3.25 shows the projection on the $Vn$-plane of trajectories in the system ($V$ and $n$ are two of the variables in the four-dimensional system). With no bias current ($I_{bias}$), there is a stable fixed point present. In this situation, injection of a single stimulus pulse produces an action potential. As a bias current is injected, one has a saddle-node bifurcation at $I_{bias} \approx 8.03$ $\mu A/cm^2$. Just beyond this saddle-node bifurcation (Figure 3.25A), there are two stable structures present, a stable fixed point and a stable limit cycle (the outer closed curve) that produces spontaneous firing of the membrane, as well as one unstable structure, an unstable limit cycle (the inner closed curve). As $I_{bias}$ is increased, the unstable limit cycle shrinks in size (Figure 3.25B,C,D), until at $I_{bias} \approx 18.56$ $\mu A/cm^2$, there is a subcritical Hopf bifurcation (see Chapter 2) in which the unstable limit cycle disappears and the stable fixed point becomes unstable. Still further increase of $I_{bias}$ leads to a shrinkage in the size of the stable limit cycle. Eventually, another Hopf bifurcation,

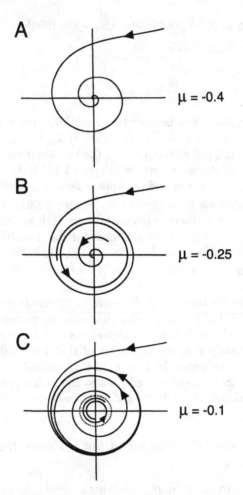

Figure 3.23. The saddle-node bifurcation of limit cycles in the two-dimensional system of ordinary differential equations given by equation 3.8. (A) $\mu = -0.4$, (B) $\mu = -0.25$, (C) $\mu = -0.1$.

which is supercritical, occurs at $I_{\text{bias}} \approx 154.5 \; \mu\text{A/cm}^2$, resulting in the disappearance of the stable limit cycle and the conversion of the unstable fixed point into a stable fixed point. Beyond this point, there is no periodic activity.

Figure 3.26 gives the bifurcation diagram for the behavior shown in Figure 3.25, computed with XPP.[†] One consequence of this diagram is that single-pulse triggering will occur only over an intermediate range of $I_{\text{bias}}$ ($8.03 \; \mu\text{A/cm}^2 < I_{\text{bias}} < 18.56 \; \mu\text{A/cm}^2$). This is due to the fact that for

---

[†]See Appendix A for an introduction to XPP.

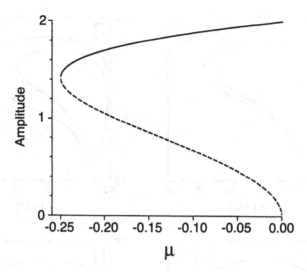

Figure 3.24. Bifurcation diagram for limit cycles in the saddle-node bifurcation of periodic orbits shown in Figure 3.23.

$I_{\text{bias}} < 8.03 \; \mu\text{A}/\text{cm}^2$ there are no limit cycles present in the system, while for $18.56 \; \mu\text{A}/\text{cm}^2 < I_{\text{bias}} < 154.5 \; \mu\text{A}/\text{cm}^2$ the fixed point is unstable, and for $I_{\text{bias}} > 154.5 \; \mu\text{A}/\text{cm}^2$, there are again no limit cycles present. There are thus, in this example, two routes by which the topology allowing annihilation and single-pulse triggering ($8.03 \; \mu\text{A}/\text{cm}^2 < I_{\text{bias}} < 18.56 \; \mu\text{A}/\text{cm}^2$) can be produced: (i) As $I_{\text{bias}}$ is increased from a very low value, there is a single saddle-node bifurcation; (ii) as $I_{\text{bias}}$ is reduced from a very high value, there are two Hopf bifurcations, the first supercritical, the second subcritical.

### 3.5.5  Hysteresis and Hard Oscillators

Another consequence of the bifurcation diagram of Figure 3.26 is that there will be hysteresis in the response to injection of a bias current. This has been investigated experimentally in the squid axon. When a ramp of current is injected into the squid axon, firing will start at a value of bias current that is higher than the value at which firing will stop as the current is ramped down (Figure 3.27). Oscillators such as that shown in Figure 3.27 that start up at large amplitude as a parameter is slowly changed are said to be "**hard**," whereas those that start up at zero amplitude (i.e., via a supercritical Hopf bifurcation) are said to be "**soft**."

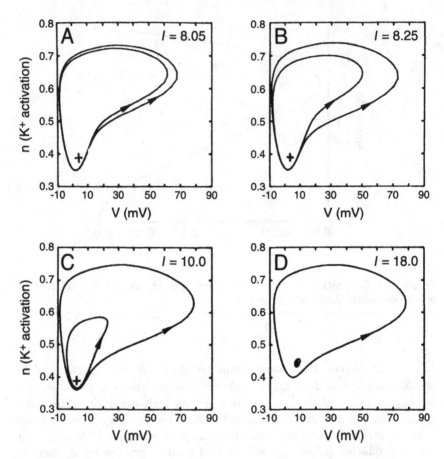

Figure 3.25. Phase-plane portrait of Hodgkin–Huxley equations as bias current $I_{\text{bias}}$ is changed. The convention that the normal resting potential is 0 mV is taken. Adapted from Guttman, Lewis, and Rinzel (1980).

## 3.5.6   Floquet Multipliers at the Saddle-Node Bifurcation

Let us now analyze the saddle-node bifurcation of Figure 3.23 by taking Poincaré sections and examining the resultant Poincaré first-return maps. In this case, since the system is two-dimensional, the Poincaré surface of section ($\Pi$) is a one-dimensional curve, and the Poincaré map is one-dimensional. At the bifurcation point, where a semistable orbit exists, one can see that there is a tangent or saddle-node bifurcation on the Poincaré map (Figure 3.28A). Beyond the bifurcation point, there is a stable fixed point on the map, corresponding to the stable limit cycle, and an unstable fixed point on the map, corresponding to the unstable limit cycle (Figure 3.28B). Remembering that the slope of the map at the fixed point gives the Floquet multiplier, one can appreciate that a saddle-node

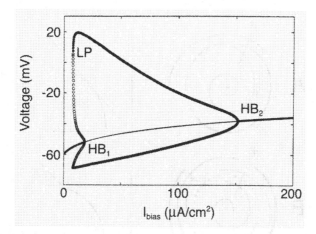

Figure 3.26. Bifurcation diagram for response of Hodgkin–Huxley equations to a bias current ($I_{\text{bias}}$), computed using **XPP**. *Thick curve*: stable fixed points; *thin curve*: unstable fixed points; *filled circles*: stable limit cycles; *unfilled circles*: unstable limit cycles.

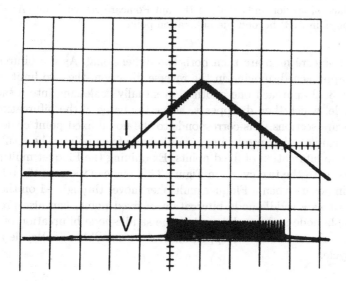

Figure 3.27. Hysteresis in the response of the squid axon to injection of a ramp of bias current. $V$ = transmembrane potential, $I$ = injected bias current. From Guttman, Lewis, and Rinzel (1980).

bifurcation occurs when a real Floquet multiplier moves through +1 on the unit circle.

When the saddle-node bifurcation of limit cycles occurs in a three-dimensional system, the stable limit cycle is a nodal cycle and the unstable limit cycle is a saddle cycle (Figure 3.29). The Poincaré plane of section

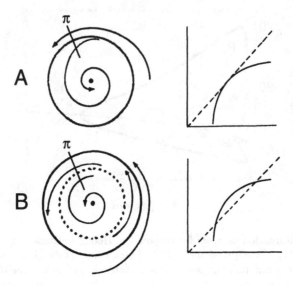

Figure 3.28. Saddle-node bifurcation in a two-dimensional system. Poincaré section (surface of section indicated by Π) and Poincaré return map: (A) at the bifurcation point, (B) beyond the bifurcation point.

and the Poincaré map are then both two-dimensional. As the bifurcation point is approached, moving in the reverse direction, the two limit cycles in Figure 3.29 approach one another, eventually coalescing into a saddle-nodal cycle, which then disappears, hence the name of the bifurcation. In the Poincaré sections, this corresponds to the nodal fixed point coalescing with the saddle point, producing a saddle-node fixed point. The result is a saddle-node bifurcation of fixed points. Examining the Floquet multipliers associated with the two cycles in Figure 3.29, one can see that the bifurcation again occurs when a Floquet multiplier moves through +1 on the unit circle. Just as a saddle-node bifurcation of fixed points can also produce an unstable node and a saddle point, the saddle-node bifurcation of limit cycles can also result in the appearance of an unstable nodal cycle and a saddle cycle.

## 3.5.7  Bistability of Periodic Orbits

We have previously considered situations in which two stable fixed points can coexist. It is also possible to have coexistence of two stable limit cycles. An example of this is shown in Figure 3.30A, in which a single cell isolated from the rabbit ventricle is driven with a train of current pulse stimuli delivered at a relatively fast rate. At the beginning of the trace, there is 1:1 synchronization between the train of stimulus pulses and the action potentials. This periodic behavior corresponds to a limit cycle in the phase

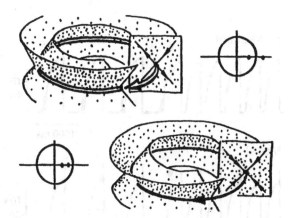

Figure 3.29. Saddle (top) and nodal (bottom) limit cycles produced by saddle-node bifurcation in a three-dimensional system. Adapted from Abraham and Shaw (1982).

space of the system. At the arrow, a single extra stimulus pulse is delivered. This extra stimulus flips the 1:1 rhythm to a 2:1 rhythm, in which every second stimulus produces only a subthreshold response. Similar behavior can be seen in an ionic model of ventricular membrane (Figure 3.30B).

There are thus two stable rhythms that coexist, with one or the other being seen, depending on initial conditions. It is also possible to flip from the 1:1 to the 2:1 rhythm by dropping pulses from the basic drive train, as well as to flip from the 2:1 rhythm back to the 1:1 rhythm by inserting an extra stimulus with the correct timing (Yehia, Jeandupeux, Alonso, and Guevara 1999). Similar results have also been described in the quiescent dog ventricle (Mines 1913) and in aggregates of spontaneously beating embryonic chick ventricular cells (Guevara, Shrier, and Glass 1990).

The existence of bistability means that hysteresis can be seen: The transition from 1:1 to 2:1 rhythm does not occur at the same driving frequency as the reverse transition from 2:1 to 1:1 rhythm. A systematic study of this phenomenon has been carried out in dog ventricle (Mines 1913), aggregates of spontaneously beating embryonic chick ventricular cells (Guevara, Shrier, and Glass 1990), frog ventricle (Hall, Bahar, and Gauthier 1999), and single rabbit ventricular cells (Yehia, Jeandupeux, Alonso, and Guevara 1999).

The bistability of Figure 3.30 implies that there are two stable limit cycles present in the phase space of the system. The simplest way in which this can occur is if there is an unstable limit cycle also present, with its stable manifold (vase-shaped surface in Figure 3.31) acting as the separatrix between the basins of attraction of the two stable limit cycles. Suppose that as a parameter is changed, there is a reverse saddle-node bifurcation of periodic cycles, destroying the unstable saddle limit cycle and one of the two stable nodal limit cycles. In that case, bistability would no longer

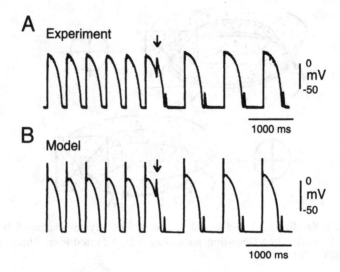

Figure 3.30. Bistability between 1:1 and 2:1 rhythms in (A) a periodically driven single cardiac cell isolated from the rabbit ventricle, (B) an ionic model of ventricular membrane. From Yehia, Jeandupeux, Alonso, and Guevara (1999).

be present, since there would be only a single limit cycle left in the phase space of the system, which would be stable, resulting in monostability.

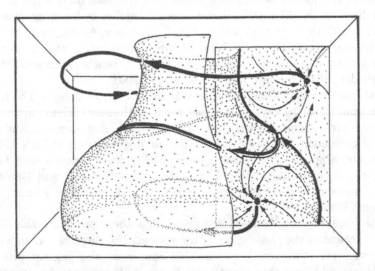

Figure 3.31. Coexistence of two stable (nodal) limit cycles, producing bistability of periodic orbits. Cycle in middle of picture is unstable (saddle) limit cycle. From Abraham and Shaw (1982).

# 3.6    Period-Doubling Bifurcation of Limit Cycles

### 3.6.1    Physiological Examples of Period-Doubling Bifurcations

Figure 3.32 shows an example of a period-doubling bifurcation in a single cell isolated from the rabbit ventricle that is subjected to periodic driving with a train of current pulses. As the interval between stimuli is decreased, the 1:1 rhythm (Figure 3.32A), in which each stimulus pulse produces an identical action potential, is replaced with an alternans or 2:2 rhythm (Figure 3.32B), in which two different morphologies of action potential are produced, which alternate in a beat-to-beat fashion. Figure 3.33 shows similar behavior in an ionic model of ventricular membrane.

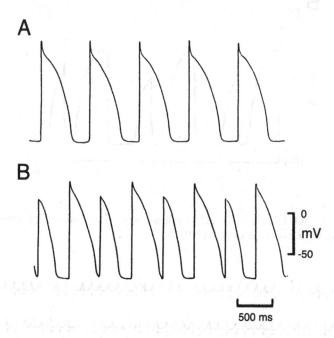

Figure 3.32. Periodic stimulation of a single rabbit ventricular cell results in (A) 1:1 or (B) 2:2 rhythm. From Guevara et al. (1989).

Another example from electrophysiology involves periodic driving of the membrane of the squid giant axon with a train of subthreshold current pulses (Figure 3.34). As the interval between pulses is increased, there is a direct transition from a 1:0 rhythm, in which there is a stereotypical subthreshold response of the membrane to each stimulus pulse, to a 2:0 response, in which the morphology of the subthreshold response alternates from stimulus to stimulus. One can also obtain responses similar to those seen in the experiments on the squid (Figure 3.34) in a reduced two-variable model, the FitzHugh–Nagumo equations (Kaplan et al. 1996).

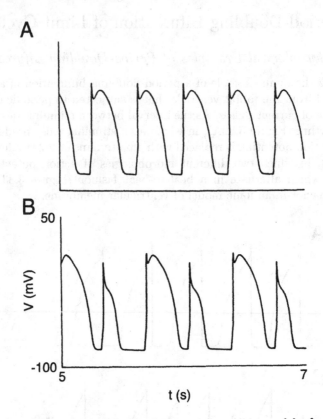

Figure 3.33. (A) 1:1 and (B) 2:2 rhythms in an ionic model of ventricular membrane. From Guevara et al. (1989).

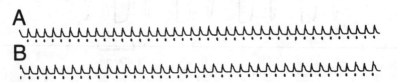

Figure 3.34. (A) 1:0 and (B) 2:0 subthreshold responses of the giant axon of the squid. From Kaplan et al. (1996).

## 3.6.2   Theory of Period-Doubling Bifurcations of Limit Cycles

In the two examples shown just above, as a parameter is changed, a periodic rhythm is replaced by another periodic rhythm of about twice the period of the original rhythm. In fact, a period-doubling bifurcation has taken place in both cases. When a period-doubling bifurcation occurs, a limit cycle reverses its stability, and in addition, a new limit cycle appears in its immediate neighborhood (Figure 3.35). This new cycle has a period that is twice as long as that of the original cycle. We have previously encountered the period-doubling bifurcation in the setting of a one-dimensional finite-difference equation (Chapter 2). In that case, a period-1 orbit is destabilized and a stable period-2 orbit produced.

Note that a period-doubled orbit cannot exist in an ordinary differential equation of dimension less than three, since otherwise, the trajectory would have to cross itself, thus violating uniqueness of solution. The trajectories and orbits shown in Figure 3.35 are thus projections onto the plane of trajectories in a three- or higher-dimensional system.

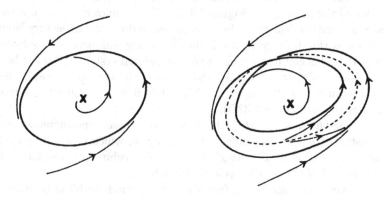

Figure 3.35. Period-doubling bifurcation of a limit cycle. From Guevara and Jongsma (1992).

A period-doubling bifurcation can be supercritical, as shown in Figure 3.35, or subcritical. Figure 3.36 gives the two corresponding bifurcation diagrams.

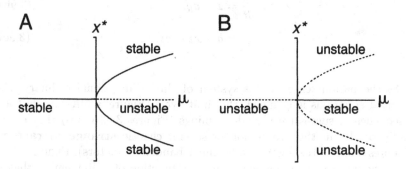

Figure 3.36. Bifurcation diagram of period-doubling bifurcation: (A) Supercritical bifurcation, (B) Subcritical bifurcation.

### 3.6.3  Floquet Multipliers at the Period-Doubling Bifurcation

A limit cycle undergoes a period-doubling bifurcation when one of its real Floquet multipliers passes through $-1$ on the unit circle (Figure 3.37A). To appreciate this fact, we must first understand how it is possible to have a negative Floquet multiplier. A negative Floquet multiplier implies, for a limit cycle in a two-dimensional system, that the slope of the return map must be negative. This means that a trajectory that has just intersected the Poincaré plane of section must next pierce the plane of section, which is a one-dimensional curve in a two-dimensional system, at a point on the other side of the limit cycle (Figure 3.37B). This cannot happen for a two-dimensional system defined in the plane, since to do so the trajectory would have to cross the limit cycle itself, thus violating uniqueness of solution. One way that this can happen in a two-dimensional system is if the orbit is a twisted orbit lying in a Möbius band. If the orbit is stable, the multiplier lies in the range $(-1, 0)$ (Figure 3.37C), while if it is unstable, it is more negative than $-1$ (Figure 3.37D).

Figure 3.38 shows a period-doubled cycle in a three-dimensional system. Also illustrated is the destabilized original cycle, which is a saddle cycle. In contrast to the case of bistability of periodic orbits (Figure 3.31), the stable manifold of the saddle cycle is twisted.

Perhaps the most interesting fact about the period-doubling bifurcation is that a cascade of such bifurcations can lead to chaotic dynamics. We have already encountered this in the setting of finite-difference equations (see Chapter 2). Figure 3.39 shows an example of this route to chaos in the much-studied Rössler equations:

$$\frac{dx}{dt} = -y - z, \tag{3.9a}$$

$$\frac{dy}{dt} = x + ay, \tag{3.9b}$$

$$\frac{dz}{dt} = b + xz - cz. \tag{3.9c}$$

As the parameter $c$ in this system of three-dimensional ordinary differential equations is changed, the limit cycle (Figure 3.39A-A) undergoes a sequence of successive period-doublings (Figures 3.39A-B-D) that eventually results in the production of several chaotic **strange attractors** (Figures 3.39A-E-H show 8-, 4-, 2- and 1-banded attractors). Figure 3.39B gives the largest Lyapunov exponent as a function of $c$. Remember that a positive Lyapunov exponent is evidence for the existence of chaotic dynamics (see Chapter 2). Figure 3.39C gives a return map extracted by plotting successive maxima of the variable $x$ for $c = 5.0$. This map is remarkably similar to the quadratic map encountered earlier (see Chapter 2).

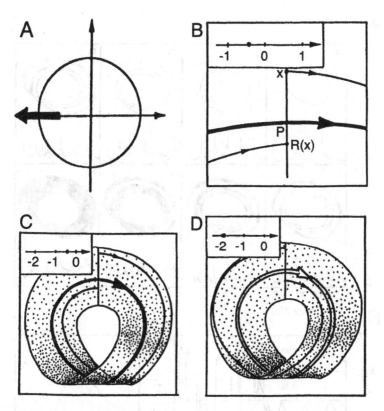

Figure 3.37. (A) Floquet diagram for period-doubling bifurcation. (B) Poincaré section of twisted cycle. (C) Stable twisted cycle. (D) Unstable twisted cycle. Panels A and B from Seydel (1994). Panels C and D adapted from Abraham and Shaw (1982).

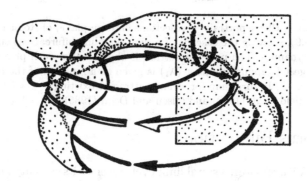

Figure 3.38. Period-doubled limit cycle in a three-dimensional system. Cycle in middle of picture is the original limit cycle, which has become destabilized. From Abraham and Shaw (1982).

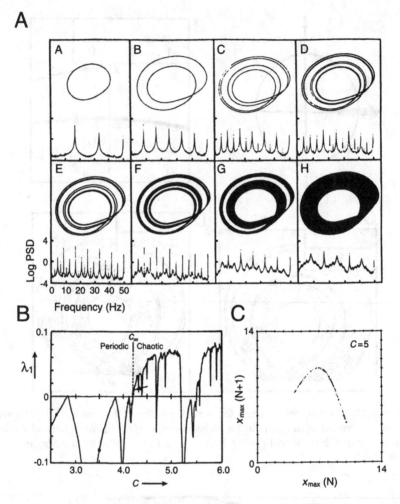

Figure 3.39. (A) Phase portraits (projections onto $xy$-plane) of Rössler equations (Equation 3.9), showing the cascade of period-doubling bifurcations culminating in chaotic dynamics. The power spectrum is shown below each phase portrait. (B) The largest Lyapunov number ($\lambda_1$) is given as a function of the bifurcation parameter, $c$. (C) Return map for strange attractor existing at $c = 5.0$. Adapted from Crutchfield et al. (1980) and Olsen and Degn (1985).

## 3.7    Torus Bifurcation

In a **torus bifurcation**, a spiral limit cycle reverses its stability and spawns a zero-amplitude torus in its immediate neighborhood, to which trajectories in the system are asymptotically attracted or repelled. The amplitude of the torus grows as the bifurcation parameter is pushed further beyond the bifurcation point. Figure 3.40 shows a supercritical torus bifurcation in a three-dimensional system. One starts off with a stable spiral limit cycle, whose Floquet multipliers are therefore a complex-conjugate pair lying within the unit circle (Figure 3.40A). Beyond the bifurcation point (Figure 3.40B), the trajectories in the system are now asymptotically attracted to orbits on the two-dimensional surface of a torus. These orbits can be either periodic or quasiperiodic. Note that the original limit cycle still exists, but that it has become an unstable spiral cycle, with a complex-conjugate pair of Floquet multipliers now lying outside of the unit circle (Figure 3.40B). A torus bifurcation thus occurs when a complex-conjugate pair of Floquet multipliers crosses the unit circle (Figure 3.40C). Figure 3.41 gives the bifurcation diagram.

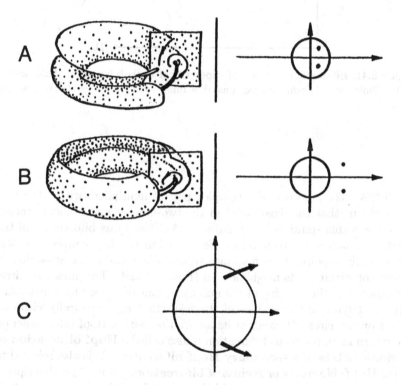

Figure 3.40. The torus bifurcation in a three-dimensional system. (A) A stable spiral limit cycle and its Floquet diagram. (B) Unstable limit cycle produced as a result of torus bifurcation. (C) Floquet multipliers crossing through unit circle at torus bifurcation point. Panels A (left) and B (left) adapted from Abraham and Shaw (1982). Panels A (right), B (right) and C adapted from Seydel (1994).

To understand why at a torus bifurcation a pair of complex-conjugate Floquet multipliers goes through the unit circle (Figure 3.40C), we must consider the significance of a Floquet multiplier being a complex number. Before the torus bifurcation occurs, when we have a stable spiral cycle (Fig-

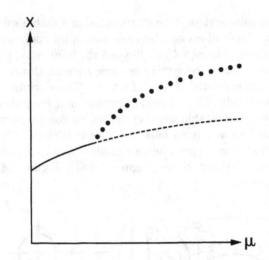

Figure 3.41. Bifurcation diagram of supercritical torus bifurcation. *Solid curve*: stable limit cycle, *dashed curve*: unstable limit cycle, *filled circles*: attracting torus.

ure 3.40A), the presence of a complex pair of multipliers within the unit circle means that the fixed point in the two-dimensional Poincaré return map is a stable spiral point (Figure 3.42A). The torus bifurcation of the orbit corresponds to a Hopf bifurcation in the map, which converts the stable spiral fixed point of the map into an unstable spiral point, spawning an **invariant circle** in its neighborhood (Figure 3.42B). This invariant circle corresponds to the piercing of the Poincaré plane of section by a quasiperiodic orbit lying in the surface of the torus that asymptotically visits all points on the circle. Because of its association with a Hopf bifurcation on the return map, the torus bifurcation is also called a **Hopf bifurcation of periodic orbits** or a **secondary Hopf bifurcation**. It is also referred to as the **Hopf–Neimark** or **Neimark bifurcation.** (Note: This description of a torus bifurcation in terms of bifurcations involving its return map is a bit simplistic, since one can have periodic as well as quasiperiodic orbits generated on the torus.) While we have illustrated the supercritical torus bifurcation above, subcritical torus bifurcations, in which an unstable limit cycle stabilizes and a repelling torus is born, also exist.

Several examples of quasiperiodic behavior have been found in biological systems. In particular, quasiperiodicity occurs naturally when one considers the weak forcing of an oscillator (see Chapter 5) or the weak interaction of two or more oscillators. In these cases, the quasiperiodic behavior arises out of a torus bifurcation (Schreiber, Dolnik, Choc, and Marek 1988).

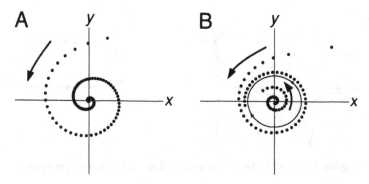

Figure 3.42. (A) Stable spiral point on Poincaré return map of original stable spiral limit cycle. (B) Invariant circle of Poincaré return map produced by Hopf bifurcation.

## 3.8   Homoclinic Bifurcation

A limit cycle is stable if all of its nontrivial Floquet multipliers lie within the unit circle. If a parameter is gradually changed, the limit cycle can lose its stability in a "local" bifurcation if and only if one or more of these multipliers crosses through the unit circle. This can happen in one of only three generic ways as a single bifurcation parameter is changed: A single real multiplier goes through +1 (saddle-node bifurcation); a single real multiplier goes through −1 (period-doubling bifurcation); or a pair of complex-conjugate multipliers crosses through the unit circle (torus bifurcation). However, there are other (nonlocal) bifurcations in which limit cycles can be created or destroyed. We now turn to consideration of one of these "global" bifurcations: the homoclinic bifurcation.

A **heteroclinic connection** is a trajectory connecting two different fixed points (thick horizontal line in Figure 3.43A). It takes an infinite amount of time to traverse the connection: The amount of time taken to leave the starting point of the connection grows without limit as one starts closer to it, while the amount of time taken to approach more closely the terminal point of the connection also grows without limit. A **heteroclinic cycle** is a closed curve made up of two or more heteroclinic connections (e.g., Figure 3.43B). To cause confusion, the term **heteroclinic orbit** has been used to denote either a heteroclinic connection or a heteroclinic cycle. We shall not use it further.

A **homoclinic orbit** is a closed curve that has a single fixed point lying somewhere along its course. Perhaps the simplest example is the case in a two-dimensional system in which the homoclinic orbit involves a saddle point (Figure 3.44A). The homoclinic orbit is formed when one of the pair of separatrices associated with the stable manifold of the saddle point coincides with one of the pair of trajectories forming its unstable manifold (Figure 3.44B). As with the heteroclinic cycle, the homoclinic orbit is of

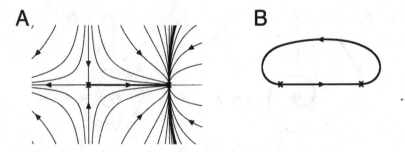

Figure 3.43. (A) Heteroclinic connection. (B) Heteroclinic cycle.

infinite period. (By continuity, it is clear that there must be some other fixed point(s) and/or limit cycle(s) present within the interior of the homoclinic orbit of Figure 3.44B.) Another type of homoclinic orbit that can occur in higher-dimensional systems is illustrated in Figure 3.44C. Here, in a three-dimensional system, the fixed point is a saddle-focus, having a pair of complex eigenvalues with positive real part, and a single real negative eigenvalue. Homoclinic orbits are not structurally stable: i.e., an infinitesimally small change in any system parameter will generally lead to their destruction, which can then result in the appearance of a periodic orbit (the homoclinic bifurcation that we shall now discuss) or chaotic dynamics (e.g., **Shil'nikov chaos**; see Guevara 1987; Guevara and Jongsma 1992; Wiggins 1988; Wiggins 1990).

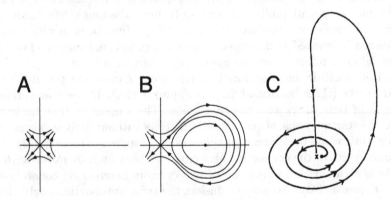

Figure 3.44. (A) Saddle point in a two-dimensional ordinary differential equation. (B) Homoclinic orbit involving a saddle point in a two-dimensional ordinary differential equation. (C) Homoclinic orbit involving a saddle focus in a three-dimensional ordinary differential equation. Panel C from Guevara and Jongsma 1992.

In one example of the homoclinic bifurcation in a two-dimensional system, there are initially two fixed points, one of which is a saddle point, and the other an unstable spiral point. There are no limit cycles present.

As a bifurcation parameter is changed, the curved trajectory associated with one of the separatrices of the unstable manifold of the saddle point approaches the trajectory associated with its stable manifold that spirals out of the unstable spiral point. At the bifurcation point, these two trajectories coincide, producing a homoclinic orbit (the closed curve starting and terminating on the saddle point). Just beyond the bifurcation point, the homoclinic orbit disappears, and is replaced by a stable limit cycle. Thus, the net result of the bifurcation is to produce a stable limit cycle, since the two preexisting fixed points remain. A homoclinic bifurcation can also result in the appearance of an unstable limit cycle (simply reverse the direction of all the arrows on the trajectories in the case described above).

There are several examples of homoclinic bifurcations in biological systems. Figure 3.45 is an example drawn from an ionic model of the sinoatrial node. As an increasingly large hyperpolarizing (positive) bias current is injected, the period of the limit cycle ($T$), which corresponds to the interval between spontaneously generated action potentials, gradually prolongs from its normal value of about 300 ms. At $I_{bias} \approx 0.39 \mu\text{A/cm}^2$, there is a homoclinic bifurcation, where $T \to \infty$. This bifurcation is heralded by a pronounced increase in the rate of growth of the period of the orbit as the bifurcation point is approached. This is a feature that distinguishes the homoclinic bifurcation from the other bifurcations we have studied so far involving periodic orbits (Hopf, saddle-node, period-doubling, and torus bifurcations). Another characteristic of the homoclinic bifurcation is that the periodic orbit appears at large (i.e., finite) amplitude. Of the other bifurcations of periodic orbits considered thus far, this feature is shared only with the saddle-node bifurcation.

## 3.9    Conclusions

A mathematical appreciation of physiological dynamics must deal with the analysis of the ways in which fixed points and oscillations can become stabilized or destabilized as parameters in physiological systems are changed. Over the years, there have been a vast number of purely experimental studies that document these sorts of dynamic changes. Many such studies are phenomenological and contain descriptions of remarkable dynamical behaviors. These papers now lie buried in dusty libraries, perhaps permanently lost to the explosion of new information technologies that has left many classical papers "off-line."

In this chapter we have illustrated some of the simplest types of bifurcations and shown biological examples to illustrate these phenomena. We have discussed three elementary one-parameter bifurcations involving fixed points alone (saddle-node, pitchfork, transcritical) as well as one two-parameter bifurcation (cusp catastrophe). We have also studied the three

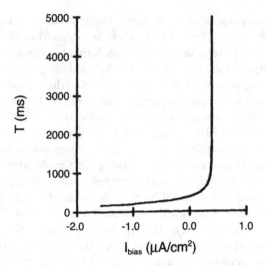

Figure 3.45. Bifurcation diagram giving period of limit cycle when a homoclinic bifurcation occurs in an ionic model of the sinoatrial node. From Guevara and Jongsma (1992).

local bifurcations of a limit cycle (saddle-node, period-doubling, torus) as well as one global bifurcation (homoclinic). Most work in physiology has centered on the local bifurcations. While there has been relatively little work on global bifurcations, one can anticipate that this will change in the future. Several other bifurcations are known to exist (Wiggins 1990), but have not yet generally shown up in modeling work on biological systems.

## 3.10    Problems

1. In a negative-feedback system, control mechanisms are present that act to reduce deviations from a set-point. For example, in a synthetic pathway with end-product inhibition, each substance is converted into a new substance, but the rate of synthesis of the initial product is a sigmoidally decreasing function of the concentration of the last product in the pathway. This problem illustrates the basic concept that an oscillation can be produced in a negative-feedback system by either increasing the gain of the negative feedback (the higher the value of $m$ in equation (3.10), the higher the gain) or by increasing the time delay; see, for example, Goodwin (1963). The oscillation arises as a consequence of a pair of complex-conjugate eigenvalues crossing the imaginary axis (a Hopf bifurcation; see Chapter 2). In this context, the multiple steps of the synthetic pathway induce a delay in the feedback. However, in other circumstances, such as the control of

blood-cell production (see Chapter 8) or the control of pupil diameter (see Chapter 9), it may be more appropriate to represent the time delay in a negative-feedback system using a time-delay differential equation.

To illustrate the properties of negative-feedback systems, we consider a simple model equation

$$\frac{dx_1}{dt} = \frac{0.5^m}{0.5^m + x_N^m} - x_1,$$

$$\frac{dx_i}{dt} = x_{i-1} - x_i, \quad i = 2, 3, \ldots, N, \tag{3.10}$$

where $m$ is a parameter (often called the Hill coefficient) controlling the steepness of the negative feedback, $N$ is the number of steps in the pathway, and $x$ and $y$ are nonnegative.

Determine the conditions for the fixed point at $x_i = 0.5$, $i = 1-N$ to be stable as a function of $m$ and $N$. In this problem if you carry through the computations until $N = 7$, the mathematics indicates the possibility for a second pair of complex eigenvalues crossing the imaginary axis as $m$ is increased. Based on your theoretical understanding of the stability of fixed points, do you expect this second bifurcation to have an effect on the observed dynamics?

An interesting project is to carry out numerical integration of this equation to compare the theoretically predicted boundaries for stability of the fixed points with the numerically computed values. Write a program in Matlab or XPP to numerically study the behavior of equation (3.10), and determine whether there is any effect on the dynamics when the second pair of eigenvalues crosses the unit circle for $N = 7$ as $m$ increases.

2. The "Brusselator" is a toy chemical reaction scheme (Prigogine and Lefever 1968) given by

$$A \longrightarrow X,$$
$$2X + Y \longrightarrow 3X,$$
$$B + X \longrightarrow Y + D,$$
$$X \longrightarrow E.$$

Given some assumptions, the dynamic equations governing these reactions are given by

$$\frac{dx}{dt} = A + Kx^2y - Bx - x = f(x, y),$$

$$\frac{dy}{dt} = -Kx^2y + Bx = g(x, y),$$

where $x$, $y$, $A$, $B$, and $K$ are nonnegative. Pick $K = 1$.

(a) Determine the fixed points of the Brusselator.

(b) Characterize the nature of the fixed points and their stability as the parameter $B$ is varied. Does a Hopf bifurcation ever occur? If so, at what value of $B$? What is the Hopf period at this value of $B$?

(c) Sketch the phase portrait of the Brusselator for various regions of the $B$ versus $A$ parameter space (don't forget that $A$ and $B$ are nonnegative).

(d) Write a `Matlab` or `XPP` program to numerically investigate the Brusselator and see how well your analytic predictions match what is seen experimentally.

3. (This problem is based on the paper by Lengyel and Epstein 1991.) Assume that two Brusselators are connected by a semipermeable membrane (that allows both $x$ and $y$ to diffuse down their concentration gradients), so that the dynamics in each compartment [1 and 2] are governed by

$$\frac{dx_1}{dt} = f(x_1, y_1) + D_x(x_2 - x_1),$$

$$\frac{dy_1}{dt} = g(x_1, y_1) + D_y(y_2 - y_1),$$

$$\frac{dx_2}{dt} = f(x_2, y_2) + D_x(x_1 - x_2),$$

$$\frac{dy_2}{dt} = g(x_2, y_2) + D_y(y_1 - y_2).$$

(a) Are there parameter values such that each Brusselator is stable, but coupling them together produces an instability?

(b) Characterize the nature of the instability as completely as possible.

(c) Modify the program you wrote for the previous problem to numerically investigate the behavior of the diffusively coupled Brusselators. How well do your analytic predictions match what you observe numerically?

## 3.11   Computer Exercises: Numerical Analysis of Bifurcations Involving Fixed Points

In these computer exercises, which involve the use of the `Auto` feature of `XPP`,[‡] we will carry out bifurcation analysis on three different simple one-dimensional ordinary differential equations.

---

[‡]See Introduction to `XPP` in Appendix A.

## Ex. 3.11-1. **Bifurcation Analysis of the Transcritical Bifurcation**

The first equation we study is

$$\frac{dx}{dt} = x(\mu - x). \tag{3.11}$$

The object of this exercise is to construct the bifurcation diagram of equation (3.11), with $x$ being the **bifurcation variable**, and $\mu$ the **bifurcation parameter**. The file **xcrit.ode** is the XPP file containing the instructions for integrating the above equation.

(a) **Finding the Fixed Points.** Start XPP and select the main XPP window (titled XPP >> xcrit.ode). After turning the bell off (in the **File** menu), start the numerical integration. You will then see a plot of the variable $x$ as a function of time. There appears to be a stable fixed point at $x = 0$.

To investigate this further, make runs from a range of initial conditions. In the **Range Integrate** menu, change **Steps** to 40, **Start** to $-2$, **End** to 2. In the plot window, one now sees the result of starting at 40 different initial conditions $x_0$ evenly spaced between $x = -2$ and $x = 2$.

The runs starting with $x_0 = 0$ and $x_0 = -1$ show that both of these points are fixed points. All initial conditions on the interval $(-1, 2]$ (in fact, $(-1, \infty)$) are attracted to the stable fixed point at $x = 0$, while those starting from $x < -1$ go off to $-\infty$. Thus, $x = -1$ is an unstable fixed point.

(b) **Effect of changing** $\mu$. Now investigate the effect of changing the bifurcation parameter $\mu$ from $-1$ (the default value assigned in xcrit.ode) to $+1$.
Has anything happened to the location of the fixed points or to their stability?
Repeat the above for $\mu = 0$.
How many fixed points are now present and what can you say about their stability?

The above method of finding fixed points and their stability using brute-force numerical integration, visual inspection of the many traces that result, and repeated change of parameter is very tedious. It makes far more sense to determine the location of a fixed point at one value of the parameter and then to follow it as the parameter is changed using some sort of continuation technique (Seydel 1988). This is exactly what Auto does.

(c) **Plotting the bifurcation diagram with Auto.** When you have opened the Auto window, change Xmin and Ymin to $-12$, and Xmax and Ymax to 12. In the AutoNum window that pops up, change Par Min to −10 and Par Max to 10 (this sets the range of $\mu$ that will be investigated). Also change NPr to 200 and Norm max to 20.

We now have to give Auto a seed from which to start. Set $\mu = -10.0$ in the XPP window Parameters, and set $x = -10.0$ in the Initial Data window. Now go back to the It's Auto man! window and select Run and click on Steady state.

A bifurcation curve with two branches, consisting of two intersecting straight lines, will now appear on the plot.
One can save a copy of the bifurcation diagram as a PostScript file to be printed out later.

(d) **Studying points of interest on the bifurcation diagram.** The numerical labels 1 to 5 appear on the plot, identifying points of interest on the diagram. In addition, Auto prints some relevant information about these labels in the xterm window from which XPP was invoked.

In this case, the point with label LAB = 1 is an endpoint (EP), since it is the starting point; the point with LAB = 2 is a branchpoint (BP), and the points with LAB = 3, 4, and 5 are endpoints of the branches of the diagram. The points lying on the parts of the branches between LAB = 2 and 5 and betwen LAB = 2 and 3 are stable, and thus are indicated by a thick line. In contrast, the other two segments between 1 and 2 and between 2 and 4 are plotted as thin lines, since they correspond to unstable fixed points. Inspect the points on the bifurcation curve by clicking on Grab in the main Auto window.

Verify that the eigenvalue lies outside of the unit circle for the first 34 points on Branch 1 of the bifurcation diagram. At point 34, the eigenvalue crosses into the unit circle, and the point becomes stable.

Ex. 3.11-2. **Bifurcation Analysis of the Saddle-Node and Pitchfork Bifurcations.** You can now invoke XPP again, carry out a few integration runs over a range of initial conditions, and obtain the bifurcation diagrams for two other files, suggestively named **saddnode.ode** and **pitchfork.ode**. The former is for the equation

$$\frac{dx}{dt} = \mu - x^2, \tag{3.12}$$

while the latter is for

$$\frac{dx}{dt} = x(\mu - x^2).$$ (3.13)

Remember that before you run Auto, you must specify a valid starting point for the continuation: Calculate and enter the steady-state value of $x$ (in the Initial Data window) appropriate to the particular value of $\mu$ being used (in the Parameters window). If "X out of bounds" error occurs, this means that $x$ has become too large or too small. Use Data Viewer window to see whether $x$ is heading toward $+\infty$ or $-\infty$.

## 3.12    Additional Computer Exercises

Additional computer exercises involving the material presented in this chapter appear at the end of Chapter 4, Dynamics of Excitable Cells, in the context of models of excitable membrane.

## 5.12 Additional Computer Exercises

# 4

# Dynamics of Excitable Cells

## Michael R. Guevara

## 4.1 Introduction

In this chapter, we describe a preparation – the giant axon of the squid – that was instrumental in allowing the ionic basis of the action potential to be elucidated. We also provide an introduction to the voltage-clamp technique, the application of which to the squid axon culminated in the Hodgkin–Huxley equations, which we introduce. Hodgkin and Huxley were awarded the Nobel Prize in Physiology or Medicine in 1963 for this work. We also provide a brief introduction to the FitzHugh–Nagumo equations, a reduced form of the Hodgkin–Huxley equations.

## 4.2 The Giant Axon of the Squid

### 4.2.1 Anatomy of the Giant Axon of the Squid

The **giant axon of the squid** is one of a pair of axons that runs down the length of the mantle of the squid in the stellate nerve (Figure 4.1). When the squid wishes to move quickly (e.g., to avoid a predator), it sends between one and eight action potentials down each of these axons to initiate contraction of the muscles in its mantle. This causes a jet of water to be squirted out, and the animal is suddenly propelled in the opposite direction. The conduction velocity of the action potentials in this axon is very high (on the order of 20 m/s), which is what one might expect for an escape mechanism. This high conduction velocity is largely due to the fact that the axon has a large diameter, a large cross-sectional area, and thus a low resistance to the longitudinal flow of current in its cytoplasm. The description of this large axon by Young in 1936 is the anatomical discovery that permitted the use of this axon in the pioneering electrophysiological work of Cole, Curtis, Marmont, Hodgkin, and Huxley in the 1940s and 1950s (see references in Cole 1968; Hodgkin 1964). The common North Atlantic squid (*Loligo pealei*) is used in North America.

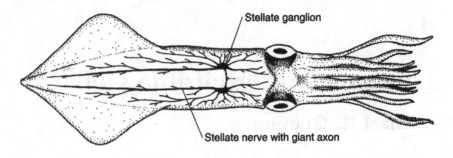

Figure 4.1. Anatomical location of the giant axon of the squid. Drawing by Tom Inoué.

### 4.2.2   Measurement of the Transmembrane Potential

The large diameter of the axon (as large as 1000 $\mu$m) makes it possible to insert an axial electrode directly into the axon (Figure 4.2A). By placing another electrode in the fluid in the bath outside of the axon (Figure 4.2B), the voltage difference across the axonal membrane (the **transmembrane potential** or **transmembrane voltage**) can be measured. One can also stimulate the axon to fire by injecting a current pulse with another set of extracellular electrodes (Figure 4.2B), producing an action potential that will propagate down the axon. This action potential can then be recorded with the intracellular electrode (Figure 4.2C). Note the afterhyperpolarization following the action potential. One can even roll the cytoplasm out of the axon, cannulate the axon, and replace the cytoplasm with fluid of a known composition (Figure 4.3). When the fluid has an ionic composition close enough to that of the cytoplasm, the action potential resembles that recorded in the intact axon (Figure 4.2D). The cannulated, internally perfused axon is the basic preparation that allowed electrophysiologists to sort out the ionic basis of the action potential fifty years ago.

The advantage of the large size of the invertebrate axon is appreciated when one contrasts it with a mammalian neuron from the central nervous system (Figure 4.4). These neurons have axons that are very small; indeed, the soma of the neuron in Figure 4.4, which is much larger than the axon, is only on the order of 10 $\mu$m in diameter.

## 4.3   Basic Electrophysiology

### 4.3.1   Ionic Basis of the Action Potential

Figure 4.5 shows an action potential in the Hodgkin–Huxley model of the squid axon. This is a four-dimensional system of ordinary differential equa-

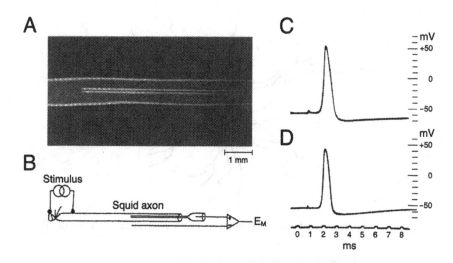

Figure 4.2. (A) Giant axon of the squid with internal electrode. Panel A from Hodgkin and Keynes (1956). (B) Axon with intracellularly placed electrode, ground electrode, and pair of stimulus electrodes. Panel B from Hille (2001). (C) Action potential recorded from intact axon. Panel C from Baker, Hodgkin, and Shaw (1961). (D) Action potential recorded from perfused axon. Panel D from Baker, Hodgkin, and Shaw (1961).

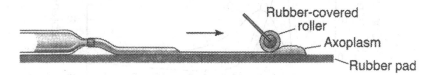

Figure 4.3. Cannulated, perfused giant axon of the squid. From Nicholls, Martin, Wallace, and Fuchs (2001).

tions that describes the three main currents underlying the action potential in the squid axon. Figure 4.5 also shows the time course of the conductance of the two major currents during the action potential. The fast inward sodium current ($I_{Na}$) is the current responsible for generating the upstroke of the action potential, while the potassium current ($I_K$) repolarizes the membrane. The leakage current ($I_L$), which is not shown in Figure 4.5, is much smaller than the two other currents. One should be aware that other neurons can have many more currents than the three used in the classic Hodgkin–Huxley description.

## 4.3.2 Single-Channel Recording

The two major currents mentioned above ($I_{Na}$ and $I_K$) are currents that pass across the cellular membrane through two different types of channels

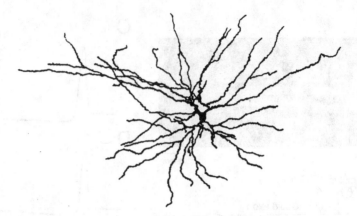

Figure 4.4. Stellate cell from rat thalamus. From Alonso and Klink (1993).

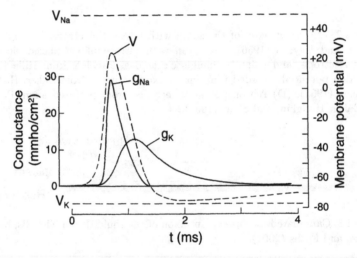

Figure 4.5. Action potential from Hodgkin–Huxley model and the conductances of the two major currents underlying the action potential in the model. Adapted from Hodgkin (1958).

lying in the membrane. The sodium channel is highly selective for sodium, while the potassium channel is highly selective for potassium. In addition, the manner in which these channels are controlled by the transmembrane potential is very different, as we shall see later. Perhaps the most direct evidence for the existence of single channels in the membranes of cells comes from the patch-clamp technique (for which the Nobel prize in Physiology or Medicine was awarded to Neher and Sakmann in 1991). In this technique, a glass microelectrode with tip diameter on the order of 1 $\mu$m is brought up against the membrane of a cell. If one is lucky, there will be only one channel in the patch of membrane subtended by the rim of the electrode,

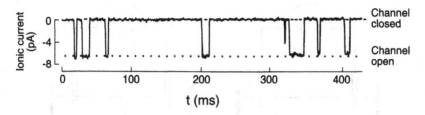

Figure 4.6. A single-channel recording using the patch-clamp technique. From Sánchez, Dani, Siemen, and Hille (1986).

and one will pick up a signal similar to that shown in Figure 4.6. The channel opens and closes in an apparently random fashion, allowing a fixed amount of current (on the order of picoamperes) to flow through it when it is in the open state.

### 4.3.3  The Nernst Potential

The concentrations of the major ions inside and outside the cell are very different. For example, the concentration of $K^+$ is much higher inside the squid axon than outside of it (400 mM versus 20 mM), while the reverse is true of $Na^+$ (50 mM versus 440 mM). These concentration gradients are set up by the sodium–potassium pump, which works tirelessly to pump sodium out of the cell and potassium into it.

A major consequence of the existence of the $K^+$ gradient is that the resting potential of the cell is negative. To understand this, one needs to know that the cell membrane is very permeable to $K^+$ at rest, and relatively impermeable to $Na^+$. Consider the thought experiment in Figure 4.7. One has a bath that is divided into two chambers by a semipermeable membrane that is very permeable to the cation $K^+$ but impermeable to the anion $A^-$. One then adds a high concentration of the salt KA into the water in the left-hand chamber, and a much lower concentration to the right-hand chamber (Figure 4.7A). There will immediately be a diffusion of $K^+$ ions through the membrane from left to right, driven by the concentration gradient. However, these ions will build up on the right, tending to electrically repel other $K^+$ ions wanting to diffuse from the left. Eventually, one will end up in electrochemical equilibrium (Figure 4.7B), with the voltage across the membrane in the steady state being given by the Nernst or equilibrium potential $E_K$,

$$E_K = \frac{RT}{zF} \ln \left( \frac{[K^+]_o}{[K^+]_i} \right), \tag{4.1}$$

where $[K^+]_o$ and $[K^+]_i$ are the external and internal concentrations of $K^+$ respectively, $R$ is the Rydberg gas constant, $T$ is the temperature in degrees Kelvin, $z$ is the charge on the ion (+1 for $K^+$), and $F$ is Faraday's constant.

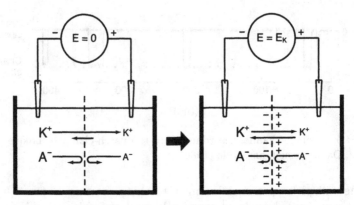

Figure 4.7. Origin of the Nernst potential.

### 4.3.4   A Linear Membrane

Figure 4.8A shows a membrane with potassium channels inserted into it. The membrane is a lipid bilayer with a very high resistance. Let us assume that the density of the channels in the membrane, their single-channel conductance, and their mean open time are such that they have a conductance of $g_K$ millisiemens per square centimeter (mS cm$^{-2}$). The current through these channels will then be given by

$$I_K = g_K(V - E_K). \tag{4.2}$$

There will thus be zero current flow when the transmembrane potential is at the Nernst potential, an inward flow of current when the transmembrane potential is negative with respect to the Nernst potential, and an outward flow of current when the transmembrane potential is positive with respect to the Nernst potential (consider Figure 4.7B to try to understand why this should be so). (An inward current occurs when there is a flow of positive ions into the cell or a flow of negative ions out of the cell.) The electrical equivalent circuit for equation (4.2) is given by the right-hand branch of the circuit in Figure 4.8B.

Now consider the dynamics of the membrane potential when it is not at its equilibrium value. When the voltage is changing, there will be a flow of current through the capacitative branch of the circuit of Figure 4.8B. The capacitance is due to the fact that the membrane is an insulator (since it is largely made up of lipids), and is surrounded on both sides by conducting fluid (the cytoplasm and the interstitial fluid). The equation of state of a capacitor is

$$Q = -CV, \tag{4.3}$$

where $Q$ is the charge on the capacitance, $C$ is the capacitance, and $V$ is the voltage across the capacitance. Differentiating both sides of this expression

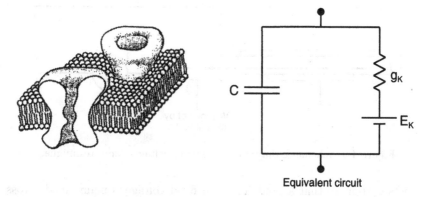

Equivalent circuit

Figure 4.8. (A) Schematic view of potassium channels inserted into lipid bilayer. Panel A from Hille (2001). (B) Electrical equivalent circuit of membrane and channels.

with respect to time $t$, one obtains

$$\frac{dV}{dt} = -\frac{I_K}{C} = -\frac{g_K}{C}(V - E_K) = -\frac{V - E_K}{\tau}, \tag{4.4}$$

where $\tau = C/g_K = R_K C$ is the **time constant** of the membrane. Here, we have also used the definition of current $I = dQ/dt$. The solution of this one-dimensional linear ordinary differential equation is

$$V(t) = E_K - (E_K - V(0))e^{-t/\tau}. \tag{4.5}$$

Unfortunately, the potassium current $I_K$ is not as simple as that postulated in equation (4.2). This is because we have assumed that the probability of the channel being open is a constant that is independent of time and voltage, and thus $g_K$ is not a function of voltage and time. This is not the case, as we shall see next.

## 4.4    Voltage-Clamping

### 4.4.1    The Voltage-Clamp Technique

While the large size of the squid axon was invaluable in allowing the transmembrane potential to be easily measured, it was really the use of this preparation in conjunction with the invention of the voltage-clamp technique that revolutionized the field. The voltage-clamp technique was pioneered by Cole, Curtis, Hodgkin, Huxley, Katz, and Marmont following the hiatus provided by the Second World War (see references in Cole 1968; Hodgkin 1964). Voltage-clamping involves placing two internal electrodes: one to measure the transmembrane potential as before, and the other to inject current (Figure 4.9). Using electronic feedback circuitry, one then

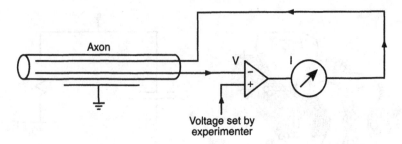

Figure 4.9. Schematic diagram illustrating voltage-clamp technique.

injects current so that a predetermined fixed voltage is maintained across the membrane. The current injected by the circuitry is then the mirror image of the current generated by the cell membrane at that potential. In addition, the effective length of the preparation is kept sufficiently short so that effects due to intracellular spread of current are mitigated: One thus transforms a problem inherently described by a partial differential equation into one reasonably well described by an ordinary differential equation.

### 4.4.2  A Voltage-Clamp Experiment

Figure 4.10B shows the clamp current during a voltage-clamp step from −65 mV to −9 mV (Figure 4.10A). This current can be broken down into the sum of four different currents: a capacitative current (Figure 4.10C), a leakage current (Figure 4.10C), a sodium current (Figure 4.10D), and a potassium current (Figure 4.10E). The potassium current turns on (**activates**) relatively slowly. In contrast, the sodium current activates very quickly. In addition, unlike the potassium current, the sodium current then turns off (**inactivates**), despite the fact that the voltage or transmembrane potential, $V$, is held constant.

### 4.4.3  Separation of the Various Ionic Currents

How do we know that the trace of Figure 4.10B is actually composed of the individual currents shown in the traces below it? This conclusion is based largely on three different classes of experiments involving ion substitution, specific blockers, and specific clamp protocols. Figure 4.11B shows the clamp current in response to a step from −65 mV to −9 mV (Figure 4.11A). Also shown is the current when all but 10% of external $Na^+$ is replaced with an impermeant ion. The difference current (Figure 4.11C) is thus essentially the sodium current. Following addition of tetrodotoxin, a specific blocker of the sodium current, only the outward $I_K$ component remains, while following addition of tetraethylammonium, which blocks potassium channels, only the inward $I_{Na}$ component remains.

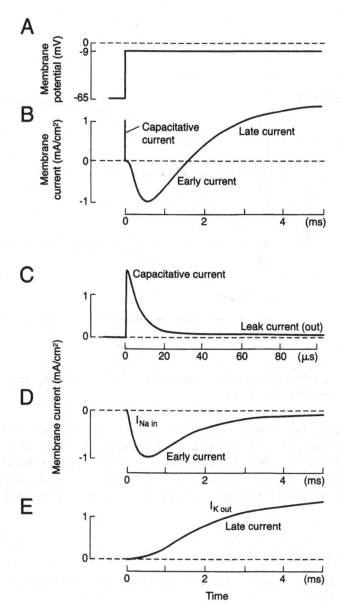

Figure 4.10. Voltage-clamp experiment on squid axon. From Nicholls, Martin, Wallace, and Fuchs (2001).

## 4.5   The Hodgkin–Huxley Formalism

### 4.5.1   Single-Channel Recording of the Potassium Current

Figure 4.12A shows a collection of repeated trials in which the voltage is clamped from $-100$ mV to $+50$ mV. The potassium channel in the

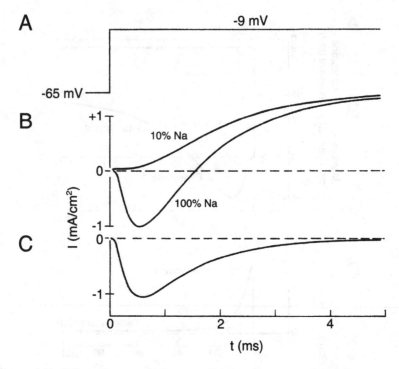

Figure 4.11. Effect of removal of extracellular sodium on voltage-clamp record. Adapted from Hodgkin (1958).

patch opens after a variable delay and then tends to stay open. Thus, the ensemble-averaged trace (Figure 4.12B) has a sigmoidal time course similar to the macroscopic current recorded in the axon (e.g., Figure 4.10E). It is thus clear that the macroscopic concept of the time course of activation is connected with the microscopic concept of the latency to first opening of a channel.

## 4.5.2    Kinetics of the Potassium Current $I_K$

The equation developed by Hodgkin and Huxley to describe the potassium current, $I_K$, is

$$I_K(V,t) = g_K(V - E_K) = \bar{g}_K[n(V,t)]^4(V - E_K), \qquad (4.6)$$

where $\bar{g}_K$ is the maximal conductance, and where $n$ is a "gating" variable satisfying

$$\frac{dn}{dt} = \alpha_n(1 - n) - \beta_n n.$$

Let us try to understand where this equation comes from.

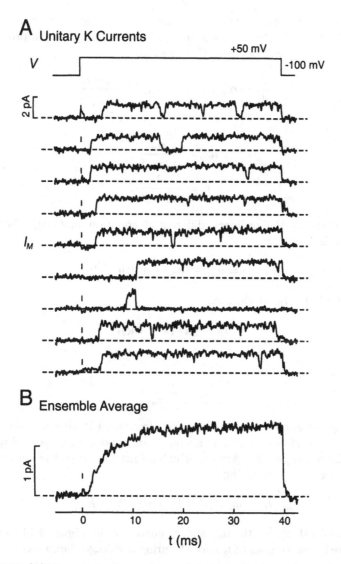

Figure 4.12. (A) Repeated voltage-clamp trials in single-channel recording mode for $I_K$. (B) Ensemble average of above recordings. Figure from F. Bezanilla.

Assume that there is one gate (the "$n$-gate") controlling the opening and closing of the potassium channel. Also assume that it follows a first-order reaction scheme

$$C \underset{\beta_n}{\overset{\alpha_n}{\rightleftharpoons}} O, \tag{4.7}$$

where the rate constants $\alpha_n$ and $\beta_n$, which are functions of voltage (but are constant at any given voltage), control the transitions between the closed

(C) and open (O) states of the gate. The variable $n$ can then be interpreted as the fraction of gates that are open, or, equivalently, as the probability that a given gate will be open. One then has

$$\frac{dn}{dt} = \alpha_n(1 - n) - \beta_n n = \frac{n_\infty - n}{\tau_n}, \tag{4.8}$$

where

$$n_\infty = \frac{\alpha_n}{\alpha_n + \beta_n},$$
$$\tau_n = \frac{1}{\alpha_n + \beta_n}. \tag{4.9}$$

The solution of the ordinary differential equation in (4.8), when $V$ is constant, is

$$n(t) = n_\infty - (n_\infty - n(0))e^{-t/\tau_n}. \tag{4.10}$$

The formula for $I_K$ would then be

$$I_K = \bar{g}_K n(V - E_K), \tag{4.11}$$

with

$$\frac{dn}{dt} = \frac{n_\infty - n}{\tau_n}, \tag{4.12}$$

where $\bar{g}_K$ is the maximal conductance. However, Figure 4.13A shows that $g_K$ has a waveform that is not simply an exponential rise, being more sigmoidal in shape. Hodgkin and Huxley thus took $n$ to the fourth power in equation (4.11), resulting in

$$I_K = \bar{g}_K n^4(V - E_K) = g_K(V - E_K). \tag{4.13}$$

Figure 4.13B shows the $n_\infty$ and $\tau_n$ curves, while Figure 4.14 shows the calculated time courses of $n$ and $n^4$ during a voltage-clamp step.

### 4.5.3  Single-Channel Recording of the Sodium Current

Figure 4.15A shows individual recordings from repeated clamp steps from $-80$ mV to $-40$ mV in a patch containing more than one sodium channel. Note that there is a variable latency to the first opening of a channel, which accounts for the time-dependent activation in the ensemble-averaged recording (Figure 4.15B). The inactivation seen in the ensemble-averaged recording is traceable to the fact that channels close, and eventually stay closed, in the patch-clamp recording.

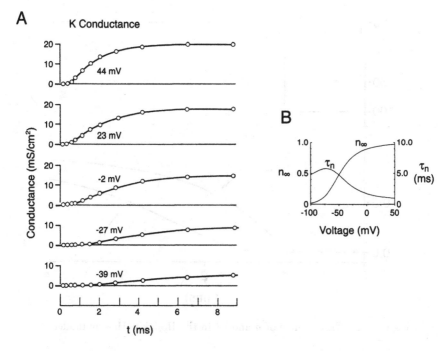

Figure 4.13. (A) Circles are data points for $g_K$ calculated from experimental values of $I_K$, $V$, and $E_K$ using equation (4.13). The curves are the fit using the Hodgkin–Huxley formalism. Panel A from Hodgkin (1958). (B) $n_\infty$ and $\tau_n$ as functions of $V$.

### 4.5.4   Kinetics of the Sodium Current $I_{Na}$

Again, fitting of the macroscopic currents led Hodgkin and Huxley to the following equation for the sodium current, $I_{Na}$,

$$I_{Na} = \bar{g}_{Na} m^3 h (V - E_{Na}) = g_{Na}(V - E_{Na}), \tag{4.14}$$

where $m$ is the **activation** variable, and $h$ is the **inactivation** variable. This implies that

$$g_{Na}(V,t) = \bar{g}_{Na}[m(V,t)]^3 h(V,t) = \frac{I_{Na}(V,t)}{(V - E_{Na})}. \tag{4.15}$$

Figure 4.16A shows that this equation fits the $g_{Na}$ data points very well.

The equations directing the movement of the $m$-gate are very similar to those controlling the $n$-gate. Again, one assumes a kinetic scheme of the form

$$C \underset{\beta_m}{\overset{\alpha_m}{\rightleftharpoons}} O, \tag{4.16}$$

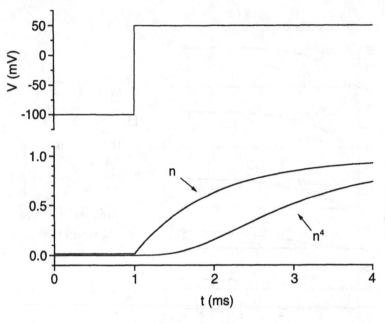

Figure 4.14. Time course of $n$ and $n^4$ in the Hodgkin–Huxley model.

where the rate constants $\alpha_m$ and $\beta_m$ are functions of voltage, but are constant at any given voltage. Thus $m$ satisfies

$$\frac{dm}{dt} = \alpha_m(1-m) - \beta_m m = \frac{m_\infty - m}{\tau_m}, \tag{4.17}$$

where

$$m_\infty = \frac{\alpha_m}{\alpha_m + \beta_m},$$

$$\tau_m = \frac{1}{\alpha_m + \beta_m}. \tag{4.18}$$

The solution of equation (4.17) when $V$ is constant is

$$m(t) = m_\infty - (m_\infty - m(0))e^{-t/\tau_m}. \tag{4.19}$$

Similarly, one has for the $h$-gate

$$C \underset{\beta_h}{\overset{\alpha_h}{\rightleftharpoons}} O, \tag{4.20}$$

where the rate constants $\alpha_h$ and $\beta_h$ are functions of voltage, but are constant at any given voltage. Thus $h$ satisfies

$$\frac{dh}{dt} = \alpha_h(1-h) - \beta_h h = \frac{h_\infty - h}{\tau_h}, \tag{4.21}$$

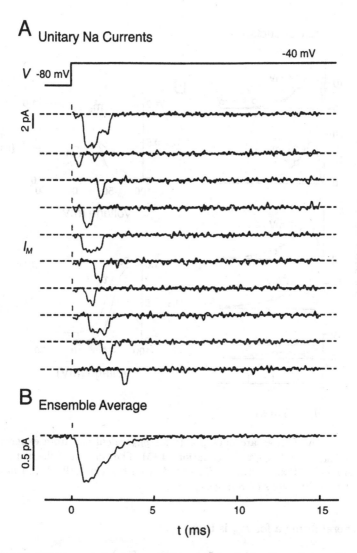

Figure 4.15. (A) Repeated voltage-clamp trials in single-channel recording mode for $I_{Na}$. (B) Ensemble average of above recordings. Figure from J.B. Patlak.

where

$$h_\infty = \frac{\alpha_h}{\alpha_h + \beta_h},$$

$$\tau_h = \frac{1}{\alpha_h + \beta_h}. \tag{4.22}$$

The solution of equation (4.21) when $V$ is constant is

$$h(t) = h_\infty - (h_\infty - h(0))e^{-t/\tau_h}. \tag{4.23}$$

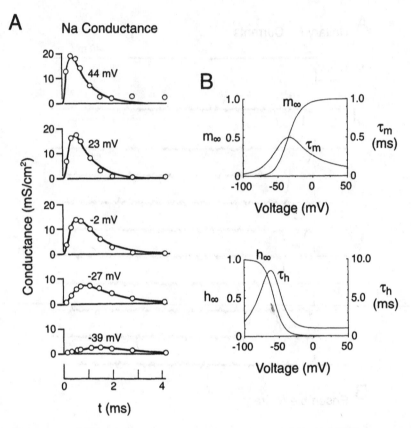

Figure 4.16. (A) Circles are data points for $g_{Na}$ calculated from experimental values of $I_{Na}$, $V$, and $E_{Na}$ using equation (4.15). The curves are the fit produced by the Hodgkin–Huxley formalism. Panel A from Hodgkin (1958). (B) $m_\infty$, $\tau_m$, $h_\infty$, and $\tau_h$ as functions of voltage.

The general formula for $I_{Na}$ is thus

$$I_{Na} = \bar{g}_{Na}m^3h(V - E_{Na}), \qquad (4.24)$$

with $m$ satisfying equation (4.17) and $h$ satisfying equation (4.21). Figure 4.16B shows $m_\infty$, $\tau_m$, $h_\infty$, and $\tau_h$, while Figure 4.17 shows the evolution of $m$, $m^3$, $h$, and $m^3h$ during a voltage-clamp step.

## 4.5.5   The Hodgkin–Huxley Equations

Putting together all the equations above, one obtains the Hodgkin–Huxley equations appropriate to the standard squid temperature of 6.3 degrees Celsius (Hodgkin and Huxley 1952). This is a system of four coupled nonlinear

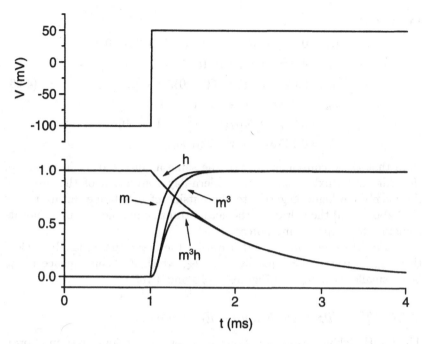

Figure 4.17. Time course of $m$, $m^3$, $h$, and $m^3 h$ in the Hodgkin–Huxley model.

ordinary differential equations,

$$\frac{dV}{dt} = -\frac{1}{C}[(\bar{g}_{Na}m^3 h(V - E_{Na}) + \bar{g}_K n^4(V - E_K)$$
$$+ \bar{g}_L(V - E_L) + I_{stim}],$$

$$\frac{dm}{dt} = \alpha_m(1 - m) - \beta_m m,$$
(4.25)

$$\frac{dh}{dt} = \alpha_h(1 - h) - \beta_h h,$$

$$\frac{dn}{dt} = \alpha_n(1 - n) - \beta_n n,$$

where

$$\bar{g}_{Na} = 120 \text{ mS cm}^{-2}, \ \bar{g}_K = 36 \text{ mS cm}^{-2}, \ \bar{g}_L = 0.3 \text{ mS cm}^{-2},$$

and

$$E_{Na} = +55 \text{ mV}, \ E_K = -72 \text{ mV}, \ E_L = -49.387 \text{ mV}, \ C = 1 \ \mu\text{F cm}^{-2}.$$

Here $I_{stim}$ is the total stimulus current, which might be a periodic pulse train or a constant ("bias") current. The voltage-dependent rate constants are given by

$$\alpha_m = 0.1(V + 35)/(1 - \exp(-(V + 35)/10)),$$
$$\beta_m = 4\exp(-(V + 60)/18),$$
$$\alpha_h = 0.07\exp(-(V + 60)/20),$$
$$\beta_h = 1/(\exp(-(V + 30)/10) + 1),$$
$$\alpha_n = 0.01(V + 50)/(1 - \exp(-(V + 50)/10)),$$
$$\beta_n = 0.125\exp(-(V + 60)/80).$$

(4.26)

Note that these equations are not the same as in the original papers of Hodgkin and Huxley, since the modern-day convention of the inside of the membrane being negative to the outside of membrane during rest is used above, and the voltage is the actual transmembrane potential, not its deviation from the resting potential.

Figure 4.18 shows $m$, $h$, and $n$ during the action potential. It is clear that $I_{Na}$ activates more quickly than $I_K$, which is a consequence of $\tau_m$ being smaller than $\tau_n$ (see Figures 4.13B and 4.16B).

### 4.5.6    The FitzHugh–Nagumo Equations

The full Hodgkin–Huxley equations are a four-dimensional system of ordinary differential equations. It is thus difficult to obtain a visual picture of trajectories in this system. In the 1940s, Bonhoeffer, who had been conducting experiments on the passivated iron wire analogue of nerve conduction, realized that one could think of basic electrophysiological properties such as excitability, refractoriness, accommodation, and automaticity in terms of a simple two-dimensional system that had a phase portrait very similar to the van der Pol oscillator (see, e.g., Figures 8 and 9 in Bonhoeffer 1948). Later, FitzHugh wrote down a modified form of the van der Pol equations to approximate Bonhoeffer's system, calling these equations the Bonhoeffer–van der Pol equations (FitzHugh 1961). FitzHugh also realized that in the Hodgkin–Huxley equations, the variables $V$ and $m$ tracked each other during an action potential, so that one could be expressed as an algebraic function of the other (this also holds true for $h$ and $n$). At about the same time as this work of FitzHugh, Nagumo et al. were working on electronic analogues of nerve transmission, and came up with essentially the same equations. These equations thus tend to be currently known as the FitzHugh–Nagumo equations and are given by

$$\frac{dx}{dt} = c\left(x - \frac{x^3}{3} + y + S(t)\right),$$
$$\frac{dy}{dt} = -\frac{(x - a + by)}{c},$$

(4.27)

where $x$ is a variable (replacing variables $V$ and $m$ in the Hodgkin–Huxley system) representing transmembrane potential and excitability, while $y$ is

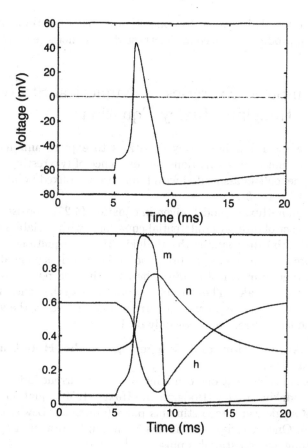

Figure 4.18. Time course of $m$, $h$, and $n$ in the Hodgkin–Huxley model during an action potential.

a variable (replacing variables $h$ and $n$ in the Hodgkin–Huxley system) responsible for refractoriness and accommodation. The function $S(t)$ is the stimulus, and $a$, $b$, and $c$ are parameters. The computer exercises in Section 4.8 explore the properties of the FitzHugh–Nagumo system in considerable detail.

## 4.6    Conclusions

The Hodgkin–Huxley model has been a great success, replicating many of the basic electrophysiological properties of the squid axon, e.g., excitability, refractoriness, and conduction speed. However, there are several discrepancies between experiment and model: For example injection of a constant bias current in the squid axon does not lead to spontaneous firing, as it does in the equations. This has led to updated versions of the Hodgkin–Huxley model being produced to account for these discrepancies (e.g., Clay 1998).

## 4.7    Computer Exercises: A Numerical Study on the Hodgkin–Huxley Equations

We will use the Hodgkin–Huxley equations to explore annihilation and triggering of limit-cycle oscillations, the existence of two resting potentials, and other phenomena associated with bifurcations of fixed points and limit cycles (see Chapter 3).

The Hodgkin–Huxley model, given in equation (4.25), consists of a four-dimensional set of ordinary differential equations, with variables $V, m, h, n$. The XPP file **hh.ode** contains the Hodgkin–Huxley equations.* The variable $V$ represents the transmembrane potential, which is generated by the sodium current (**curna** in **hh.ode**), the potassium current (**curk**), and a leak current (**curleak**). The variables $m$ and $h$, together with $V$, control the sodium current. The potassium current is controlled by the variables $n$ and $V$. The leak current depends only on $V$.

**Ex. 4.7-1. Annihilation and Triggering in the Hodgkin–Huxley Equations.**

We shall show that one can annihilate bias-current induced firing in the Hodgkin–Huxley (Hodgkin and Huxley 1952) equations by injecting a single well-timed stimulus pulse (Guttman, Lewis, and Rinzel 1980). Once activity is so abolished, it can be restarted by injecting a strong enough stimulus pulse.

(a) **Nature of the Fixed Point.** We first examine the response of the system to a small perturbation, using direct numerical integration of the system equations. We will inject a small stimulus pulse to deflect the state point away from its normal location at the fixed point. The way in which the trajectory returns to the fixed point will give us a clue as to the nature of the fixed point (i.e., node, focus, saddle, etc.). We will then calculate eigenvalues of the point to confirm our suspicions.

Start up XPP using the source file **hh.ode**.

The initial conditions in **hh.ode** have been chosen to correspond to the fixed point of the system when no stimulation is applied. This can be verified by integrating the equations numerically

---

*See Introduction to XPP in Appendix A.

(select Initialconds and then Go from the main XPP window).
Note that the transmembrane potential $(V)$ rests at $-60$ mV,
which is termed the **resting potential**.

Let us now investigate the effect of injecting a stimulus pulse.
Click on the Param button at the top of the main XPP window.
In the Parameters window that pops up, the parameters
tstart, duration, and amplitude control the time at which the
current-pulse stimulus is turned on, the duration of the pulse,
and the amplitude of the pulse, respectively (use the ∨, ∨∨, ∧,
and ∧∧ buttons to move up and down in the parameter list).
When the amplitude is positive, this corresponds to a **depolarizing** pulse, i.e., one that tends to make $V$ become more positive.
(Be careful: This convention as to sign of stimulus current is
reversed in some papers!) Conversely, a negative amplitude corresponds to a **hyperpolarizing** pulse.

Change amplitude to 10. Make an integration run by clicking
on Go in this window.

You will see a nice action potential, showing that the stimulus
pulse is **suprathreshold**.

Decrease amplitude to 5, rerun the integration, and notice the
subthreshold response of the membrane (change the scale on the
$y$-axis to see better if necessary, using Viewaxes from the main
XPP window). The damped oscillatory response of the membrane
is a clue to the type of fixed point present. Is it a node? a saddle?
a focus? (see Chapter 2).

Compute the eigenvalues of the fixed point by selecting Sing
Pts from the main XPP window, then Go, and clicking on YES
in response to Print eigenvalues? Since the system is four-
dimensional, there are four eigenvalues. In this instance, there
are two real eigenvalues and one complex-conjugate pair. What
is the significance of the fact that all eigenvalues have negative
real part? Does this calculation of the eigenvalues confirm your
guess above as to the nature of the fixed point? Do the numerical values of the eigenvalues (printed in the main window
from which XPP was originally invoked) tell you anything about
the frequency of the damped subthreshold oscillation observed
earlier (see Chapter 2)? Estimate the period of the damped oscillation from the eigenvalues.

Let us now investigate the effect of changing a parameter in the equations.

Click on **curbias** in **Parameters** and change its value to $-7$.
This change now corresponds to injecting a constant **hyperpolarizing** current of 7 $\mu A/cm^2$ into the membrane (see Chapter 4).
Run the integration and resize the plot window if necessary using **Viewaxes**.
After a short transient at the beginning of the trace, one can see that the membrane is now resting at a more hyperpolarized potential than its original resting potential of $-60$ mV.
How would you describe the change in the qualitative nature of the subthreshold response to the current pulse delivered at $t = 20$ ms?
Use **Sing Pts** to recalculate the eigenvalues and see how this supports your answer.

Let us now see the effect of injecting a constant **depolarizing** current.

Change **curbias** to 5, and make an integration run.
The membrane is now, of course, resting, depolarized to its usual value of $-60$ mV.
What do you notice about the damped response following the delivery of the current pulse?
Obtain the eigenvalues of the fixed point.
Try to understand how the change in the numerical values of the complex pair explains what you have just seen in the voltage trace.

(b) **Single-Pulse Triggering.** Now set **curbias** to 10 and remove the stimulus pulse by setting **amplitude** to zero. Carry out an integration run. You should see a dramatic change in the voltage waveform. What has happened?

Recall that $n$ is one of the four variables in the Hodgkin–Huxley equations. Plot out the trajectory in the $(Vn)$-plane: Click on **Viewaxes** in the main XPP window and then on 2D; then enter **X-axis:V, Y-axis:n, Xmin:-80, Ymin:0.3, Xmax:40, Ymax:0.8**.
You will see that the limit cycle (or, more correctly, the projection of it onto the $(Vn)$-plane) is asymptotically approached by the trajectory. If we had chosen as our initial conditions a point exactly on the limit cycle, there would have been no such transient present. You might wish to examine the projection

of the trajectory onto other two-variable planes or view it in a three-dimensional plot (using Viewaxes and then 3D, change the Z-axis variable to $h$, and enter 0 for Zmin and 1 for Zmax). Do not be disturbed by the mess that now appears! Click on Window/Zoom from the main XPP window and then on Fit. A nice three-dimensional projection of the limit-cycle appears.

Let us check to see whether anything has happened to the fixed point by our changing curbias to 10. Click on Sing pts, Go, and then YES.

XPP confirms that a fixed point is still present, and moreover, that it is still stable.

Therefore, if we carry out a numerical integration run with initial conditions close enough to this fixed point, the trajectory should approach it asymptotically in time.

To do this, click on Erase in the main XPP window, enter the equilibrium values of the variables $V, m, h, n$ as displayed in the bottom of the Equilibria window into the Initial Data menu, and carry out an integration run.

You will probably not notice a tiny point of light that has appeared in the 3D plot. To make this more transparent, go back and plot $V$ vs. $t$, using Viewaxes and 2D.

Our calculation of the eigenvalues of the fixed point and our numerical integration runs above show that the fixed point is stable at a depolarizing bias current of 10 $\mu A/cm^2$. However, remember that our use of the word stable really means **locally** stable; i.e., initial conditions in a sufficiently small neighborhood around the fixed point will approach it. In fact, our simulations already indicate that this point cannot be globally stable, since there are initial conditions that lead to a limit cycle. One can show this by injecting a current pulse.

Change amplitude from zero to 10, and run a simulation. The result is the startup of spontaneous activity when the stimulus pulse is injected at $t = 20$ ms (**"single-pulse triggering"**). Contrast with the earlier situation with no bias current injected.

(c) **Annihilation.** The converse of single-pulse triggering is **annihilation**. Starting on the limit cycle, it should be possible to terminate spontaneous activity by injecting a stimulus that would put the state point into the basin of attraction of the fixed point. However, one must choose a correct combination of stimulus "strength" (i.e., amplitude and duration) and timing. Search for and find a correct combination (Figure 3.20 will give you a hint as to what combination to use).

(d) **Supercritical Hopf Bifurcation.** We have seen that injecting a depolarizing bias current of 10 $\mu$A/cm$^2$ allows annihilation and single-pulse triggering to be seen.

Let us see what happens as this current is increased.
Put `amplitude` to zero and make repeated simulation runs, changing `curbias` in steps of 20 starting at 20 and going up to 200 $\mu$A/cm$^2$.
What happens to the spontaneous activity? Is there a bifurcation involved?

Pick one value of bias current in the range just investigated where there is no spontaneous activity.
Find the fixed point and determine its stability using `Sing pts`. Will it be possible to trigger into existence spontaneous activity at this particular value of bias current?
Conduct a simulation (i.e., a numerical integration run) to back up your conclusion.

(e) **Auto at Last!** It is clear from the numerical work thus far that there appears to be no limit-cycle oscillation present for bias current sufficiently small or sufficiently large, but that there is a stable limit cycle present over some intermediate range of bias current (our simulations so far would suggest somewhere between 10 and 160 $\mu$A/cm$^2$). It would be very tedious to probe this range finely by carrying out integration runs at many values of the parameter `curbias`, and in addition injecting pulses of various amplitudes and polarities in an attempt to trigger or annihilate activity. The thing to do here is to run `Auto`, man!

Open `Auto` by selecting `File` and then `Auto` from the main `XPP` window. This opens the main `Auto` window (`It's Auto man!`). Select the `Parameter` option in this window and replace the default choice of `Par1 (blockna)` by `curbias`. In the `Axes` window in the main `Auto` menu, select `hI-lo`, and then, in the resulting `AutoPlot` window, change `Xmin` and `Ymin` to 0 and −80, respectively, and `Xmax` and `Ymax` to 200 and 20, respectively.
These parameters control the length of the $x$- and $y$-axes of the bifurcation diagram, with the former corresponding to the bifurcation variable (bias current), and the latter to one of the four possible bifurcation variables $V, m, n, h$ (we have chosen $V$).
Invoke `Numerics` from the main `Auto` window, change `Par Max` to 200 (this parameter, together with `Par Min`, which is set to zero, sets the range of the bifurcation variable (`curbias`) that will be investigated).
Also change `Nmax`, which gives the maximum number of points

that `Auto` will compute along one bifurcation branch before stopping, to 500.

Also set NPr to 500 to avoid having a lot of labeled points. Set `Norm Max` to 150.

Leave the other parameters unchanged and return to the main `Auto` window.

Click on the main `XPP` window (*not* the main `Auto` window), select `ICs` to bring up the `Initial Data` window, and click on `default` to restore our original initial conditions ($V, m, h$, and $n$ equal to $-59.996$, $0.052955$, $0.59599$, and $0.31773$, respectively). Also click on `default` in the `Parameters` window. Make an integration run.

Click on `Run` in the main `Auto` window (`It's Auto man!`) and then on `Steady State`. A branch of the bifurcation diagram appears, with the numerical labels 1 to 4 identifying points of interest on the diagram. In this case, the point with label LAB = 1 is an endpoint (EP), since it is the starting point; the points with LAB = 2 and LAB = 3 are Hopf-bifurcation points (HB), and the point with LAB = 4 is the endpoint of this branch of the bifurcation diagram (the branch ended since `Par Max` was attained). The points lying on the parts of the branch between LAB = 1 and 2 and between 3 and 4 are plotted as thick lines, since they correspond to stable fixed points. The part of the curve between 2 and 3 is plotted as a thin line, indicating that the fixed point is unstable.

Click on `Grab` in the main `Auto` window. A new line of information will appear along the bottom of the main `Auto` window. Using the → key, move sequentially through the points on the bifurcation diagram. Verify that the eigenvalues (plotted in the lower left-hand corner of the main `Auto` window) all lie within the unit circle for the first 47 points of the bifurcation diagram. At point 48 a pair of eigenvalues crosses through the unit circle, indicating that a Hopf bifurcation has occurred. A transformation has been applied to the eigenvalues here, so that eigenvalues in the left-hand complex plane (i.e., those with negative real part) now lie within the unit circle, while those with positive real part lie outside the unit circle.

Let us now follow the limit cycle created at the Hopf bifurcation point (Pt = 47, LAB = 2), which occurs at a bias current of $18.56$ $\mu A/cm^2$. Select this point with the → and ← keys. Then press the `<Enter>` key and click on `Run`. Note that the menu has changed. Click on `Periodic`. You will see a series of points (the second branch of the bifurcation diagram) gradually being computed and then displayed as circles.

The open circles indicate an unstable limit cycle, while the filled circles indicate a stable limit cycle. The set of circles lying above the branch of fixed points gives the maximum of $V$ on the periodic orbit at each value of the bias current, while the set of points below the branch of fixed points gives the minimum. The point with LAB = 5 at a bias current of 8.03 $\mu$A/cm$^2$ is a limit point (LP) where there is a saddle-node bifurcation of periodic orbits (see Chapter 3).

Are the Hopf bifurcations sub- or supercritical (see Chapter 2)?

For the periodic branch of the bifurcation diagram, the Floquet multipliers are plotted in the lower left-hand corner of the screen. You can examine them by using Grab (use the <Tab> key to move quickly between labeled points). Note that for the stable limit cycle, all of the nontrivial multipliers lie within the unit circle, while for the unstable cycle, there is one that lies outside the unit circle. At the saddle-node bifurcation of periodic orbits, you should verify that this multiplier passes through +1 (see Chapter 3).

After all of this hard work, you may wish to keep a PostScript copy of the bifurcation diagram (using file and Postscript from the main Auto window). You can also keep a working copy of the bifurcation diagram as a diskfile for possible future explorations (using File and Save diagram). Most importantly, you should sit down for a few minutes and make sure that you understand how the bifurcation diagram computed by Auto explains all of the phenomenology that you obtained from the pre-Auto (i.e., numerical integration) part of the exercise.

(f) **Other Hodgkin–Huxley Bifurcation Parameters.** There are many parameters in the Hodgkin–Huxley equations. You might try constructing bifurcation diagrams for any one of these parameters. The obvious ones to try are $gna, gk, gleak, ena, ek,$ and $eleak$ (see equation (4.25)).

(g) **Critical Slowing Down.** When a system is close to a saddle-node bifurcation of periodic orbits, the trajectory can spend an arbitrarily long time traversing the region in a neighborhood of where the semistable limit cycle will become established at the bifurcation point. Try to find evidence of this in the system studied here.

**Ex. 4.7-2. Two Stable Resting Potentials in the Hodgkin–Huxley Equations.** Under certain conditions, axons can demonstrate two stable resting potentials (see Figure 3.1 of Chapter 3). We will compute the bifurcation diagram of the fixed points using the modified Hodgkin–Huxley equations in the XPP file **hh2sss.ode**. The bifurcation parameter is now the external potassium concentration **kout**.

(a) **Plotting the bifurcation diagram.** This time, let us invoke Auto as directly as possible, without doing a lot of preliminary numerical integration runs.

Make a first integration run using the source file **hh2sss.ode**. The transient at the beginning of the trace is due to the fact that the initial conditions correspond to a bias current of zero, and we are presently injecting a hyperpolarizing bias current of $-18$ $\mu A/cm^2$.

Make a second integration run using as initial conditions the values at the end of the last run by clicking on `Initialconds` and then `Last`. We must do this, since we want to invoke Auto, which needs a fixed point to get going with the continuation procedure used to generate a branch of the bifurcation diagram. Start up the main `Auto` window. In the `Parameter` window enter kout for `Par1`. In the `Axes` window, enter 10, $-100$, 400, and 50 for `Xmin`, `Ymin`, `Xmax`, and `Ymax`, respectively. Click on `Numerics` to obtain the `AutoNum` window, and set `Nmax:2000`, `NPr:2000`, `Par Min:10`, `Par Max:400`, and `Norm Max:150`. Start the computation of the bifurcation diagram by clicking on `Run` and `Steady state`.

(b) **Studying points of interest on the bifurcation diagram.** The point with label LAB = 1 is an endpoint (EP), since it is the starting point; the points with LAB = 2 and LAB = 3 are limit points (LP), which are points such as saddle-node and saddle-saddle bifurcations, where a real eigenvalue passes through zero (see Chapter 3).

The point with LAB = 4 is a Hopf-bifurcation (HB) point that we will study further later. The endpoint of this branch of the bifurcation diagram is the point with LAB = 5 (the branch ended since `Par Max` was attained).

The points lying on the parts of the branch between LAB = 1 and 2 and between 4 and 5 are plotted as thick lines, since they correspond to stable fixed points. The parts of the curve between 2 and 3 and between 3 and 4 are plotted as thin lines, indicating that the fixed points are unstable.

Note that there is a range of kout over which there are two stable fixed points (between the points labeled 4 and 2).

Let us now inspect the points on the curve.
Click on `Grab`.
Verify that the eigenvalues all lie within the unit circle for the first 580 points of the bifurcation diagram. In fact, all eigenvalues are real, and so the point is a stable node.

At point 580 (LAB = 2) a single real eigenvalue crosses through the unit circle at +1, indicating that a limit point or turning point has occurred. In this case, there is a saddle-node bifurcation of fixed points.

Between points 581 and 1086 (LAB = 2 and 3 respectively), there is a saddle point with one positive eigenvalue. The stable manifold of that point is thus of dimension three, and separates the four-dimensional phase space into two disjoint halves.

At point 1086 (LAB = 3), a second eigenvalue crosses through the unit circle at +1, producing a saddle point whose stable manifold, being two-dimensional, no longer divides the four-dimensional phase space into two halves. The two real positive eigenvalues collide and coalesce somewhere between points 1088 and 1089, then split into a complex-conjugate pair.

Both of the eigenvalues of this complex pair cross and enter the unit circle at point 1129 (LAB = 4). Thus, the fixed point becomes stable. A reverse Hopf bifurcation has occurred, since the eigenvalue pair is purely imaginary at this point (see Chapter 2). Press the `<Esc>` key to exit from the `Grab` function.

Keep a copy of the bifurcation diagram on file by clicking on `File` and `Save Diagram` and giving it the filename **hh2sss.ode.auto**.

(c) **Following Periodic Orbit Emanating from Hopf Bifurcation Point.** A Hopf bifurcation (HB) point has been found by `Auto` at point 1129 (LAB = 4). Let us now generate the periodic branch emanating from this HB point.
Load the bifurcation diagram just computed by clicking on `File` and then on `Load diagram` in the main `Auto` window. The bifurcation diagram will appear in the plot window.
Reduce `Nmax` to 200 in `Numerics`, select the HB point (LAB = 4) on the diagram, and generate the periodic branch by clicking on `Run` and then `Periodic`.
How does the periodic branch terminate? What sort of bifurca-

tion is involved (see Chapter 3)?

Plotting the period of the orbit will help you in figuring this out. Do this by clicking on **Axes** and then on **Period**. Enter 50, 0, 75, and 200 for **Xmin**, **Ymin**, **Xmax**, and **Ymax**.

(d) **Testing the results obtained in Auto.** Return to the main XPP menu and, using direct numerical integration, try to see whether the predictions made above by **Auto** (e.g., existence of two stable resting potentials) are in fact true.

Compare your results with those obtained in the original study (Aihara and Matsumoto 1983) on this problem (see e.g., Figure 3.7).

Ex. 4.7-3. **Reduction to a Three-Dimensional System.** It is difficult to visualize what is happening in a four-dimensional system. The best that the majority of us poor mortals can handle is a three-dimensional system. It turns out that much of the phenomenology described above will be seen by making an approximation that reduces the dimension of the equations from four to three. This involves removing the time-dependence of the variable $m$, making it depend only on voltage.

Exit XPP and copy the file **hh.ode** to a new file **hhminf.ode**. Then edit the file **hhminf.ode** so that $m^3$ is replaced by $m_\infty{}^3$ in the code for the equation for **curna**: That is, replace the line,

    curna(v) = blockna*gna*m∧3*h*(v-ena)

with

    curna(v) = blockna*gna*minf(v)∧3*h*(v-ena)

Run XPP on the source file **hhminf.ode** and view trajectories in 3D.

## 4.8 Computer Exercises: A Numerical Study on the FitzHugh–Nagumo Equations

In these computer exercises, we carry out numerical integration of the FitzHugh–Nagumo equations, which are a two-dimensional system of ordinary differential equations.

The objectives of the first exercise below are to examine the effect of changing initial conditions on the time series of the two variables and on the phase-plane trajectories, to examine nullclines and the direction field in the phase-plane, to locate the stable fixed point graphically and numerically, to determine the eigenvalues of the fixed point, and to explore the concept of excitability. The remaining exercises explore a variety of phenomena characteristic of an excitable system, using the FitzHugh–Nagumo equations as a prototype (e.g., refractoriness, anodal break excitation, recovery of latency and action potential duration, strength–duration curve, the response to periodic stimulation). While detailed keystroke-by-keystroke instructions are given for the first exercise, detailed instructions are not given for the rest of the exercises, which are of a more exploratory nature.

## The FitzHugh–Nagumo Equations

We shall numerically integrate a simple two-dimensional system of ordinary differential equations, the FitzHugh–Nagumo equations (FitzHugh 1961). As described earlier on in this chapter, FitzHugh developed these equations as a simplification of the much more complicated-looking four-dimensional Hodgkin–Huxley equations (Hodgkin and Huxley 1952) that describe electrical activity in the membrane of the axon of the giant squid. They have now become the prototypical example of an excitable system, and have been used as such by physiologists, chemists, physicists, mathematicians, and other sorts studying everything from reentrant arrhythmia in the heart to controlling chaos. The FitzHugh–Nagumo equations are discussed in Section 4.5.6 in this chapter and given in equation (4.27). (See also Kaplan and Glass 1995, pp. 245–248, for more discussion on the FitzHugh–Nagumo equations.)

The file **fhn.ode** is the XPP file[†] containing the FitzHugh–Nagumo equations, the initial conditions, and some plotting and integrating instructions for XPP. Start up XPP using the source file **fhn.ode**.

Ex. 4.8-1. **Numerical Study of the FitzHugh–Nagumo equations.**

(a) **Time Series.** Start a numerical integration run by clicking on Initialconds and then Go. You will see a plot of the variable $x$ as a function of time.

Notice that $x$ asymptotically approaches its equilibrium value of about 1.2.

Examine the transient at the start of the trace: Using Viewaxes and 2D, change the ranges of the axes by entering 0 for Xmin, 0 for Ymin, 10 for Xmax, 2.5 for Ymax. You will see that the range of $t$ is now from 0 to 10 on the plot.

Now let us look at what the other variable, $y$, is doing. Click on Xi vs t and enter $y$ in the box that appears at the top of the XPP window. You will note that when $y$ is plotted as a function of $t$, $y$ is monotonically approaching its steady-state value

---

[†]See Introduction to XPP in Appendix A.

of about $-0.6$.

We thus know that there is a stable fixed point in the system at $(x, y) \approx (1.2, -0.6)$. You will be able to confirm this by doing a bit of simple algebra with equation (4.27).

(b) **Effect of Changing Initial Conditions.** The question now arises as to whether this is the only stable fixed point present in the system. Using numerical integration, one can search for multiple fixed points by investigating what happens as the initial condition is changed in a systematic manner.

Go back and replot $x$ as a function of $t$.
Rescale the axes to 0 to 20 for $t$ and 0 to 2.5 for $x$.
Click on the ICs button at the top of the XPP window. You will see that $x$ and $y$ have initial conditions set to 2, as set up in the file **fhn.ode**. Replace the the initial condition on $x$ by $-1.0$.
When you integrate again, you will see a second trace appear on the screen that eventually approaches the same asymptotic or steady-state value as before.
Continue to modify the initial conditions until you have convinced yourself that there is only one fixed point in the system of equation (4.27).

Note: Remember that if the plot window gets too cluttered with traces, you can erase them (**Erase** in main XPP window).

You might also wish to verify that the variable $y$ is also approaching the same value as before ($\approx -0.6$) when the initial conditions on $x$ and $y$ are changed.

Reset the initial conditions to their default values.
In finding the equilibria (by using **Sing pts** in XPP), you will see that the fixed point is stable and lies at $(x^\star, y^\star) = (1.1994, -0.62426)$.
While XPP uses numerical methods to find the roots of $dx/dt = 0$ and $dy/dt = 0$ in equation (4.27), you can very easily verify this from equation (4.27) with a bit of elementary algebra.

(c) **Trajectory in the $(xy)$ phase plane.** We have examined the time series for $x$ and $y$. Let us now look at the trajectories in the $(xy)$ phase plane.

Change the axis settings to to X-axis:x, Y-axis:y, Xmin:-2.5, Ymin:-2.5, Xmax:2.5, Ymax:2.5, and make another integration run. The path followed by the state-point of the system (the trajectory) will then be displayed.

The computation might proceed too quickly for you to see the direction of movement of the trajectory: However, you can figure this out, since you know the initial condition and the location of the fixed point.

Try several different initial conditions (we suggest the $(x, y)$ pairs $(-1, -1)$, $(-1,2)$, $(1,-1)$, $(1,1.5)$).

An easy way to set initial conditions is to use the Mouse key in the Initialconds menu.

You have probably already noticed that by changing initial conditions, the trajectory takes very different paths back to the fixed point.

To make this clear graphically, carry out simulations with the trajectory starting from an initial condition of $(x_0, y_0) = (1.0, -0.8)$ and then from $(1.0, -0.9)$. In fact, it is instructive to plot several trajectories starting on a line of initial conditions at $x = 1$ (using Initialconds and then Range) with $y$ changing in steps of 0.1 between $y = -1.5$ and $y = 0$. Note that there is a critical range of $y_0$, in that for $y_0 > -0.8$, the trajectory takes a very short route to the fixed point, while for $y_0 < -0.9$, it takes a much longer route.

Use Flow in the Dir.field/flow menu to explore initial conditions in a systematic way. What kind of fixed point is present (e.g., node, focus, saddle)?

(d) **Nullclines and the Direction Field.** The above results show that a small change in initial conditions can have a dramatic effect on the resultant evolution of the system (do not confuse this with "sensitive dependence on initial conditions").

To understand this behavior, draw the nullclines and the direction fields (do this both algebraically and using XPP).

What is the sign of $dx/dt$ in the region of the plane above the $x$-nullcline? below the $x$-nullcline? How about for the $y$-nullcline? You can figure this out from equation (4.27), or let XPP do it for you.

By examining this direction field (the collection of tangent vectors), you should be able to understand why trajectories have the shape that they do.

Try running a few integrations from different initial conditions set with the mouse, so that you have a few trajectories superimposed on the vector field.

(e) **Excitability.** The fact that small changes in initial conditions can have a large effect on the resultant trajectory is responsible for the property of **excitability** possessed by the FitzHugh–

Nagumo equations. Let us look at this directly.

Click on **Viewaxes** and then 2D. Set X-axis:t, Y-axis:x, Xmin:0.0, Ymin:-2.5, Xmax:20.0, Ymax:2.5. Set $x = 1.1994$ and $y = -0.62426$ as the new initial conditions and start an integration run.

The trace is a horizontal line, since our initial conditions are now at the fixed point itself: the transient previously present has evaporated. Let us now inject a stimulus pulse.

In the **Param(eter)** window, enter

amplitude=-2.0, tstart=5.0, and duration=0.2.

We are thus now set up to inject a depolarizing stimulus pulse of amplitude 2.0 and duration 0.2 time units at $t = 5.0$. We shall look at the trace of the auxiliary variable $v$, which was defined to be equal to $-x$ in **fhn.ode**. We need to look at $-x$, since we will now identify $v$ with the transmembrane potential.

Using **Xi vs t** to plot $v$ vs. $t$, you will see a "voltage waveform" very reminiscent of that recorded experimentally from an excitable cell, i.e., there is an action potential with a fast upstroke phase, followed by a phase of repolarization, and a hyperpolarizing afterpotential.

Change the pulse amplitude to $-1.0$ and run a simulation. The response is now subthreshold.

Make a phase-plane plot (with the nullclines) of the variables $x$ and $y$ for both the sub- and suprathreshold responses, to try to explain these two responses.

Explore the range of amplitude between 0.0 and $-2.0$. The concept of an effective "threshold" should become clear.

Calculate the eigenvalues of the fixed point using the **Sing pts** menu. The eigenvalues appear in the window from which XPP was invoked.

Does the nature of the eigenvalues agree with the fact that the subthreshold response resembled a damped oscillation?

Predict the period of the damped oscillation from the eigenvalues and compare it with what was actually seen.

(f) **Refractoriness.** Modify the file **fhn.ode** so as to allow two successive suprathreshold stimulus pulses to be given (it might be wise to make a backup copy of the original file). Investigate how the response to the second stimulus pulse changes as the interval between the two pulses (the coupling interval) is changed. The Heaviside step-function heav$(t - t_0)$, which is available in XPP, is 0 for $t < t_0$ and 1 for $t > t_0$. Is there a coupling interval below which one does not get an action potential?

(g) **Latency.** The latency (time from onset of stimulus pulse to upstroke of action potential) increases as the interval between two pulses is decreased. Plot latency as a function of coupling interval. Why does the latency increase with a decrease in the coupling interval?

(h) **Action Potential Duration.** The action potential duration (time between depolarization and repolarization of the action potential) decreases as the coupling interval decreases. Why is this so?

(i) **Strength–Duration Curve: Rheobase and Chronaxie.** It is known from the earliest days of neurophysiology that a pulse of shorter duration must be of higher amplitude to produce a suprathreshold response of the membrane. Plot the threshold pulse amplitude as a function of pulse duration. Try to explain why this curve has the shape that it has. The minimum amplitude needed to provoke an action potential is termed the *rheobase*, while the shortest possible pulse duration that can elicit an action potential is called *chronaxie*.

(j) **Anodal Break Response.** Inject a hyperpolarizing pulse (i.e., one with a positive `amplitude`). Start with a pulse amplitude of 5 and increase in increments of 5. Can you explain why an action potential can be produced by a hyperpolarizing stimulus ("anodal break response")?

(k) **Response to Periodic Stimulation.** Investigate the various rhythms seen in response to periodic stimulation with a train of stimulus pulses delivered at different stimulation frequencies. The response of the FitzHugh–Nagumo equations to a periodic input is very complicated, and has not yet been completely characterized.

(l) **Automaticity.** Another basic property of many excitable tissues is automaticity: the ability to spontaneously generate action potentials. Set `a = 0` and `tmax = 50` (click on `Numerics` and then `Total`) and run `fhn.ode`. What has changing the parameter `a` done? What is the attractor now? What has happened to the stability of the fixed point? What kind of bifurcation is involved? Changing `a` systematically in the range from 0 to 2 might assist you in answering these questions. How does the shape of the limit cycle change as `a` is changed?

(m) **Phase-Locking.** Explore the response of the FitzHugh–Nagumo oscillator to periodic stimulation with a periodic train of current pulses. Systematically change the frequency and amplitude of the pulse train. In another exercise (Section 5.9), phase-locking is studied in an oscillator that is simple enough to allow reduction of the dynamics to consideration of a one-dimensional map. What is the dimension of the map that would result from reduction of the FitzHugh–Nagumo case to a map? Can you find instances where the response of the FitzHugh–Nagumo oscillator is different from that of the simpler oscillator?

**Ex. 4.8-2. Two Stable Fixed Points in the FitzHugh–Nagumo Equations.** In the standard version of the FitzHugh–Nagumo equations, there is only one fixed point present in the phase-space of the system.

Since the $x$-isocline is cubic, the possibility exists for there to be three fixed points in the FitzHugh–Nagumo equations.

Try to find combinations of the parameters $a$, $b$, and $c$ such that two stable fixed points coexist ("bistability").

How would you have to change the FitzHugh–Nagumo equations to obtain more than two stable fixed points (i.e., multistability)?

# 5

# Resetting and Entraining Biological Rhythms

## Leon Glass

## 5.1 Introduction

Biological rhythms are ubiquitous. Their periods of oscillation range from fractions of a second to a year. Independent of the period of the oscillation, and the precise mechanism underlying the generation of the oscillation, certain underlying mathematical concepts are broadly applicable. Appropriate stimuli delivered to the oscillators usually induce a resetting of the oscillation, so that the timing of the oscillation will be different from what it would have been if the stimulus were not delivered. Occasionally, a stimulus delivered during an oscillation will terminate the oscillation, or lead to a different oscillation. Determining the response of oscillators to perturbations administered at different phases of the cycle can give important information about the oscillator, and also may be useful in determining the behavior of the oscillator in the fluctuating environment. It is likely that in every branch of biology in which oscillations are observed, there is a literature analyzing the oscillations from the idiosyncratic perspective of the particular discipline. Yet, from a mathematical perspective there is a commonality of ideas and approaches (Pavlidis 1973; Guckenheimer 1975; Kawato and Suzuki 1978; Kawato 1981; Winfree 2000; Guevara, Glass, and Shrier 1981; Glass and Winfree 1984; Winfree 1987; Glass and Mackey 1988).

Resetting can be measured experimentally by delivering a stimulus to an oscillating system and determining the resulting dynamics. By delivering stimuli at different phases of an oscillation and with different magnitudes, the underlying dynamical system generating the oscillation can be probed. I give a dramatic clinical example to illustrate the approach. Figure 5.1 shows an example of a stimulus delivered by an electrode directly in a person's heart during the course of a serious cardiac arrhythmia, ventricular tachycardia (Josephson et al. 1993). The different traces represent the simultaneously recorded activity from several different sites both on the body surface and also in the heart itself. The sharp deflections represent the times when waves of electrical activation pass a given location. In the top panel, the stimulus was delivered at 270 ms following an activation recorded in the right ventricular apex, and in the bottom panel, the stimulus was delivered 260 ms after an activation from the right ventricular apex. The effects were completely different. A pulse at 270 ms reset the rhythm, whereas when a pulse was delivered 10 ms earlier, the abnormal tachycardia was abolished and the original rhythm was reestablished. Cardiologists often take this type of result to indicate that the original rhythm was generated by an excitation traveling in a reentrant circuit, in which the excitation repeatedly circulates like a toy train going around a circular track. From a mathematical perspective, we are led to inquire exactly what one can infer about underlying physiological mechanisms based on experimental data concerning resetting.

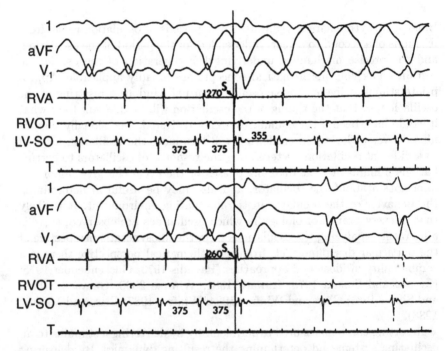

Figure 5.1. Resetting and annihilation of ventricular tachycardia. In each panel the traces labeled 1, aVF, $V_1$ are surface electrocardiograms, and the other traces are from the right ventricular apex (RVA), right ventricular outflow tract (RVOT), and the site of origin of the tachycardia in the left ventricle (LV-SO). In the top panel a stimulus (S) delivered 270 ms after a complex in the RVA resets the tachycardia, whereas in the lower panel, a stimulus delivered 260 ms after a complex in the RVA annihilates the tachycardia. From Josephson, Callans, Almendral, Hook, and Kleiman (1993).

Stimuli need not be delivered as single isolated pulses, but can also be delivered as periodic trains. In general, the observed rhythms will depend on the frequency and amplitude of the periodic pulse train. In some cases regular rhythms are set up, whereas in other cases, there are complex aperiodic rhythms. A formidable experimental challenge is to determine experimentally the dynamics as a function of stimulation parameters. Another issue is whether we can predict the effects of periodic stimulation based on knowledge about the resetting induced by a single stimulus.

Because of our knowledge about the mathematical properties of equations generating oscillations, we have expectations concerning the results of resetting and entrainment experiments. The presence of stable oscillations in mathematical models of oscillations enables us to make theoretical predictions concerning resetting and entrainment experiments. Since predictions may be quite different for oscillations generated by different mechanisms, knowledge about the results of resetting and entrainment experiments may be helpful in determining underlying mechanisms.

In this chapter, I will summarize the application of these ideas in idealized situations as well as in concrete experimental settings. I will also mention recent advances and open problems.

## 5.2  Mathematical Background

### 5.2.1  W-Isochrons and the Perturbation of Biological Oscillations by a Single Stimulus

Since biological oscillations often have "stable" periods and amplitudes (coefficient of variation on the order of 3%), it is usual to associate the oscillation with a stable limit cycle in some appropriate nonlinear theoretical model of the oscillation (Winfree 2000). Recall from Chapter 2 that a **stable limit cycle** is a periodic solution of a differential equation that is attracting in the limit of $t \to \infty$ for all points in a neighborhood of the limit cycle. Say that the period of the oscillation is $T_0$. We will designate a particular event to be the fiducial event, designated as phase, $\phi = 0$. The **phase** at any subsequent time $t > 0$ is defined to be $\phi = t/T_0$ (mod 1). The phase here is defined to lie between 0 and 1; to convert it to radians multiply it by $2\pi$.

The set of all initial conditions that attract to the limit cycle in the limit $t \to \infty$ is called the **basin of attraction** of the limit cycle. Let $x(t)$ be on a limit cycle at time $t$ and $y(t)$ be in the basin of attraction of the limit cycle. Denote the distance between $a$ and $b$ by $d[a, b]$. Let the phase of $x$ at $t = 0$ be $\phi$. Then if in the limit $t \to \infty$,

$$d[x(t), y(t)] = 0,$$

the **latent** or **asymptotic phase** of $y(t)$ is also $\phi$. We say that $y(t)$ is on the same **W-isochron** as $x(t)$.

The development of the concept of W-isochrons and the recognition of their significance is due to Winfree (2000). Many important mathematical results concerning W-isochrons were established by Guckenheimer (1975), who considered dynamical systems in $n$-dimensional Euclidean space. He proved the existence of isochrons and showed that every neighborhood of every point on the frontier of the basin of attraction of a limit cycle intersects every W-isochron. Moreover, the dimension of the frontier of the basin of attraction is greater than or equal to $n-2$.

We now consider the effects of perturbations delivered to the biological oscillation. Assume that a perturbation delivered to an oscillation at phase $\phi$ shifts the oscillation to the latent phase $g(\phi)$. The function $g(\phi)$ is called the **phase resetting curve**. The following **continuity theorem** summarizes important aspects of the effects of perturbations on limit cycle oscillations in ordinary differential equations (Guckenheimer 1975) and partial differential equations (Gedeon and Glass 1998). *If a perturbation delivered at any phase of a limit cycle oscillation leaves the state point in the basin of attraction of the asymptotically stable limit cycle, then the resetting curves characterizing the effects of the stimuli will be continuous.* In general, the phase resetting curve $g(\phi)$ is a circle map $g : S^1 \to S^1$.

Circle maps can be continuous or discontinuous. Continuous circle maps can be characterized by their **(topological) degree** or **winding number**. The degree of a continuous circle map measures the number of times the latent phase $g(\phi)$ wraps around the unit circle as $\phi$ goes around the circle once. For example, for oscillations associated with stable limit cycle oscillations in differential equations and for very weak perturbations in general, $g(\phi) \approx \phi$ and the degree is 1. In many instances, as Winfree discusses (Winfree 2000), the degree of the resetting curve is 0 when the stimulation is strong. If the degree of the resetting curve is 1 for weak stimuli and 0 for strong stimuli, there must be an intermediate stimulus (or stimuli) that will perturb the system outside of the basin of attraction of the limit cycle, though whether the limit cycle is eventually reestablished depends on whether the stimulus perturbs the system to the basin of attraction of another stable attractor. Similarly, if the resetting curve is discontinuous, there must be a stimulus phase or range of stimulus phases that will perturb the system outside of the basin of attraction of the limit cycle (Gedeon and Glass 1998).

These abstract notions are directly related to experiment. The phase resetting curve can be measured experimentally. Assume once again that the marker event of an oscillation is defined as $t = 0$, $\phi = 0$. Assume that in response to a perturbation delivered at phase $\phi$, marker events recur at successive times $T_1(\phi), T_2(\phi), \ldots, T_n(\phi)$. Let us assume that for all $j$ sufficiently large, the limit cycle is asymptotically approached, so that $T_j(\phi) - T_{j-1}(\phi) = T_0$, where $T_0$ is the control cycle length.

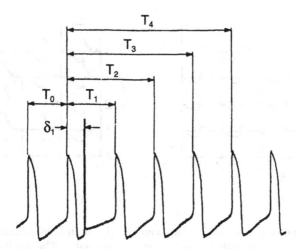

Figure 5.2. Resetting the intrinsic rhythm in a spontaneously beating aggregate of cells from embryonic chick heart. A single stimulus delivered at a phase $\phi = \delta_1/T_0$ leads to a resetting of the oscillation. The time from the action potential before the stimulus to the $j$th action potential after the stimulus is designated $T_j$. The reestablishment of an oscillation with the same amplitude and period as before the stimulus is evidence for a stable limit cycle oscillation in this preparation. From Zeng, Glass, and Shrier (1992).

Figure 5.2 shows the effects of a single stimulus delivered to a spontaneously beating aggregate of cells from embryonic chick heart. There is a rapid reestablishment of the original oscillation. This is experimental evidence that the rhythm is being generated by a stable limit cycle oscillation.

Figure 5.3 shows the results of a resetting experiment in an aggregate of cells from embryonic chick heart. A single stimulus is delivered at different phases of the oscillation. The panel on the left is typical of weak stimulation, and the panel on the right is typical of strong stimulation.

The phase resetting curve can be determined from the data in Figure 5.3. It is given by

$$g(\phi) = \phi - \frac{T_j(\phi)}{T_0} \pmod 1. \tag{5.1}$$

Winfree (2000) gives many examples of resetting biological oscillators. The degree of the experimentally measured phase resetting curve is usually 1 or 0, though in some cases it was discontinuous (Winfree 2000). Though most are not much bothered by discontinuities in resetting experiments, understanding their origin is a challenge (Glass and Winfree 1984).

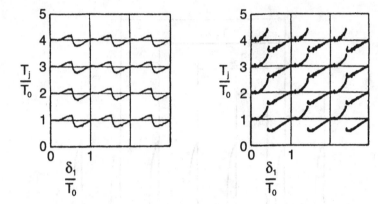

Figure 5.3. Resetting curves derived from an experiment in which a single stimulus is delivered to spontaneously beating heart cell aggregates. The results are triple plotted. A stimulus of 13 nA gives weak resetting, and a stimulus of 26 nA gives strong resetting. The time from the action potential before the stimulus to the $j$th action potential after the stimulus is plotted as a function of the phase of the stimulus. From Zeng, Glass, and Shrier (1992).

### 5.2.2   Phase Locking of Limit Cycles by Periodic Stimulation

The earliest studies drawing a connection between the resetting and the entrainment of limit cycles involved the computation of the effects of periodic stimulation on a stretch receptor (Perkel et al. 1964) and the entrainment of circadian rhythms (Pavlidis 1973). This connection can be developed mathematically using the concept of the W-isochron introduced in the last section (Glass and Mackey 1988).

In general, the effect of a single stimulus is to shift an oscillator from one W-isochron to a new W-isochron. Consequently, it is possible to define a one-dimensional map that relates the phase of an oscillation before a stimulus to the phase of an oscillation before the following stimulus. Iteration of such a map enables prediction of the dynamics during periodic stimulation. Although iteration of a one-dimensional map determined from resetting experiments can offer excellent insight into dynamics observed in real systems (Perkel et al. 1964; Guevara et al. 1981), there are many underlying assumptions. First, it is necessary to assume that the stimulation does not change the properties of the oscillation, so that the same resetting curve that is found using single pulses is also applicable under periodic stimulation. In addition, it is necessary to assume that the resetting induced by a single isolated stimulus is the same as the resetting induced by a single stimulus at the same phase delivered during a periodic train of stimuli. Even if the properties of the oscillation are not changed by the periodic stimulation protocol, there are at least two different reasons why knowing the effects of a single stimulus would not be adequate to predict the results of a periodic train of stimuli. First, it is possible that the oscillation did not relax back to its asymptotic attractor before the next stimulus was applied. Such an effect would be particularly important if there were slow relaxation to the limit cycle oscillation following a stimulus. In addition, for limit cycles that are associated with circulating oscillations in excitable media in space, the effects of a single stimulus might be blocked by a propagating wave associated with the oscillation, whereas a pulse during a periodic train of stimuli might penetrate the circuit and lead to resetting of the circulating wave.

From the assumptions above, we can derive an appropriate one-dimensional map to predict the effects of periodic stimulation with period $t_s$ of a limit cycle with intrinsic period $T_0$. Call $\phi_n$ the phase of stimulus $n$. Then, if the phase resetting curve is $g(\phi_n)$, the effects of periodic stimulation are given by

$$\phi_{n+1} = g(\phi_n) + \tau \pmod{1} \equiv f(\phi_n, \tau), \tag{5.2}$$

where $\tau = t_s/T_0$. Starting from an initial condition $\phi_0$ we generate the sequence of points $\phi_1, \phi_2, \ldots, \phi_n$.

The sequence $\{\phi_n\}$ is well-defined, provided no stimulus results in a resetting to a point outside the basin of attraction of the limit cycle. If $\phi_p = \phi_0$ and $\phi_n \neq \phi_0$ for $1 \leq n < p$, where $n$ and $p$ are positive integers, there is a periodic cycle of period $p$. A periodic cycle of period $p$ is stable if

$$\left| \frac{\partial f^p(\phi_0)}{\partial \phi} \right| = \prod_{n=0}^{p-1} \left| \frac{\partial f}{\partial \phi} \right|_{\phi_n} < 1. \tag{5.3}$$

The **rotation number**, $\rho$, gives the average increase in $\phi$ per iteration. Calling

$$\Delta_{n+1} = g(\phi_n) + \tau - \phi_i, \tag{5.4}$$

we have

$$\rho = \limsup_{N \to \infty} \frac{1}{N} \sum_{n=1}^{N} \Delta_i. \tag{5.5}$$

Stable periodic orbits are associated with **phase locking**. In $p : m$ phase locking, there is a periodic orbit consisting of $p$ stimuli and $m$ cycles of the oscillator leading to a rotation number $m/p$. For periodically forced oscillators neither the periodicity nor the rotation number alone is adequate to characterize the dynamics.

To illustrate the application and limitations of this basic theory we consider two simple theoretical models: the Poincaré oscillator and excitation circulating in a two-dimensional reentrant circuit.

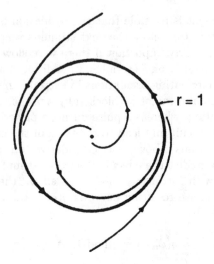

Figure 5.4. The phase plane portrait for the Poincaré oscillator. From Glass and Mackey (1988).

## 5.3    The Poincaré Oscillator

We first illustrate these concepts in a very simple ordinary differential equation that has been used extensively as a theoretical model in biology. Since this prototypical example of a nonlinear oscillation was first used by Poincaré as an example of stable oscillations, it has been called the Poincaré oscillator (Glass and Mackey 1988).

The Poincaré oscillator has been considered many times as a model of biological oscillations (Winfree 2000; Guevara and Glass 1982; Hoppensteadt and Keener 1982; Keener and Glass 1984; Glass and Mackey 1988; Glass and Sun 1994). The model has uncanny similarities to experimental data and has been useful as a conceptual model to think about the effects of periodic stimulation of cardiac oscillators.

The Poincaré oscillator is most conveniently written in a polar coordinate system where $r$ is the distance from the origin and $\phi$ is the angular coordinate; see Chapter 2. The equations are written

$$\frac{dr}{dt} = kr(1 - r),$$
$$\frac{d\phi}{dt} = 2\pi, \tag{5.6}$$

where $k$ is a positive parameter. Starting at any value of $r$, except $r = 0$, there is an evolution until $r = 1$. The parameter $k$ controls the relaxation rate. The phase, $\phi = 0$, corresponds to the upstroke of the action potential or the onset of the contraction.

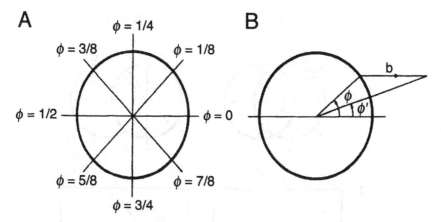

Figure 5.5. (a) Isochrons in the Poincaré oscillator. (b) A stimulus is assumed to induce a horizontal translation $b$. From Glass and Mackey (1988).

We show the phase plane portrait in Figure 5.4. Since the rate of change of $\phi$ is not a function of $r$, the isochrons are open sets lying along radii of the coordinate system. In this case the frontier of the basin of attraction of the limit cycle is the origin. The dimension of the frontier of the isochrons is 0, which is $\geq (n - 2)$ (see p. 126), Figure 5.5.

We assume that perturbations are modeled by a horizontal translation to the right by a distance $b$, Figure 5.5. In the experimental setting, perturbations are an electrical stimulus that depolarizes the membrane. A stimulus induces (after a delay) a new action potential if it is delivered in the latter part of the cycle.

This theoretical model facilitates analytical work because of its comparatively simple analytical form. The phase resetting curve, $g(\phi)$, is readily computed and is given by

$$g(\phi) = \frac{1}{2\pi} \arccos \frac{\cos 2\pi\phi + b}{(1 + b^2 + 2b\cos 2\pi\phi)^{1/2}} \ (\text{mod } 1). \qquad (5.7)$$

In computations using equation (5.7), in evaluating the arccosine function, take $0 < \phi_i' < 0.5$ for $0 < \phi_i < 0.5$, and $0.5 < \phi_i' < 1$ for $0.5 < \phi_i < 1$.

In Figure 5.6, I plot the perturbed cycle length and the phase resetting curve for the Poincaré oscillator.

The effects of periodic stimulation can now be computed by application of equations (5.2) and (5.7). The geometry of the locking zones is very complicated; a partial representation is shown in Figure 5.7. Here I summarize several important properties. For further details the original references (Guevara and Glass 1982; Keener and Glass 1984; Glass and Sun 1994) should be consulted.

There are symmetries in the organization of the locking zones as originally derived in Guevara and Glass (1982). The symmetries are:

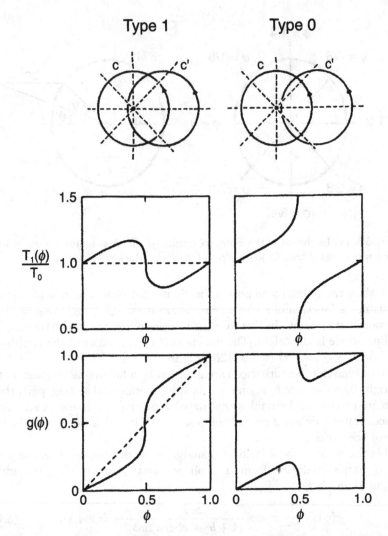

Figure 5.6. Perturbed cycle length and phase resetting curves for the Poincaré oscillator for weak, Type 1 (left panels) and strong, Type 0 (right panels) stimuli. From Glass and Winfree (1984).

- *Symmetry 1.* Assume that there is a stable period $p$ cycle with fixed points $\phi_0, \phi_1, \ldots, \phi_{p-1}$ for $\tau = 0.5 - \delta$, $0 < \delta < 0.5$, associated with $p : m$ phase locking. Then for $\tau = 0.5 + \delta$, there will be a stable cycle of period $p$ associated with a $p : p - m$ phase locking ratio. The $p$ fixed points are $\psi_0, \psi_1, \ldots, \psi_{p-1}$ where $\psi_i = 1 - \phi_i$.

- *Symmetry 2.* Assume that there is a stable period $p$ cycle with fixed points $\phi_0, \phi_1, \ldots, \phi_{p-1}$ for $\tau = \delta$, $0 < \delta < 1.0$, associated with $p : m$ phase locking. Then for $\tau = \delta + k$, where $k$ is a positive integer,

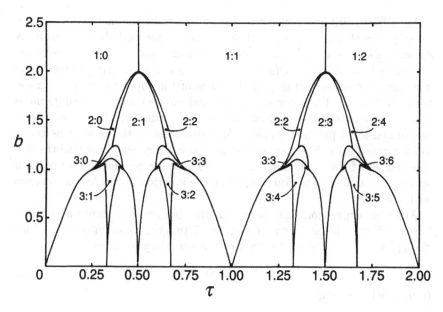

Figure 5.7. Locking zones for the Poincaré oscillator. Adapted from Glass and Sun (1994).

there will be a stable cycle of period $p$ associated with a $p : m + pk$ phase locking ratio. The $p$ fixed points are $\psi_0, \psi_1, \ldots, \psi_{p-1}$ where $\psi_i = 1 - \phi_i$.

These symmetries can be confirmed in Figure 5.7.

I now summarize main features of the organization of the locking zones. The topology of $g(\phi)$ changes at $b = 1$, and this has profound effects on the organization of the locking zones.

**Case: $0 \le b < 1$.** The map is an invertible differentiable map of the circle (Arnold 1983). An **Arnold tongue of rotation number** $m/p$ is defined as the union of values in parameter space for which there is unique attracting $p : m$ phase locking for all initial conditions. For invertible differentiable maps of the circle of the form in equation (5.2), if there is $p : m$ phase locking for $\tau$ and $p' : m'$ phase locking for $\tau'$, then there exists a value $\tau < \tau* < \tau'$, leading to $p + p' : m + m'$ phase locking. Usually, the range of values of $\tau$ associated with a given Arnold tongue covers an open interval in parameter space. For a given set of parameters the rotation number is unique. If it is rational, there is phase locking, and if it is irrational, there is **quasiperiodicity**. The organization of phase locking zones for $0 \le b < 1$ shown in Figure 5.7 for $b < 1$ is typical, and is called the **classic Arnold tongue structure**. The periodic orbits lose stability via a tangent bifurcation.

**Case: $1 < b$.** The map now has two local extrema. For any set of parameter values there is no longer necessarily a unique attractor. It is possible to have **bistability** in which there exist two stable attractors for a given set of parameter values. The attractors are either periodic or chaotic. A **superstable cycle** is a cycle containing a local extremum. Such cycles are guaranteed to be stable. One way to get a good geometric picture of the structure of the zones is to plot the locus of the superstable cycles in the parameter space. The structure of bimodal interval maps and circle maps has been well studied and shows complex cascades of bifurcations in the two-dimensional parameter space. As $b$ decreases in this zone, new phase locking zones arise; however, almost all these zones disappear into the discontinuities of the circle map at $b = 1$. There are accumulation points of an infinite number of periodic orbits at the junction of the Arnold tongues with the line $b = 1$.

Analytic expressions for some of the bifurcations can be derived. For $0 < b < 1$ the stability is lost by a tangent bifurcation for which $\partial \phi_{n+1}/\partial \phi_n = 1$. This implies that at the boundary we have

$$b + \cos 2\pi \phi_0 = 0,$$

from which we compute

$$b = |\sin 2\pi \tau|. \tag{5.8}$$

The fixed point at the stability boundary is at

$$\phi_0 = \tau + \frac{1}{4}, \text{ for } 0 < \tau < \frac{1}{4},$$

and

$$\phi_0 = \tau + \frac{3}{4}, \text{ for } \frac{3}{4} < \tau < 1.$$

For $1 < b < 2$ stability of the period-1 fixed point is lost by a period-doubling bifurcation for which $\partial \phi_{n+1}/\partial \phi_n = -1$. From this we compute that at the boundary we have

$$2 + b^2 + 3b \cos 2\pi \phi_0 = 0.$$

Carrying through the trigonometry we find the stability boundary

$$b = \sqrt{4 - 3\sin^2 2\pi \tau}. \tag{5.9}$$

The fixed point at the boundary is given by

$$\phi_0 = \tau + \frac{1}{2\pi} \sin^{-1} \sqrt{\frac{4 - b^2}{3b^2}}.$$

It is not generally appreciated that in this system there can be changes in the rotation number without a change in periodicity (Guevara and Glass 1982). For example, for $2 < b$ as $\tau$ increases with $b$ fixed there is a change from 1:0 phase locking to 1:1 phase locking along the line $\tau = 0.5$.

The analysis above assumes instantaneous relaxation back to the limit cycle. Although this idealization is clearly not obeyed in real systems (Zeng,

Glass, and Shrier 1992), it can nevertheless provide a good approximation to the dynamics of real systems, particularly for the case in which the stimulation frequency is roughly comparable to the intrinsic period of the limit cycle.

We now consider the consequences of a finite relaxation time in the Poincaré oscillator (Glass and Sun 1994). We again assume that a stimulus is schematically represented by a horizontal translation of magnitude $b$; Figure 5.5. The stimulus takes point $(r, \phi_n)$ to point $(r'_n, \phi'_n)$, where

$$r'_n = (r_n^2 + b^2 + 2br_n \cos 2\pi\phi_n)^{1/2},$$
$$\phi'_n = \frac{1}{2\pi} \arccos \frac{r_n \cos 2\pi\phi_n + b}{r'_n}. \tag{5.10}$$

Following the stimulus, the equations of motion take over, so that by direct integration, we find that immediately before stimulus $(n+1)$ delivered at a time $\tau$ after the first stimulus, we have

$$r_{n+1} = \frac{r'_n}{(1 - r'_n)\exp(-k\tau) + r'_n},$$
$$\phi_{n+1} = \phi'_n + \tau \pmod 1. \tag{5.11}$$

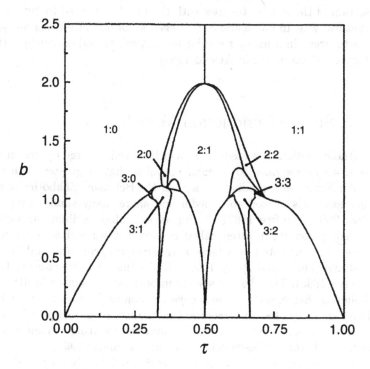

Figure 5.8. Locking zones for periodically stimulated Poincaré oscillator with finite relaxation times, $k = 10$. Adapted from Glass and Sun (1994).

An important difference is present in the organization of locking zones; even for low stimulation amplitudes the classic Arnold tongue structure described earlier does **not** apply. This fact does not seem to be widely appreciated. Even for low–amplitude stimulation, for any amplitude and frequency of stimulation, there will always be a period-1 orbit. In contrast, in the infinite relaxation limit, for $b < 1$, inside the Arnold tongues associated with locking of period $p \neq 1$, there is no period-1 cycle. The existence of period-1 cycles follows immediately from an application of the Brouwer fixed point theorem (Guillemin and Pollack 1975, p. 65). Consequently, the result is also applicable to a broad class of periodically stimulated oscillators and excitable systems, provided there is a sufficient contraction for large excursions from the limit cycle (Glass and Sun 1994). Of course, the period-1 cycle is not always stable, so that in experimental work, it will often appear as though the classic Arnold tongue structure is being observed. Subsequent to publication of this result I found a similar result in Levinson (1944). The result deserves to be better known.

Finite relaxation to the limit cycle will also destroy the symmetries in the infinite relaxation case. Moreover, the fine details of the locking zones change in subtle ways not yet well understood. For example, the points of accumulation of an infinite number of locking zones, which occur at the intersection of the Arnold tongues with the line $b = 1$, need to "unfold" in some natural way. In Glass and Sun (1994), we observe that this unfolding appears to occur in a manner similar to that envisioned earlier by Arnold (see Figure 153 on p. 312 in Arnold 1983).

## 5.4   A Simple Conduction Model

An excitable medium is a medium in which there is a large excursion from steady state in response to a small stimulus that is greater than some threshold. Nerve cells, cardiac tissue, and the Belousov–Zhabotinsky reaction are examples of excitable media and share many similar properties (Winfree 2000; Winfree 1987). A ring of excitable medium can support a circulating excitation, often called a reentrant wave. Reentrant waves have been demonstrated in a large number of experimental and theoretical systems (Quan and Rudy 1990b; Rudy 1995; Courtemanche, Glass, and Keener 1993). They have a special importance to human health, since it is believed that many cardiac tachyarrhythmias (abnormally fast heart rhythms) are associated with reentrant mechanisms (see Chapter 7). There is a large cardiological literature that involves the resetting and entrainment of cardiac arrhythmia (Josephson et al. 1993; Gilmour 1990).

The previous section dealt with the resetting of a highly simplified model containing a stable limit cycle oscillation in a nonlinear ordinary differential equation. In this section we describe a highly simplified model to illustrate some of the main features as well as the subtleties that arise in the analysis of resetting and entrainment of periodic reentrant waves in nonlinear partial differential equations.

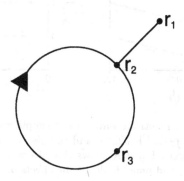

Figure 5.9. A simple model for reentrant excitation. The excitation travels at a fixed velocity around the ring and along the tail. From Glass, Nagai, Hall, Talajic, and Nattel (2002).

We assume that a wave circulates on a one-dimensional ring to which a "tail" has been added; Figure 5.9. The basic cycle length $T_0$ is given by

$$T_0 = \frac{L}{c},$$

where $L$ is the circumference of the ring and $c$ is the velocity of propagation. At any point on the ring, for a time interval of $R$ after passage of the wave, the tissue is refractory. Otherwise, the medium is excitable. A stimulus delivered during the refractory period has no effect, whereas a stimulus delivered during the excitable period will generate waves propagating into the excitable medium. In the current presentation, I consider only the resetting as measured from a single site, which might be the same or different from the stimulation site. The current discussion is based on the analysis in Glass et al. (2002), which should be consulted for more details.

First, assume that the stimulus and recording site are both directly on the ring. We select this point as a fiducial point and assume that the circulating wave crosses the fiducial point at time $t_0$. The phase is $\phi(t) = (t - t_0)/T_0$ (mod 1). The ring is parameterized by an angular coordinate, $\theta$. We set $\theta = 0$ at the fiducial point so that the angular position of the wave around the ring at time $t$ is $\theta(t) = \phi(t)$. This example has been set up so that the location of the wave on the ring is the same as the phase of the oscillation. This example is an interesting contrast to the Poincaré oscillator in which the trajectory was a closed circular path in the two-dimensional phase space. In this example, the trajectory is a closed circular path in two-dimensional physical space.

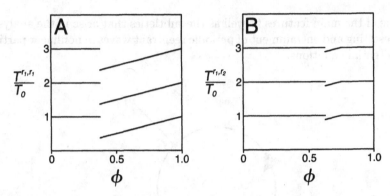

Figure 5.10. Resetting a reentrant wave in the simple geometry shown in Figure 5.9. The refractory period is $3T_0/8$, and the distance from $r_1$ to $r_2$ is $\frac{1}{4}$ the circumference of the circle. Panel A shows the effects of stimulation and measurement from point $r_2$, and panel B shows the effects of stimulation from point $r_1$ and measurement from point $r_2$.

If the stimulus is delivered during the refractory period, then it has no effect. If the stimulus is delivered outside the refractory time, then it will induce two waves, one traveling in the opposite (antidromic) and the other in the same direction (orthodromic) to the original wave. The antidromic wave will collide with it and be annihilated, whereas the orthodromic wave will continue to propagate, leading to a resetting of the original rhythm. The perturbed cycle length $T_j(\phi(t))$ is

$$\frac{T_j(\phi(t))}{T_0} = \begin{cases} j - (1 - \phi(t)), & 1 > \phi(t) > \frac{R}{T_0}, \\ j, & \frac{R}{T_0} > \phi(t) > 0. \end{cases} \qquad (5.12)$$

Using equation (5.1) we obtain the phase transition curve

$$g(\phi(t)) = \begin{cases} 0, & 1 > \phi(t) > \frac{R}{T_0}, \\ \phi(t), & \frac{R}{T_0} > \phi(t) > 0. \end{cases} \qquad (5.13)$$

The time interval during which the stimulus resets the rhythm is called the **excitable gap**. The excitable gap is $G = T_0 - R$.

Iteration of equation (5.2) is easily carried out. Provided the stimulation period falls in the range $R < t_s < T_0$, there is a stable fixed point on the period-1 map, associated with entrainment of the reentrant excitation to the periodic stimulation. The phase of the fixed point is $t_s/T_0$.

Now consider the effect of stimulating the excitation from a point $r_1$ off the ring that lies at a distance $l$ from the fiducial point $r_2$ on the ring and on the same radius as $r_2$; see Figure 5.9. Measurements are carried out at the point $r_2$. Because the conduction from the reentrant pathway can collide with the excitation from the stimulating electrode before it resets the reentrant excitation, the range of phases over which resetting is observed is reduced. The perturbed cycle length is

$$\frac{T_j(\phi(t))}{T_0} = \begin{cases} j - 1 + \phi(t) + \frac{l}{cT_0}, & 1 - \frac{l}{cT_0} > \phi(t) > \frac{R}{T_0} + \frac{l}{cT_0}, \\ j, & \text{otherwise,} \end{cases} \qquad (5.14)$$

and the associated resetting curve becomes

$$g(\phi(t)) = \begin{cases} 1 - \frac{l}{cT_0}, & 1 - \frac{l}{cT_0} > \phi(t) > \frac{R}{T_0} + \frac{l}{cT_0}, \\ \phi(t), & \text{otherwise.} \end{cases} \qquad (5.15)$$

The range of values over which resetting occurs due to a single pulse decreases as the distance of the stimulus from the ring increases. Specifically, the excitable gap is $(T_0 - R - 2l/c)$. From this it follows that for $l > c(T_0 - R)/2$, there is no resetting. If the phase transition curve is used to predict the effects of periodic stimulation, then one theoretically predicts that there will be 1:1 entrainment for stimulation periods in the range $T_0 > t_s > R + 2l/c$. However, this is not correct, and in this case the resetting curve can no longer be used to predict the effects of periodic stimulation. The reason for this is that the collisions between the wave from the stimulus and the reentrant wave lead to a reduced range of values for the resetting. During periodic stimulation at a rate $t_s$ in the range $T_0 > t_s > R$, the collisions between the waves originating from the periodic stimulation and the reentrant wave will occur successively closer to the reentrant circuit and the waves originating from the periodic forcing will eventually penetrate the reentrant circuit and entrain the reentrant wave (Krinsky and Agladze 1983; Biktashev 1997). Therefore, the theoretical prediction of the entrainment zone based on the resetting curve will underestimate the range of values leading to entrainment by a value of $2l/c$ for $l < c(T_0 - R)/2$. Thus, one can estimate the distance of a stimulus from the reentrant circuit by multiplying the discrepancy between the predicted and observed high-frequency boundaries of the 1:1 locking by the velocity of propagation of the wave, provided the stimulus is not very distant from the reentrant circuit.

Just as the Poincaré oscillator captures important aspects of nonlinear oscillators generated by ordinary differential equations, the current model captures some important aspects of dynamics in nonlinear partial differential equations that support reentrant excitation. The resetting curves shown in Figure 5.10 are discontinuous, similar to resetting curves observed from stimulation of the nonlinear FitzHugh–Nagumo equation of a pulse circulating on a one-dimensional ring (Glass and Josephson 1995; Nomura and Glass 1996) or a two-dimensional annulus (Glass, Nagai, Hall, Talajic, and Nattel 2002). By the continuity theorem discussed in Section 5.2.1 (p. 126), the observation of a discontinuity in the resetting curves has an important implication: that there should exist stimulation parameters that would lead to a transition so that the system no longer displays a single reentrant circulating wave (Glass and Josephson 1995; Gedeon and Glass 1998; Glass, Nagai, Hall, Talajic, and Nattel 2002). This type of annihilation is observed in numerical studies. Further, the observation of the annihilation of ventricular tachycardia by a single pulse, shown in Figure 5.1, is consistent with this theoretical concept and also the theoretical interpretation that the ventricular tachycardia in the patient whose record is displayed in Figure 5.1 was generated by a reentrant excitation.

## 5.5    Resetting and Entrainment of Cardiac Oscillations

The computational machinery outlined above can be applied in practical situations. I will very briefly recount work from our group, and give references to more complete descriptions.

Extensive studies of the effects of single and periodic stimulation on spontaneously beating aggregates of embryonic chick heart cells have been carried out by Michael Guevara, Wanzhen Zeng, and Arkady Kunysz working in Alvin Shrier's laboratory at McGill University. The objective has been to determine the phase resetting behavior under single stimuli and to apply these results to compute the effects of periodic stimulation (Guevara et al. 1981; Guevara et al. 1986; Glass et al. 1983; Glass et al. 1984; Glass et al. 1987; Guevara et al. 1988; Zeng et al. 1990; Kowtha et al. 1994).

The results of the studies on the entrainment of the spontaneously beating aggregates of heart cells are summarized in Figure 5.11, which shows the different main locking regions using numerical iteration of experimentally determined resetting curves using the methods described above and shows examples of representative rhythms. The main findings of the experimental studies are:

- There are many different phase locking regions. For low to moderate stimulation amplitudes, the largest zones that can be readily observed in every experiment, are 1:1, 1:2, 3:2, 2:1, 3:1, 2:3. In addition, other zones corresponding to rational ratios $p : m$, where $p$ and $m$ are 4 or less, can usually be observed near the theoretically predicted region in Figure 5.11.

- For several different sets of stimulation amplitude and frequency there are aperiodic dynamics (Guevara, Glass, and Shrier 1981). There is a particular zone, using moderate stimulation amplitude and frequencies slightly less than the intrinsic frequency, that leads to period-doubling bifurcations and deterministic chaos. In this region, plots of $\phi_{i+1}$ as a function of $\phi_i$ based on the experimental data are approximately one-dimensional with characteristic shape associated with one-dimensional maps that give chaotic dynamics.

As a second example, I briefly recount more recent results from experiments in which reentrant excitation in an animal was subjected to single

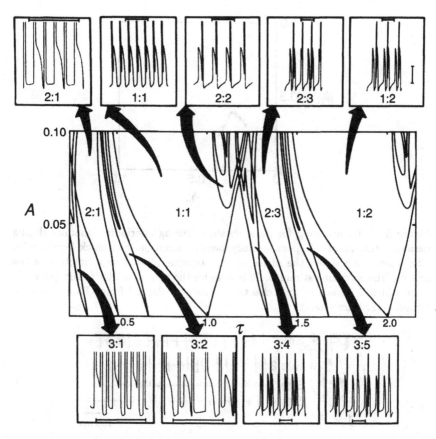

Figure 5.11. Locking zones for periodically stimulated heart cell aggregates. The computations are based on experimentally measured resetting curves. The time bar is 1 sec. Adapted from Glass, Guevara, and Shrier (1987).

and periodic stimuli carried out in the laboratory of Stanley Nattel at the Montreal Heart Institute; see Glass, Nagai, Hall, Talajic, and Nattel (2002), which should be consulted for full details.

Experiments were carried out in anesthetized dogs. One of the upper chambers of the heart, the right atrium, had incisions made that result in the establishment of a circulating excitation similar to the clinical arrhythmia atrial flutter. During the course of this rhythm, both single and periodic stimuli could be delivered from different places on the heart's surface, and the dynamics could be measured at other sites on the heart's surface. Figure 5.13 shows a schematic diagram of the surgery and the observed activity during the tachycardia.

During periodic stimulation the activity could be entrained in a 1:1 fashion over a broad range of stimulation periods ranging from 115 ms to 145 ms; Figure 5.14. However, the experimental determination of the reset-

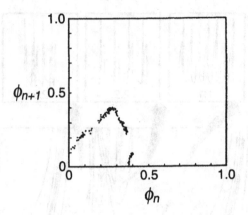

Figure 5.12. Return map for data obtained during aperiodic dynamics during periodic stimulation of spontaneously beating aggregates of chick hearts cells. The return map shows the phase of one stimulus plotted as a function of the phase of the preceding stimulus. This form for the map is similar to the quadratic map, which is known to give chaotic dynamics. Adapted from Glass, Guevara, Bélair, and Shrier (1984).

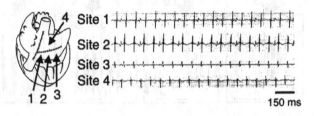

Figure 5.13. Schematic diagram of surgical procedures that led to the establishment of a reentrant rhythm in a canine model of human atrial flutter and electrical recordings from different sites in the heart during atrial flutter. From Glass, Nagai, Hall, Talajic, and Nattel (2002).

ting curve followed by the application of the theory in Section 5.2.2 led to the conclusion that 1:1 entrainment should be possible over a more limited range of stimulation periods from 130 ms to 145 ms. The failure of the resetting curve to predict the range of entrainment in this example likely arises because the stimulus is some distance away from the circuit responsible for the reentrant tachycardia.

## 5.6   Conclusions

Single stimuli reset or annihilate stable nonlinear oscillations. Periodic stimuli delivered during the course of nonlinear oscillations can lead to a wide range of different behaviors including quasiperiodicity, entrainment, chaos,

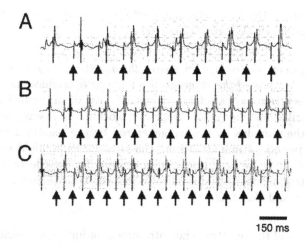

Figure 5.14. Experimental studies of the entrainment of atrial flutter with stimulation at site 4 and recording from site 2 in Figure 5.13. A. $t_s = 145$ ms, $t_s/T_0 = 0.97$; B. $t_s = 130$ ms, $t_s/T_0 = 0.0.87$; C. $t_s = 115$ ms, $t_s/T_0 = 0.77$. From Glass, Nagai, Hall, Talajic, and Nattel (2002).

and annihilation. Although the origin of all these different behaviors can be interpreted using the iteration of low-dimensional maps, our studies have shown that a number of different factors that exist in the real world tend to limit our ability to predict the effects of periodic stimulation using iteration of low-dimensional maps.

- *Stimulation can change the properties of the oscillation.* Although in numerical models, time constants are often very fast and the ongoing activity does not modify system properties, in experiments and in clinical circumstances the parameters of cardiac oscillations are modified under the rapid repetitive activity. Thus, the idealizations that the properties of oscillators are not affected by the stimulation need to be modified for a more complete analysis. Though such modifications have been implemented occasionally (Kunysz, Glass, and Shrier 1997; Vinet 1999), the simplicity of the one-dimensional circle map as a model for periodically forced biological oscillators is lost in the process, and the resulting models, though more realistic, are often less esthetically pleasing.

- *There can be slow relaxation times to the limit cycle.* Following a stimulus, the limit cycle is not necessarily reestablished prior to the time of the next stimulus (Zeng, Glass, and Shrier 1992). At fast stimulation rates, the finite relaxation time to the limit cycle may make it difficult to predict the effects of periodic stimulation based on resetting curves. Even in the Poincaré oscillator, when the relaxation time is finite, there are still only limited results on the fine

structure of the dynamics as a function of the stimulation frequency
and amplitude (Glass and Sun 1994).

- *For systems distributed in space, the resetting may depend on the
  spatial location of the stimulus.* For limit cycle oscillations associated
  with reentrant excitation in space, the failure of an excitation to reset
  the oscillation may result from the blocking of a wave originating
  from the stimulus by a wave generated by the reentrant excitation.
  Since periodic stimulation from the same location might entrain the
  oscillation, there would be a failure to predict the effects of periodic
  stimulation using the information about resetting (Glass, Nagai, Hall,
  Talajic, and Nattel 2002).

To understand the resetting and entrainment of limit cycle oscillations in
biological systems, there needs to be a mix of theory and experiment. Experimental studies of the effects of single and multiple stimuli delivered to
biological oscillators yield beautiful data in which the timing of key events
can be measured over time. Since the rhythms that are observed depend
on the parameters of the stimulation (amplitude of stimuli, frequency of
stimuli, number of stimuli, initial phase of stimuli), systematic studies are
essential. This chapter has sketched out the basic mathematics that I believe is essential for understanding these experiments, and has given some
examples where the methods have yielded unexpected results.

## 5.7   Acknowledgments

This chapter reviews work carried out with a number of colleagues. I have
benefited greatly from these collaborations and give particular thanks to
Michael Guevara, Alvin Shrier, Arthur T. Winfree, Zeng Wanzhen, Jiong
Sun, Arkady Kunysz, Taishin Nomura, Stanley Nattel, Yoshihiko Nagai.
This material has been presented at a number of summer schools held in a
variety of locales including Salt Lake City, Montreal, Saõ Paulo, Maribor,
and Warsaw. I appreciate the feedback and interactions from the students
which have helped me understand this work. The work has been supported
by the Natural Sciences and Engineering Council of Canada, the Canadian
Heart and Stroke Foundation, the Mathematics of Information Technology
and Complex Systems Centre of Excellence (Canada), and the Research
Resource for Complex Physiologic Signals funded by the National Institutes
of Health, USA. Some of current chapter is based on an earlier presentation
of the same material in (Glass 1997).

# 5.8  Problems

Analysis of the following problems will facilitate understanding and may be useful for those who wish to do further research on these topics.

1. Suppose that you observe a biological oscillation with a period of oscillation that is quite stable; say there is a coefficient of variation of about 3%. Also assume that you have found a stimulus that can reset the oscillation, as measured by a marker event of the oscillation. For example, the stimulus might be a current pulse delivered to a neural or cardiac oscillation, and the marker event would be the onset of an action potential. How can the new phase induced by the stimulus be measured experimentally?

2. Annihilation of a limit cycle oscillation by a single stimulus can be fatal, if the oscillation is essential for life, or helpful, if the oscillation is a dangerous arrhythmia. Discuss the circumstances under which knowledge about the resetting curves of biological systems can be used to predict whether annihilation of a stable limit cycle oscillation using a single stimulus is possible. Critically discuss and contrast the postulated mechanisms for annihilation of limit cycles presented in Jalife and Antzelevitch (1979), Paydarfar and Buerkel (1995), and Glass and Josephson (1995).

3. The Poincaré oscillator provides a simple model that is amenable to significant algebraic analysis. The text gives the algebraic expression for the resetting curve, equation (5.7), and also the boundaries of the 1:1 locking regions in equations (5.8) and (5.9). Derive these equations.

4. Extending the results in the problem above to the finite relaxation case is difficult; the current state of the art is in Glass and Sun (1994). The problem is to analyze the loss of stability of the period-1 fixed points in the Poincaré oscillator with finite relaxation times as a function of the stimulation strength $b$ and the relaxation $k$. It is necessary to keep in mind that there can be more than one fixed point for given parameter values, and that the initial condition is important. Good results would merit publication.

5. Experimental project. Select any biological or physical oscillation that can be reset using brief pulsatile stimuli. Determine the phase resetting curves for a range of stimuli and amplitudes, and the phase locking for a range of amplitudes and periods. Can the resetting curves be used to predict the effects of periodic stimulation? Careful studies would merit publication.

## 5.9    Computer Exercises: Resetting Curves for the Poincaré Oscillator

One of the simplest models of a limit cycle oscillation is the Poincaré oscillator. The equations for this model are

$$\frac{dr}{dt} = kr(1 - r),$$

$$\frac{d\phi}{dt} = 2\pi, \qquad\qquad (5.16)$$

where $k$ is a positive parameter. Starting at any value of $r$, except $r = 0$, there is an evolution until $r = 1$. The parameter $k$ controls the relaxation rate. In these exercises, we consider the relaxation in the limit $k \to \infty$.

### Software

There are 2 Matlab* programs you will use for this exercise:

**resetmap(b)** This program computes the resetting curve (new phase versus old phase) for a stimulus strength **b**. The output is a matrix with 102 columns and 2 arrays. The first array is the old phase ranging from 0 to 1. There are two points just less than and just greater than $\phi = 0.5$. These points are needed especially for the case where $b > 1$. The second array is the new phase.

**poincare(phizero,b,tau,niter)** This program does an iteration of the periodically stimulated Poincaré oscillator, where **phizero** is the initial phase, **b** is the stimulation strength, **tau** is the period of the stimulation, and **niter** is the number of iterations. It is valid for $0 < \tau < 1$. The output consists of two arrays:

The first array (called **phi** in the following) is a listing of the successive phases during the periodic stimulation.
The second array (called **beats** in the following) is a listing of the number of beats that occur between successive stimuli.

### How to Run the Programs

- To compute the resetting curve for $b = 1.10$, type

        [phi,phiprime]=resetmap(1.10);

- To plot out the resetting curve just computed, type

        plot(phi,phiprime,'*')

---

*See Introduction to Matlab in Appendix B.

- To simulate periodic stimulation of the Poincaré oscillator, type

    ```
    [phi,beats]=poincare(.3,1.13,0.35,100);
    ```

    This will generate two time series of 100 iterates from an initial condition of $\phi = 0.3$, with $b = 1.13$ and $\tau = 0.35$. The array **phi** is the successive phases during the stimulation. The array **beats** is the number of beats between stimuli.

- To display the output as a return map, type

    ```
    plot(phi(2:99),phi(3:100),'*')
    ```

    This plots out the successive phases of each stimulus as a function of the phase of the preceding stimulus. The points lie on a one-dimensional curve. The dynamics in this case are chaotic. In fact, what is observed here is very similar to what is actually observed during periodic stimulation of heart cell aggregates described in the first chapter.

- To display the number of beats between stimuli, type

    ```
    plot(beats,'*')
    ```

- The rotation number gives the ratio between the number of beats and the number of stimuli during a stimulation. This is the average number of beats per stimulus. To compute the rotation number, type

    ```
    sum(beats)/length(beats)
    ```

## Exercises

Ex. 5.9-1. **Compute resetting curves for varying values of b.** Use the program **resetmap** to compute the resetting curves for several values of $b$ in the range from 0 to 2. In particular, determine the value of $b$ at which the topology of the resetting curve changes. **Note:** The value $b = 1$ is a singular value, and the program does not work for that value. You could try to compute the analytic form for $b = 1$, or you could consult (Keener and Glass 1984) for the surprising answer.

Ex. 5.9-2. **Test the periodicity of iterates of phi.** Use the program **poincare** to compute the succesive iterates of **phi** for different values of $(b, \tau)$ and use the program **testper** (see lab 2.8) to determine whether or not the successive iterates of **phi** are periodic. You might wish to modify the programs so that they loop through several values of $(b, \tau)$. Stable periodic points correspond to stable patterns of entrainment between the stimulus and the oscillator.

   (a) Refer to Figure 5.7. Select values of $b$ and $\tau$ that are expected to give 1:1, 2:1, 3:2, 2:2 phase locking and try to confirm that these behaviors are in fact observed in the simulations. In doing this, you should realize that the rotation number is given by the ratio of the number of action potentials to the number of stimuli.

(b) Find values for which there are different asymptotic behaviors depending on the initial condition.

(c) Find values of $b$ and $\tau$ that give quasiperiodicity. How many iterates do you need to carry out to convince yourself that the behavior is quasiperiodic rather that periodic with a long period? Choose a value of $b > 0$.

(d) Find a period-doubling route to chaos.

**Ex. 5.9-3. Dynamics over the $(b,\tau)$ plane.** (Hard): Determine the dynamics over the $(b, \tau)$ parameter plane and draw a diagram with the results. You should get the diagram in Figure 5.7.

**Ex. 5.9-4. Dynamics of 2-D Poincaré oscillator with finite relaxation time.** (Research level). Consider the two-dimensional Poincaré oscillator, equation (5.16), with finite relaxation time ($k$ is finite). Investigate the dynamics of this equation. What has been found out so far is in (Glass and Sun 1994).

Any good results on the following questions merit a research publication:

(a) What are the dynamics where the period-1 orbit becomes unstable? You need to get some analytic results giving particular consideration to the presence of subcritical and supercritical Hopf bifurcations.

(b) How many stable orbits can exist simultaneously? Describe the different stable periodic orbits for some subset of parameter space.

(c) For what range of parameter values are there chaotic dynamics?

(d) Give an analytic proof of chaos in this example and explore the routes to chaos.

Do you agree that it is important to understand this example as well as the ionic mechanisms of heart cell aggregates to understand the effects of periodic stimulation of the aggregates?

# 6

# Effects of Noise on Nonlinear Dynamics

## André Longtin

## 6.1  Introduction

The influence of noise on nonlinear dynamical systems is a very important area of research, since all systems, physiological and other, evolve in the presence of noisy driving forces. It is often thought that noise has only a blurring effect on the evolution of dynamical systems. It is true that that can be the case, especially for so-called "observational" or "measurement" noise, as well as for linear systems. However, in nonlinear systems with dynamical noise, i.e., with noise that acts as a driving term in the equations of motion, noise can drastically modify the deterministic dynamics. For example, the hallmark of nonlinear behavior is the bifurcation, which is a qualitative change in the phase space motion when the value of one or more parameter changes. Noise can drastically modify the dynamics of a deterministic dynamical system. It can make the determination of bifurcation points very difficult, even for the simplest bifurcations. Noise can shift bifurcation points or induce behaviors that have no deterministic counterpart, through what are known as noise-induced transitions (Horsthemke and Lefever 1984). The combination of noise and nonlinear dynamics can also produce time series that are easily mistakable for deterministic chaos. This is especially true in the vicinity of bifurcation points, where the noise has its greatest influence.

This chapter considers these issues starting from a basic level of description: the stochastic differential equation. It discusses sources of noise, and shows how noise, or "stochastic processes," can be coupled to deterministic differential equations. It also discusses analytical tools to deal with stochastic differential equations, as well as simple methods to numerically integrate such equations. It then focuses in greater detail on trying to pinpoint a Hopf bifurcation in a real physiological system, which will lead to the notion of a noise-induced transition. This system is the human pupil light reflex. It has been studied by us and others both experimentally and theoretically. It is also the subject of Chapter 9, by John Milton; he has been involved in the work presented here, along with Jelte Bos (who carried out some of the experiments at the Free University of Amsterdam) and Michael Mackey. This chapter then considers noise-induced firing in excitable systems, and how noise interacts with deterministic input such as a sine wave to produce various forms of "stochastic phase locking."

Noise is thought to arise from the action of a large number of variables. In this sense, it is usually understood that noise is high-dimensional. The mathematical analysis of noise involves associating a random variable with the high-dimensional physical process causing the noise. For example, for a cortical cell receiving synaptic inputs from ten thousand other cells, the ongoing synaptic bombardment may be considered as a source of current noise. The firing behavior of this cell may then be adequately described by assuming that the model differential equations governing the excitability of this cell (e.g., Hodgkin–Huxley-type equations) are coupled to a random variable describing the properties of this current noise.

Although we may consider these synaptic inputs as noise, the cell may actually make more sense of it than we can, such as in temporal and spatial coincidences of inputs. Hence, one person's noise may be another person's information: It depends ultimately on the phenomena you are trying to understand. This explains in part why there has been such a thrust in the last decades to discover simple low-dimensional deterministic laws (such as chaos) governing observed noisy fluctuations.

From the mathematical standpoint, noise as a random variable is a quantity that fluctuates aperiodically in time. To be a useful quantity to describe the real world, this random variable should have well-defined properties that can be measured experimentally, such as a distribution of values (a density) with a mean and other moments, and a two-point correlation function. Thus, although the variable itself takes on a different set of values every time we look at it or simulate it (i.e., for each of its "realizations"), its statistical and temporal properties remain constant. The validity of these assumptions in a particular experimental setting must be properly assessed, for example by verifying that certain stationarity criteria are satisfied.

One of the difficulties with modeling noise is that in general, we do not have access to the noise variable itself. Rather, we usually have access to a state variable of a system that is perturbed by one or more sources of noise. Thus, one may have to begin with assumptions about the noise and its coupling to the dynamical state variables. The accuracy of these assumptions can later be assessed by looking at the agreement of the predictions of the resulting model with the experimental data.

## 6.2   Different Kinds of Noise

There is a large literature on the different kinds of noise that can arise in physical and physiological systems. Excellent references on this subject can be found in Gardiner (1985) and Horsthemke and Lefever (1984). An excellent reference for noise at the cellular level is the book by DeFelice (1981). These books provide background material on thermal noise (also known as Johnson–Nyquist noise, fluctuations that are present in any system due to its temperature being higher than absolute zero), shot noise (due to the motion of individual charges), $1/f^\alpha$ noise (one-over-$f$ noise or flicker noise, the physical mechanisms of which are still the topic of whole conferences), Brownian motion, Ornstein–Uhlenbeck colored noise, and the list goes on.

In physiology, an important source of noise consists of conductance fluctuations of ionic channels, due to the (apparently) random times at which they open and close. There are many other sources of noise associated with channels (DeFelice 1981). In a nerve cell, noise from synaptic events can be more important than the intrinsic sources of noise such as conductance fluctuations. Electric currents from neighboring cells or axons are a form of noise that not only affects recordings through measurement noise but also a cell's dynamics. There are fluctuations in the concentrations of ions and other chemicals forming the milieu in which the cells live. These fluctuations may arise on a slower time scale than the other noises mentioned up to now.

The integrated electrical activity of nerve cells produces the electroencephalogram (EEG) pattern with all its wonderful classes of fluctuations. Similarly, neuromuscular systems are complex connected systems of neurons, axons, and muscles, each with its own sources of noise. Fluctuations in muscle contraction strength are dependent to a large extent on the firing patterns of the motorneurons that drive them. All these examples should convince you that the modeling of noise requires a knowledge of the basic physical processes governing the dynamics of the variables we measure.

There is another kind of distinction that must be applied, namely, that between observational, additive, and multiplicative noise (assuming a stationary noise). In the case of observational noise, the dynamical system evolves deterministically, but our measurements on this system are contaminated by noise. For example, suppose a one-dimensional dynamical system is governed by

$$\frac{dx}{dt} = f(x, \mu), \tag{6.1}$$

where $\mu$ is a parameter. Then observational noise corresponds to the measurement of $y(t) \equiv x(t) + \xi(t)$, where $\xi(t)$ is the observational noise process. The measurement $y$, but not the evolution of the system $x$, is affected by the presence of noise. While this is often an important source of noise with which analyses must contend, and the simplest to deal with mathematically, it is also the most boring form of noise in a physical system: It does not give rise to any new effects.

One can also have additive sources of noise. These situations are characterized by noise that is independent of the precise state of the system:

$$\frac{dx}{dt} = f(x, \mu) + \xi(t) \,. \tag{6.2}$$

In other words, the noise is simply added to the deterministic part of the dynamics. Finally, one can have multiplicative noise, in which the noise is dependent on the value of one or many state variables. Suppose that $f$ is separable into a deterministic part and a stochastic part that depends on $x$:

$$\frac{dx}{dt} = h(x) + g(x)\xi(t) \,. \tag{6.3}$$

We then have the situation where the effect of the noise term will depend on the value of the state variable $x$ through $g(x)$. Of course, a given system may have one or more noise sources coupled in one or more of these ways. We will discuss such sources of noise in the context of our experiments on the pupil light reflex and our study of stochastic phase locking.

## 6.3  The Langevin Equation

Modeling the precise effects of noise on a dynamical system can be very difficult, and can involve a lot of guesswork. However, one can already gain insight into the effect of noise on a system by coupling it additively to **Gaussian white noise**. This noise is a mathematical construct that approximates the properties of many kinds of noise encountered in experimental situations. It is Gaussian distributed with zero mean and autocorrelation $\langle \xi(t)\xi(s) \rangle = 2D\delta(t - s)$, where $\delta$ is the Dirac *delta* function. The quantity $D \equiv \sigma^2/2$ is usually referred to as the **intensity of the Gaussian white noise** (the actual intensity with the autocorrelation scaling used in our example is $2D$). Strictly speaking, the variance of this noise is infinite, since it is equal to the autocorrelation at zero lag (i.e. at $t = s$). However, its intensity is finite, and $\sigma$ times the square root of the time step will be the standard deviation of Gaussian random numbers used to numerically generate such noise (see below).

The Langevin equation refers to the stochastic differential equation obtained by adding Gaussian white noise to a simple first-order linear dynamical system with a single stable fixed point:

$$\frac{dx}{dt} = -\alpha x + \xi(t) \,. \tag{6.4}$$

This is nothing but equation (6.2) with $f(x, \mu) = -\alpha x$, and $\xi(t)$ given by the Gaussian white noise process we have just defined. The stochastic process $x(t)$ is also known as Ornstein–Uhlenbeck noise with correlation time $1/\alpha$, or as lowpass-filtered Gaussian white noise. A noise process that is not white noise, i.e. that does not have a delta-function auto-correlation, is called "colored noise". Thus, the exponentially correlated Ornstein–Uhlenbeck noise is a colored noise. The probability density of this process is given by a Gaussian with zero mean and variance $\sigma^2/2\alpha$ (you can verify this using the Fokker–Planck formalism described below). In practical work, it is important to distinguish between this variance, and the intensity $D$ of the Gaussian white noise process used to produce it. It is in fact common to plot various quantities of interest for a stochastic dynamical system as a function of the intensity $D$ of the white noise used in that dynamical system, no matter where it appears in the equations.

The case in which the deterministic part is nonlinear yields a **nonlinear Langevin equation**, which is the usual interesting case in mathematical physiology. One can simulate a nonlinear Langevin equation with deterministic flow $h(x, t, \mu)$ and a coefficient $g(x, t, \mu)$ for the noise process using various stochastic numerical integration methods of different orders of precision (see Kloeden, Platen, and Schurz 1991 for a review). For example, one can use a simple "stochastic" Euler–Maruyama method with fixed time step $\Delta t$ (stochastic simulations are much safer with fixed step methods). Using the definition of Gaussian white noise $\xi(t)$ as the derivative of the Wiener process $W(t)$ (the Wiener process is also known as "Brownian motion"), this method can be written as

$$x(t + \Delta t) = x(t) + \Delta t \, h(x, t, \mu) + g(x, t, \mu)\Delta W_n \,, \qquad (6.5)$$

where the $\{\Delta W_n\}$ are "increments" of the Wiener process. These increments can be shown to be independent Gaussian random variables with zero mean and standard deviation given by $\sigma\sqrt{\Delta t}$. Because the Gaussian white noise $\xi(t)$ is a function that is nowhere differentiable, one must use another kind of calculus to deal with it (the so-called stochastic calculus; see Horsthemke and Lefever 1984; Gardiner 1985). One consequence of this fact is the necessity to exercise caution when performing a nonlinear change of variables on stochastic differential equations: The "Stratonovich" stochastic calculus obeys the laws of the usual deterministic calculus, but the "Ito" stochastic calculus does not. One thus has to associate an Ito or Stratonovich interpretation with a stochastic differential equation before performing coordinate changes. Also, for a given stochastic differential equation, the properties of the random process $x(t)$ such as its moments will depend on which calculus is assumed. Fortunately, there is a simple transformation between the Ito and Stratonovich forms of the stochastic differential equation. Further, the properties obtained with both calculi are identical when the noise is additive.

It is also important to associate the chosen calculus with a proper integration method (Kloeden, Platen, and Schurz 1991). For example, an explicit Euler–Maruyama scheme is an Ito method, so numerical results with this method will agree with any theoretical results obtained from an analysis of the Ito interpretation of the stochastic differential equation, or of its equivalent Stratonovich form. In general, the Stratonovich form is best suited to model "real colored noise" and its effects in the limit of vanishing correlation time, i.e. in the limit where colored noise is allowed to become white after the calculation of measurable quantities.

Another consequence of the stochastic calculus is that in the Euler–Maruyama numerical scheme, the noise term has a magnitude proportional to the square root of the time step, rather than to the time step itself. This makes this method an "order $\frac{1}{2}$" method, which converges more slowly than the Euler algorithm for deterministic differential equations. Higher-order methods are also available (Fox, Gatland, Roy, and Vemuri 1988; Mannella and Palleschi 1989; Honeycutt 1992); some are used in the computer exercises associated with this chapter and with the chapter on the pupil light reflex (Chapter 9). Such methods are especially useful for stiff stochastic problems, such as the Hodgkin–Huxley or FitzHugh–Nagumo equations with stochastic forcing, where one usually uses an adaptive method in the noiseless case, but is confined to a fixed step method with noise. Stochastic simulations usually require multiple long runs ("realizations": see below) to get good averaging, and higher-order methods are useful for that as well.

The Gaussian random numbers $\Delta W_n$ are generated in an uncorrelated fashion, for example by using a pseudorandom number generator in combination with the Box–Müller algorithm. Such algorithms must be "seeded," i.e., provided with an initial condition. They will then output numbers with very small correlations between themselves. A simulation that uses Gaussian numbers that follow one initial seed is called a realization. In certain problems, it is important to repeat the simulations using $M$ different realizations of $N$ points (i.e., $M$ with different seeds). This performs an average of the stochastic differential equation over the distribution of the random variable. It also serves to reduce the variance of various statistical quantities used in a simulation (such as power spectral amplitudes). In the case of very long simulations, it also avoids problems associated with the finite period of the random number generator.

A stochastic simulation yields a different trajectory for each different seed. It is possible also to describe the action of noise from another point of view, that of probability densities. One can study, for example, how an ensemble of initial conditions, characterized by a density, propagates under the action of the stochastic differential equation. One can study also the probability density of measuring the state variable between $x$ and $x + dx$ at a given time. The evolution of this density is governed by a deterministic partial differential equation in the density variable $\rho(x, t)$, known as the Fokker–Planck equation. In one dimension, this equation is

$$\frac{\partial \rho}{\partial t} = \frac{1}{2} \frac{\partial^2 \left[ g^2(x) \rho \right]}{\partial x^2} - \frac{\partial \left[ h(x) \rho \right]}{\partial x}. \tag{6.6}$$

Setting the left-hand side to zero and solving the remaining ordinary differential equation yields the asymptotic density for the stochastic differential equation, $\rho^* = \rho(x, \infty)$. This corresponds to the probability density of finding the state variable between $x$ and $x + dx$ once transients have died out, i.e., in the long-time limit. Since noisy perturbations cause transients, this long-time density somehow characterizes not only the deterministic attractors, but also the noise-induced transients around these attractors. For simple systems, it is possible to calculate $\rho^*$, and sometimes even $\rho(x, t)$ (this can always be done for linear systems, even with time-dependent coefficients). However, for nonlinear problems, in general one can at best approximate $\rho^*$. For delay differential equations such as the one studied in the next section, the Fokker–Planck formalism breaks down. Nevertheless, it is possible to calculate $\rho^*$ numerically, and even understand some of its properties analytically (Longtin, Milton, Bos, and Mackey 1990; Longtin 1991a; Guillouzic, L'Heureux, and Longtin 1999).

A simple example of a nonlinear Langevin equation is

$$\frac{dx}{dt} = x - x^3 + \xi(t), \tag{6.7}$$

which models the overdamped noise-driven motion of a particle in a bistable potential. The deterministic part of this system has three fixed points, an unstable one at the origin and stable ones at $\pm 1$. For small noise intensity $D$, the system spends a long time fluctuating on either side of the origin before making a switch to the other side, as shown in Figure 6.1. Increasing the noise intensity increases the frequency of the switches across the origin. At the same time, the asymptotic probability density broadens around the stable points, and the probability density in a neighborhood of the (unstable) origin increases; the vicinity of the origin is thus stabilized by noise. One can actually calculate this asymptotic density exactly for this system using the Fokker–Planck formalism (try it! the answer is $\rho(x) = C \exp\left[(x^2 - x^4)/2D\right]$, where $C$ is a normalization constant). Also, because this is a one-dimensional system with additive noise, the maxima of the density are always located at the same place as for the deterministic case. The maxima are not displaced by noise, and no new maxima are created; in other words, there are no noise-induced states. This is not always the case for multiplicative noise, or for additive or multiplicative noise in higher dimensions.

## 6.4   Pupil Light Reflex: Deterministic Dynamics

We illustrate the effect of noise on nonlinear dynamics by first considering how noise alters the behavior of a prototypical physiological control system. The pupil light reflex, which is the focus of Chapter 9, is a negative feedback control system that regulates the amount of light falling on the retina. The

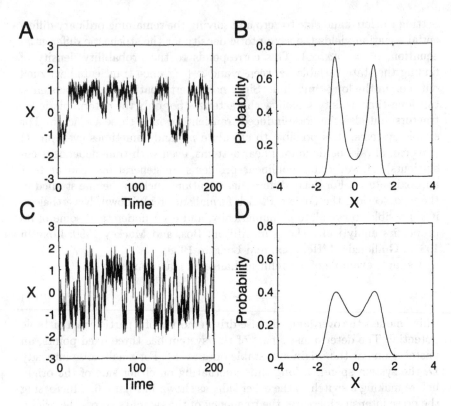

Figure 6.1. Realizations of equation (6.7) at (A) low noise intensity $D = 0.5$ and (C) high noise intensity $D = 1.0$. The corresponding normalized probability densities are shown in (B) and (D) respectively. These densities were obtained from 30 realizations of 400,000 iterates; the integration time step for the stochastic Euler method is 0.005.

pupil is the hole in the middle of the colored part of the eye called the iris. If the ambient light falling on the pupil increases, the reflex response will contract the iris sphincter muscle, thus reducing the area of the pupil and the light flux on the retina. The delay between the variation in light intensity and the variation in pupil area is about 300 msec. A mathematical model for this reflex is developed in Chapter 9. It can be simplified to the following form:

$$\frac{dA}{dt} = -\alpha A + \frac{c}{1 + \left[\frac{A(t-\tau)}{\theta}\right]^n} + k, \tag{6.8}$$

where $A(t)$ is the pupil area, and the second term on the right is a sigmoidal negative feedback function of the area at a time $\tau = 300$ msec in the past. Also, $\alpha, \theta, c, k$ are constants, although in fact, $c$ and $k$ fluctuate noisily. The parameter $n$ controls the steepness of the feedback around the fixed point,

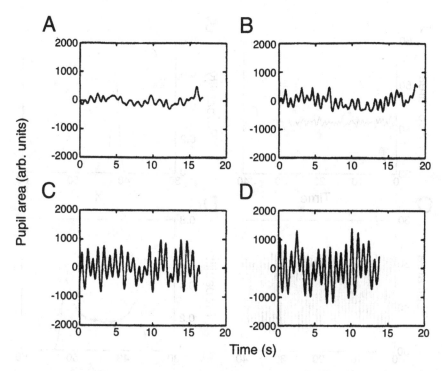

Figure 6.2. Experimental time series of pupil area measured on a human subject for four different values of the feedback gain. Gain values are (A) 1.41, (B) 2.0, (C) 2.82, and (D) 4.0. From Longtin (1991b).

which is proportional to the feedback gain. If $n$ is increased past a certain value $n_o$, or the delay $\tau$ past a critical delay, the single stable fixed point will become unstable, giving rise to a stable limit cycle (supercritical Hopf bifurcation). It is possible to artificially increase the parameter $n$ in an experimental setting involving humans (Longtin, Milton, Bos, and Mackey 1990). We would expect that under normal operating conditions, the value of $n$ is sufficiently low that no periodic oscillations in pupil area are seen. As $n$ is increased, the deterministic dynamics tell us that the amplitude of the oscillation should start increasing proportionally to $\sqrt{n - n_o}$. Experimental data are shown in Figure 6.2, in which the feedback gain, proportional to $n$, increases from panel A to D.

What is apparent in this system is that noisy oscillations are seen even at the lowest value of the gain; they are seen even below this value (not shown). In fact, it is difficult to pinpoint a qualitative change in the oscillation waveform as the gain increases. Instead, the amplitude of the noisy oscillation simply increases, and in a sigmoidal fashion rather than a square root fashion. This is not what the deterministic model predicts. It is possible that the aperiodic fluctuations arise through more complicated dynamics

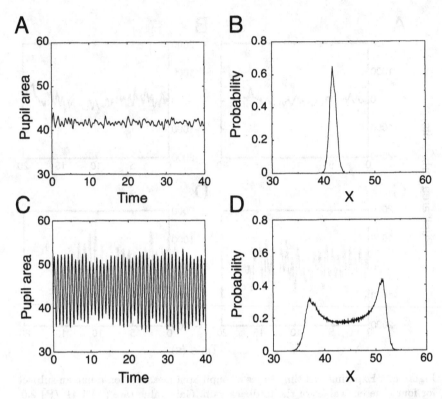

Figure 6.3. Characterization of pupil area fluctuations obtained from numerical simulations of equation (6.8) with multiplicative Gaussian colored noise on the parameter $k$; the intensity is $D = 15$, and the noise correlation time is $\alpha^{-1} = 1$. The bifurcation parameter is $n$; a Hopf bifurcation occurs (for $D = 0$) at $n = 8.2$. (A) Realization for $n = 4$; the corresponding normalized probability density is shown in (B). (C) and (D) are the same as, respectively, (A) and (B) but for $n = 10$. The densities were computed from 10 realizations, each of duration equal to 400 delays.

(e.g., chaos) in this control system. However, such dynamics are not present in the deterministic model for any combination of parameters and initial conditions. In fact, there are only two globally stable solutions, either a fixed point or a limit cycle.

Another possibility is that noise is present in this reflex, and what we are seeing is the result of noise driving a system in the vicinity of a Hopf bifurcation. This would not be surprising, since the pupil has a well-documented source of fluctuations known as pupillary hippus. It is not known what the precise origin of this noise is, but the following section will show that we can test for certain hypotheses concerning its nature. Incorporating noise into the model can in fact produce fluctuations that vary similarly to those in Figure 6.2 as the feedback gain is increased, as we will now see.

## 6.5   Pupil Light Reflex: Stochastic Dynamics

We can explain the behaviors seen in Figure 6.2 if noise is incorporated into our model. One can argue, based on the known physiology of this system (see Chapter 9), that noise enters the reflex pathway through the parameters $c$ and $k$, and causes fluctuations about their mean values $\bar{c}$ and $\bar{k}$, respectively. In other words, we can suppose that $c = \bar{c} + \eta(t)$, i.e., that the noise is multiplicative, or $k = \bar{k} + \eta(t)$, i.e., that the noise is additive, or both. This noise represents the fluctuating neural activity from many different areas of the brain that connect to the Edinger–Westphal nucleus, the neural system that controls the parasympathetic drive of the iris. It also is meant to include the intrinsic noise at the synapses onto this nucleus and elsewhere in the reflex arc.

Without noise, equation (6.8) undergoes a supercritical Hopf bifurcation as the gain is increased via the parameter $n$. We have investigated both the additive and multiplicative noise hypotheses by performing stochastic simulations of equation (6.8). The noise was chosen to be Ornstein–Uhlenbeck noise with a correlation time of one second (Longtin, Milton, Bos, and Mackey 1990). Some results are shown in Figure 6.3, where a transition from low amplitude fluctuations to more regular high-amplitude fluctuations is seen as the feedback gain is increased. Results are similar with noise on either $c$ or $k$. Even before the deterministic bifurcation, oscillations with roughly the same period as the limit cycle that appears at the bifurcation are excited by the noise. Increasing the gain just makes them more prominent: In fact, there is no actual bifurcation when noise is present, only a graded appearance of oscillations.

## 6.6   Postponement of the Hopf Bifurcation

We now discuss the problem of pinpointing a Hopf bifurcation in the presence of noise. This is a difficult problem not only for the Hopf bifurcation, but for other bifurcations as well (Horsthemke and Lefever 1984). From the time series point of view, noise causes fluctuations on the deterministic solution that exists without noise. However, and this is the more interesting effect, it can also produce noisy versions of behaviors that occur nearby in parameter space for the noiseless system. For example, as we have seen in the previous section, near a Hopf bifurcation the noise will produce a mixture of fixed-point and limit cycle solutions. In the most exciting examples of the effect of noise on nonlinear dynamics, even new behaviors having no deterministic counterpart can be produced.

The problem of pinpointing a bifurcation in the presence of noise arises because there is no obvious qualitative change in dynamics from the time series point of view, in contrast with the deterministic case. The definition of a bifurcation as a qualitative change in the dynamical behavior when a parameter is varied has to be modified for a noisy system. There is usually more than one way of doing this, depending on which **order parameter** one chooses, i.e., which aspect of the dynamics one focuses on; the location of the bifurcation may also depend on this choice of order parameter.

In the case of the pupil light reflex near a Hopf bifurcation, it is clear from Figure 6.3 that noise causes oscillations even though the deterministic behavior is a fixed point. The noise is simply revealing the behavior beyond the Hopf bifurcation. It is as though noise causes the bifurcation parameter to fluctuate across the deterministic bifurcation. This is a useful way to visualize the effect of noise, but it may be misleading, since the parameter need not fluctuate across the bifurcation point to see a mixture of behaviors below and beyond this point. One can thus say, from the time series point of view, that noise advances the bifurcation point, since (noisy) oscillations are seen where, deterministically, a fixed point should be seen. Further, one can compare features of the noisy oscillation in time with, for example, the same features predicted by a model (see below).

This analysis has its limitations, however, because power spectral (or autocorrelation) measures of the strength of the oscillatory component of the time series do not exhibit a qualitative change as parameters (including noise strength) vary. Rather, for example, the peak in the power spectrum associated with the oscillation simply increases as the underlying deterministic bifurcation is approached or the noise strength is increased. In other words, there is no bifurcation from the spectral point of view. Also, this point of view does not necessarily give a clear picture of the behavior beyond the deterministic bifurcation. For example, can one say that the fixed point, which is unstable beyond the deterministic bifurcation, is stabilized by noise? In other words, does the system spend more time near the fixed point than without noise? This can be an important piece of information about the behavior of a real control system (see also Chapter 8 on cell replication and control).

There are measures that reveal a bifurcation in the noisy system. One measure is based on the computation of invariant densities for the solutions. In other words, let the solution run long enough so that transients have disappeared, and then build a histogram of values of the solution. It is better to repeat this process for many realizations of the stochastic process in order to obtain a smooth histogram.

In the deterministic case, this will produce two qualitatively different densities, depending on whether $n$ is below or above the deterministic Hopf bifurcation point $n_o$. If it is below, then the asymptotic solution is a fixed point, and the density is a delta function at this fixed point: $\rho^*(x) = \delta(x - x^*)$. If $n > n_o$, the solution is approximately a sine wave, for which the density is

$$\rho^*(x) = \frac{1}{\pi A \cos\left[\arcsin(x/A)\right]},$$  (6.9)

where $A$ is the amplitude of the sine wave.

When the noise intensity is greater than zero, the delta function gets broadened to a Gaussian distribution, and the density for the sine wave gets broadened to a smooth double-humped or bimodal function. Examples are shown in Figure 6.3 for two values of the feedback parameter $n$. It is possible then to define the bifurcation in the stochastic context by the transition from unimodality to bimodality (Horsthemke and Lefever 1984). The distance between the peaks can serve as an order parameter for this transition (different order parameters can be defined, as in the physics literature on phase transitions). It represents in some sense the mean amplitude of the fluctuations.

We have found that the transition from a unimodal to a bimodal density occurs at a value of $n$ greater than $n_o$ (Longtin, Milton, Bos, and Mackey 1990). In this sense, the bifurcation is postponed by the noise, with the magnitude of the postponement being proportional to the noise intensity. In certain simple cases (although not yet for the delay-differential equation studied here), it is possible to analytically approximate the behavior of the order parameter with noise. This allows one to predict the presence of a postponement, and to relate this postponement to certain model parameters, especially those governing the nonlinear behavior. A postponement does not imply that there are no oscillations if $n < n_p$, where $n_p$ is the extrapolated bifurcation point for the noisy case (it is very time-consuming to numerically determine this point accurately). As we have seen, when there is noise near a Hopf bifurcation, oscillations are present. However, a postponement does imply that if $n > n_o$, *the presence of noise stabilizes the fixed point*. In other words, the system spends more time near the fixed point with noise than without noise. This is why the density for the stochastic differential equation fills in between the two peaks of the deterministic distribution given in equation (6.9).

One can try to pinpoint the bifurcation in the pupil data by computing such densities at different values of the feedback gain. The result is shown in Figure 6.4 for the data used for Figure 6.2. Even for the highest value of gain, there are clearly oscillations, and the distribution still appears unimodal. However, this is not to say that it is unimodal. A problem arises because a large number of simulated data points (two orders of magnitude more than experimentally available) are needed to properly measure the order parameter, i.e., the distance between the two peaks of the probability density. The bifurcation is not seen from the density point of view in this system with limited data sets and large amounts of noise (the higher the noise, the more data points are required). The model does suggest, however, that a postponement can be expected from the density point of view; in particular, the noisy system spends more time near the fixed point than without noise, even though oscillations occur. Further, the mean, moments, and other features of these densities could be compared to those obtained from time series of similar duration generated by models, in the hope of better understanding the dynamics underlying such noisy experimental systems.

This lack of resolution to pinpoint the Hopf bifurcation motivated us to validate our model using other quantities, such as the mean and relative standard deviation of the amplitude and period fluctuations as gain increases (Longtin, Milton, Bos, and Mackey 1990). That study showed that for noise intensity $D = 15$ and a noise correlation time around one, these quantities have similar values in the experiments and the simulations. This strengthens our belief that stochastic forces are present in this system. Interestingly, our approach of investigating fluctuations across a bifurcation (supercritical Hopf in this case) allows us to amplify the noise in the system, in the sense that it is put under the magnifying glass. This is because noise has a strong effect on the dynamics of a system in the vicinity of a bifurcation point, since there is loss of linear stability at this point (neither the fixed point nor the zero-amplitude oscillation is attracting).

Finally, there is an interesting theoretical aspect to the postponements. We are dealing here with a first-order differential-delay equation. Noise-induced transitions such as postponements are not possible with additive noise in one-dimensional ordinary differential equations (Horsthemke and Lefever 1984). But our numerical results show that in fact, additive noise-induced transitions are possible in a first-order delay-differential equation. The reason behind this is that while the highest-order derivative is one, the delay-differential equation is infinite-dimensional, since it evolves in a functional space (an initial function must be specified). More details on these theoretical aspects of the noise-induced transitions can be found in Longtin (1991a).

## 6.7   Stochastic Phase Locking

The nervous system has evolved with many sources of noise, acting from the microscopic ion channel scale up to the macroscopic scale of the EEG activity. This is especially true for cells that transduce physical stimuli into neuroelectrical activity, since they are exposed to environmental sources of noise, as well as to intrinsic sources of noise such as ionic channel conductance fluctuations, synaptic fluctuations, and thermal noise. Traditionally, sources of noise in sensory systems, such as the senses of audition and touch, have been perceived as a nuisance. For example, they have been thought to limit our aptitude for detecting or discriminating between stimuli.

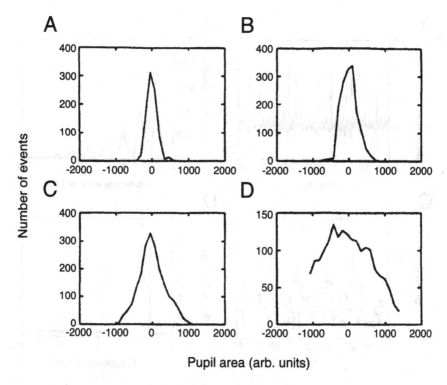

Figure 6.4. Densities corresponding to the time series shown in Figure 6.2 (more data were used than are shown in Figure 6.2). From Longtin (1991b).

In the past decades, there have been studies that revealed a more constructive role for neuronal noise. For example, noise can increase the dynamic range of neurons by linearizing their stimulus–response characteristics (see, e.g., Spekreijse 1969; Knight 1972; Treutlein and Schulten 1985). In other words, noise smoothes out the abrupt increase in mean firing rate that occurs in many neurons as the stimulus intensity increases; this abruptness is a property of the deterministic bifurcation from nonfiring to firing behavior. Noise also makes detection of weak signals possible (see, e.g., Hochmair-Desoyer, Hochmair, Motz, and Rattay 1984). And noise can stabilize systems by postponing bifurcation points, as we saw in the previous section (Horsthemke and Lefever 1984).

In this section, we focus on a special kind of firing behavior exhibited by many kinds of neurons across many different sensory modalities. In general terms, it can be referred to as "stochastic phase locking," but in more specific terms it is known as "skipping." An overview of physiological examples of stochastic phase locking in neurons can be found in Segundo, Vibert, Pakdaman, Stiber, and Martinez (1994). Figure 6.5 plots the membrane potential versus time for a model cell driven by a sinusoidal stimulus

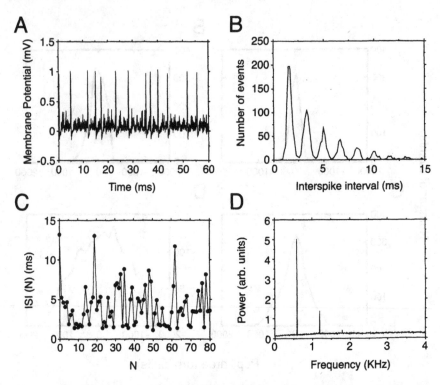

Figure 6.5. Numerical simulation of the FitzHugh–Nagumo equations in the sub-threshold regime, in the presence of noise and sinusoidal forcing. (A) Time series of the membrane potential. (B) Interspike interval histogram obtained from 100 realizations of this stochastic system yielding a total of 2048 intervals. (C) An aperiodic sequence of interspike intervals (ISI). (D) Power spectrum of the spike train, averaged over 100 realizations.

and noise (the model is the FitzHugh–Nagumo equations; see below). The sharp upstrokes superimposed on the "noisy" oscillation are action potentials. The stimulus period here is long in comparison to the action potential duration. The feature of interest is that while the spikes are phase locked to the stimulus, they do not occur at every cycle of the stimulus. Instead, a seemingly random *integer* number of periods of the stimulus are "skipped" between any two successive spikes, thus the term "skipping." This is shown in the interspike interval histogram in Figure 6.5B. The peaks in this interspike interval histogram line up with the integer multiples of the driving period ($T_o = 1.67$ msec). The lack of periodicity in the firing pattern can be inferred from Figure 6.5C, where the interval value is plotted as a function of the interval number (80 intervals are plotted).

In practice, one usually does not have access to the membrane voltage itself, since the sensory cells or their afferent nerve fibers cannot be impaled by a microelectrode. Instead, one has a sequence of interspike intervals from which the mechanisms giving rise to signal encoding and skipping must be inferred.

In the rest of this chapter, we describe some mechanisms of skipping in sensory cells, as well as the potential significance of such firing patterns for sensory information processing. We discuss the phenomenology of skipping patterns, and then describe efforts to model these patterns mathematically. We describe the stochastic resonance effect in this context, and discuss its origins. We also discuss skipping patterns in the context of "bursting" firing patterns. We consider the relation of noise-induced firing to linearization by noise. We also show how noise can alter the shape of tuning curves, and end with an outlook onto interesting issues for future research.

## 6.8 The Phenomenology of Skipping

A firing pattern in which cycles of a stimulus are skipped is a common occurrence in physiology. For example, this behavior underlies $p : m$ phase locking seen in cardiac and other excitable cells, i.e., firing patterns with $m$ responses to $p$ cycles of the stimulus. The main additional properties here are that the phase locking pattern is aperiodic, and remains qualitatively the same as stimulus characteristics are varied. In other words, abrupt changes between patterns with different phase locking ratios are not seen under "skipping" conditions. For example, as the amplitude of the stimulus increases, the skipping pattern remains aperiodic, but there is a higher incidence of short skips rather than long skips.

A characteristic interspike interval histogram for a skipping pattern is shown in Figure 6.5B, and again in Figure 6.6A for in a bursty P-type electroreceptor of a weakly electric fish. The stimulus in this latter case is a 660 Hz oscillatory electric field generated by the fish itself (its "electric organ discharge"). It is modulated by food particles and other objects and fish, and the 660 Hz carrier along with its modulations are read by the receptors in the skin of the fish. This electrosensory system is used for electrolocation and electrocommunication. The interspike interval histogram in Figure 6.6A again consists of a set of peaks located at integer multiples of the driving period. Note from Figure 6.6B that there is no apparent periodicity in the interval sequence. The firing patterns of electroreceptors were first characterized in Scheich, Bullock, and Hamstra Jr (1973). These receptors are known are P-units or "probability coders," since it is thought that their probability of firing is proportional to, and thus encodes, the instantaneous amplitude of the electric organ discharge. Hence this probability, determined by various parameters including the intensity of noise sources acting in the receptor, is an important part of the neuronal code in this system.

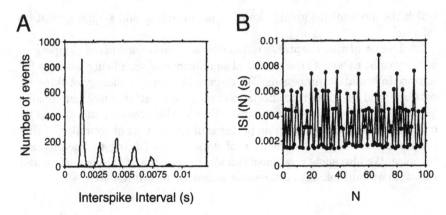

Figure 6.6. Interspike interval histogram and sequence of interspike intervals (ISI) measured from a primary afferent fiber of an electroreceptor of the weakly electric fish *Apteronotus leptorhynchus*. The stimulus frequency is 660 Hz, and is generated by the fish itself. Data provided courtesy of Joe Bastian, U. Oklahoma at Norman.

Another classic example of skipping is found in mammalian auditory fibers. Rose, Brugge, Anderson, and Hind (1967) show that skipping patterns occur at frequencies from below 80 Hz up to 1000 Hz and beyond in a single primary auditory fiber of the squirrel monkey. For all amplitudes, the modes in the interspike interval histogram line up with the integer multiples of the stimulus period, and there is a mode centered on each integer between the first and last visible modes. At low frequencies, multiple firings can occur in the preferred part of the stimulus cycle, and thus a peak corresponding to very short intervals is also seen. At frequencies beyond 1000 Hz, the first peak is usually missing due to the refractory period of the afferent fiber; in other words, the cell cannot recover fast enough to fire spikes one millisecond apart. Nevertheless, phase locking persists as evidenced by the existence of other modes. This is true for electroreceptors as well (Chacron, Longtin, St-Hilaire, and Maler 2000).

The degree of phase locking in all cases is evident from the width of the interspike interval histogram peaks: Sharp peaks correspond to a high degree of phase locking, i.e., to a narrow range of phases of the stimulus cycle during which firing preferentially occurs. As amplitude increases, the multimodal structure of the interspike interval histogram is still present, but the intervals are more concentrated at low values. In other words, the higher the intensity, the lower the (random) integer number of cycles skipped between firings. What is astonishing is that these neurons are highly "tunable" across such a broad range of stimulus parameters, with the modes always lining up with multiples of the driving period. There are many examples of skipping in other neurons, sensory and otherwise, e.g., in mechanoreceptors and thermoreceptors (see Longtin 1995; Segundo, Vibert, Pakdaman,

Stiber, and Martinez 1994; Ivey, Apkarian, and Chialvo 1998 and references therein).

Another important characteristic of skipping is that the positions of the peaks vary smoothly with stimulus frequency, and the envelope of the interspike interval histogram varies smoothly with stimulus amplitude. This is different from the phase-locking patterns governed, for example, by phase-resetting curves leading to an Arnold tongue structure as stimulus amplitude and period are varied. We will see that a plausible mechanism for skipping involves the combination of noise with subthreshold dynamics, although suprathreshold mechanisms exist, as we will see below (Longtin 1998). In fact, we have recently found (Chacron, Longtin, St-Hilaire, and Maler 2000) that suprathreshold periodic forcing of a leaky integrate-and-fire model with voltage and threshold reset can produce patterns close to those seen in electroreceptors of the nonbursty type (interspike interval histogram similar to that in Figure 6.6A, except that the first peak is missing). We focus below on the subthreshold scenario in the context of the FitzHugh–Nagumo equations, with one suprathreshold example as well.

## 6.9  Mathematical Models of Skipping

The earliest analytical/numerical study of skipping was performed by Gerstein and Mandelbrot (1964). Cat auditory fibers recorded during auditory stimulation with periodic "clicks" of noise (at frequencies less than 100 clicks/sec) showed skipping behavior. Gerstein and Mandelbrot were interested in reproducing experimentally observed spontaneous interspike interval histograms using "random walks to threshold models" of neuron firing activity. In one of their simulations, they were able to reproduce the basic features of the interspike interval histogram in the presence of the clicks by adding a periodically modulated drift term to their random walk model. The essence of these models is that the firing activity is entirely governed by noise plus a constant and/or a periodically modulated drift. A spike is associated with the crossing of a fixed threshold by the random variable.

Since this early study, there have been other efforts aimed at understanding the properties of neurons driven by periodic stimuli and noise. The following examples have been excerpted from the large literature on this subject. French et al. (1972) showed that noise breaks up patterns of phase locking to a periodic signal, and that the mean firing rate is proportional to the amplitude of the signal. Glass et al. (1980) investigated an integrate-and-fire model of neural activity in the presence of periodic forcing and noise. They found unstable zones with no phase locking, as well as quasi-periodic dynamics and firing patterns with stochastic skipped beats. Keener et al. (1981) were able to analytically investigate the dynamics of phase locking in a leaky integrate-and-fire model without noise. Alexander et al. (1990) studied phase locking phenomena in the FitzHugh–Nagumo model of a neuron in the excitable regime, again without noise. There have also been studies of noise-induced limit cycles in excitable cell models like the Bonhoeffer–van der Pol equations (similar to the FitzHugh–Nagumo model), but in the absence of periodic forcing (Treutlein and Schulten 1985).

The past two decades have seen a revival of stochastic models of neural firing in the context of skipping. Hochmair-Desoyer et al. (1984) have looked at the influence of noise on firing patterns in auditory neurons using models such as the FitzHugh–Nagumo equations, and shown that it can alter the tuning curves (see below). This model generates real action potentials with a refractory period. It also has many other behaviors that are found in real neurons, such as a resonance frequency. It is a suitable model, however, only when an action potential is followed by a hyperpolarizing after-potential, i.e., the voltage goes below the resting potential after the spike, and slowly increases towards it. It is also a good simplified model to study certain neural behaviors qualitatively; better models exist (they are usually more complex) and should be used when quantitative agreement between theory and experiment is sought. The study of the FitzHugh–Nagumo model in Longtin (1993) was motivated by the desire to understand how stochastic resonance could occur in real neurons, as opposed to bistable systems where the concept had been confined. In fact, the FitzHugh–Nagumo model has a cubic nonlinearity, just as does the standard quartic bistable system in equation (6.7); however, it has an extra degree of freedom that serves to reset the system after the threshold for spiking is crossed.

We illustrate here the behavior of the FitzHugh–Nagumo model with simultaneous stimulation by a periodic signal and by noise. The latter can be interpreted as either synaptic noise, or signal noise, or conductance fluctuations (although the precise modeling of such fluctuations is better done with conductance-based models such as Hodgkin–Huxley-type models). The model equations are (Longtin 1993)

$$\epsilon \frac{dv}{dt} = v(v - a)(1 - v) - w + \eta(t), \tag{6.10}$$

$$\frac{dw}{dt} = v - dw - b - r \sin \beta t, \tag{6.11}$$

$$\frac{d\eta}{dt} = -\lambda \eta + \lambda \xi(t). \tag{6.12}$$

The variable $v$ is the fast voltage-like variable, while $w$ is a recovery variable. Also, $\xi(t)$ is a zero-mean Gaussian white additive noise, which is lowpass filtered to produce an Ornstein–Uhlenbeck-type additive noise denoted by $\eta$. The autocorrelation of the white noise is $\langle \xi(t)\xi(s) \rangle = 2D\delta(t - s)$; i.e., it is delta-correlated. The parameter $D$ is the intensity of the white noise. This Ornstein–Uhlenbeck noise is Gaussian and exponentially correlated, with a correlation time (i.e., the $1/e$ time) of $t_c = \lambda^{-1}$. The periodic signal of amplitude $r$ and frequency $\beta$ is added here to the recovery variable $w$ as in Alexander, Doedel, and Othmer (1990), yielding qualitatively similar dynamics as in the case in which it is added to the voltage equation (after proper adjustment of the amplitude; see Longtin 1993). The periodic forcing should be added to the voltage variable when the period of stimulation is smaller than the refractory period of the action potential.

The parameter regime used to obtain the results in Figure 6.5 can be understood as follows. In the absence of periodic stimulation, one would see a smooth unimodal interspike interval histogram (close to a gamma-type distribution) governed by the interaction of the two-dimensional FitzHugh–Nagumo dynamics with noise. The periodic stimulus thus carves peaks out of this "background" distribution of the interspike interval histogram. If the noise is turned off and the stimulus is turned on, there would be no firings whatsoever. This is a crucial point: The condition for obtaining skipping with the tunability properties described in the previous section is that *the deterministic dynamics must be subthreshold*. This feature can be controlled by the parameter $b$, which sets the proximity of the resting potential (i.e., the single stable fixed point) to the threshold. In fact, this dynamical system goes through a supercritical Hopf bifurcation at $b_H = 0.35$. It can also be controlled by a constant current that could be added to the left hand side of the first equation.

Figure 6.7 contrasts the subthreshold ($r = 0.2$) and suprathreshold ($r = 0.22$) behavior in the FitzHugh–Nagumo system. In the subthreshold case, the noise is essential for firings to occur: No intervals are obtained when the noise intensity, $D$, is zero. For $r = 0.22$ and $D = 0$, only one kind of interval is obtained, namely, that corresponding to the period of the deterministic limit cycle. For $D > 0$, the limit cycle is perturbed by the noise, and sometimes comes close to but misses the separatrix: No action potential is generated during one or more cycles of the stimulus. In the subthreshold case, one also sees skipping behavior. At higher noise intensities, the interspike interval histograms hardly differ, and thus we cannot tell from such an interspike interval histogram whether the system is suprathreshold or subthreshold. This distinction can be made by varying the noise level as illustrated in this figure. In the subthreshold case, the mean of the distribution will always move to lower intervals as $D$ increases, although this is not true for the suprathreshold case.

There have also been numerous modeling studies based on noise and sinusoidally forced integrate-and-fire-type models (see, e.g., Shimokawa, Pakdaman, and Sato 1999; Bulsara, Elston, Doering, Lowen, and Lindenberg 1996; and Gammaitoni, Hänggi, Jung, and Marchesoni 1998 and references therein). Other possible generic dynamical behavior might lurk behind this form of phase locking. In fact, the details of the phase locking, and of the physiology of the cells, are important in determining which specific dynamics are at work. Subthreshold chaos might be involved (Kaplan,

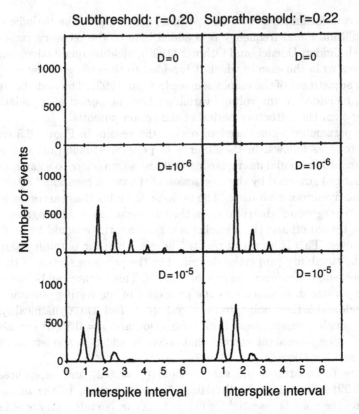

Figure 6.7. Comparison of interspike interval histograms with increasing noise intensity in the subthreshold regime (left panels) and suprathreshold regime (right panels).

Clay, Manning, Glass, Guevara, and Shrier 1996), although probably with noise as well if one seeks smooth interspike interval histograms with symmetric modes and without missing modes between the first and the last (as with the data in Rose, Brugge, Anderson, and Hind 1967; see Longtin 1998). An example of these effects of noise on the multimodal interspike interval histograms produced by the chaos (with pulsatile forcing) is shown in Figure 6.8. In other systems, chaos may be the main player, as in the experimental system (pulsatile stimulation of squid axons) considered in Kaplan et al. (1996).

In Figure 10 of Longtin (1993), a "chaos" hypothesis for skipping was investigated using the FitzHugh–Nagumo model with sinusoidal forcing (instead of pulses as in Kaplan, Clay, Manning, Glass, Guevara, and Shrier 1996). Without noise, this produced subthreshold chaos, as described in Kaplan, Clay, Manning, Glass, Guevara, and Shrier (1996), although clearly, when spikes occur (using some criterion for graded responses) these are "suprathreshold" responses to this "subthreshold chaos"; in other words,

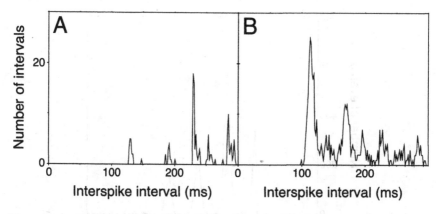

Figure 6.8. Interspike interval histogram from the FitzHugh–Nagumo system $dv/dt = v(v - 0.139)(1 - v) - w + I + \eta(t)$, $dw/dt = 0.008(v - 2.54w)$, where $I$ consists of rectangular pulses of duration 1.0 msec and height 0.28, repeated every 28.5 msec. Each histogram is obtained from one realization of $5 \times 10^7$ time steps. The Ornstein–Uhlenbeck noise $\eta(t)$ has a correlation time of 0.001 msec. (A) Noise intensity $D = 0$. (B) $D = 2.5 \times 10^{-5}$.

the chaos could just as well be referred to as "suprathreshold." In this FitzHugh–Nagumo chaos case, some features of the Rose et al. data could be reproduced, but others not. For example, multimodal histograms were found. But the individual peaks had more "internal" structure than seen in the data (including electroreceptor and mechanoreceptor data); they were not aligned very well with the integer multiples of the driving period; and some peaks were missing, as in the case of pulsatile forcing shown in Figure 6.8. Further, the envelope did not have the characteristic exponential decay (past the second mode) seen in the Rose et al. data (which is what is expected for uncorrelated intervals). Additive dynamical noise on top of this chaos did a better job at reproducing these qualitative features, at least for the parameters explored (Longtin 1998). The modes of the interspike interval histogram were still a bit lopsided, and the envelopes were different from those of the data. Interestingly, solutions that "look chaotic" often end up on periodic orbits after a long while. A bit of noise would probably keep these solutions bouncing around irregularly.

The other reason that some stochastic component may be a necessary ingredient is the smoothness observed in the transitions between interspike interval histograms as stimulation parameters are changed. Changing period or amplitude in the chaotic models leads to sometimes abrupt changes in the multimodal structure (and some peaks just keep on missing). Noise induced firing with deterministically subthreshold dynamics does produce the required smooth transitions in the proper sequence seen in Rose et al. (1967). Note, however, that in certain parameter ranges, it is possible to get multimodal histograms with suprathreshold forcing. This is shown

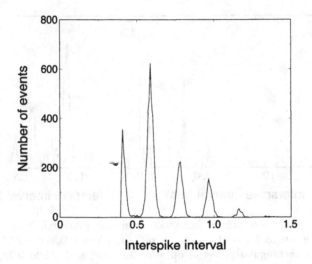

Figure 6.9. Interspike interval histogram from the FitzHugh–Nagumo system with fast sinusoidal forcing $\beta = 32$. Other parameters are $a = 0.5$, $b = 0.15$, $d = 1$, $\epsilon = 0.005$, $I = 0.04$ and $r = 0.06$. The histogram is obtained from 10 realizations of 500,000 time steps.

in Figure 6.9, for which the forcing frequency is high, and the deterministic solution is a periodic 3:1 solution.

All this discussion does not exclude the possibility that chaos alone (e.g., with other parameters or in a more refined model) might give the right picture for this kind of data, or that deterministic chaos or periodic phase locking combined with noise might give it as well. Only good intracellular data can ultimately settle the issue of the origin of the stochastic phase locking, and provide an explanation for the smooth skipping patterns.

## 6.10   Stochastic Resonance

The notion that skipping neurons in the subthreshold regime rely on noise to fire is interesting from the point of view of signal processing. In order to transmit information about the stimulus (the input) to a neuron in its spike train (the output), noise must be present. Without noise, there are no firings, and with too much noise, we expect to see a very noisy output with again no information (or very little) about the stimulus. Hence, there must be a noise value for which information about the stimulus is optimally transmitted to the output. In other words, starting from zero noise, adding noise will increase the signal-to-noise ratio, and an optimal noise level can be found where the signal-to-noise ratio peaks. This is indeed the case in the FitzHugh–Nagumo model studied above. This effect, in which the signal-to-noise ratio is optimal for some intermediate noise intensity, is known as **stochastic resonance**. It has been studied for over a decade, usually in bistable systems. It had been studied theoretically in bistable neurons (Bulsara, Jacobs, Zhou, Moss, and Kiss 1991), and predicted to occur in real neurons (Longtin, Bulsara, and Moss 1991). Thereafter, it was studied theoretically in an excitable system (Longtin 1993; Chialvo and Apkarian 1993; Chapeau-Blondeau, Godivier, and Chambet 1996) and a variety of other systems (Gammaitoni, Hänggi, Jung, and Marchesoni 1998), and shown to occur in real systems (see, e.g., Douglass, Wilkens, Pantazelou, and Moss 1993; Levin and Miller 1996). It is one of many constructive roles for noise discovered in recent decades (Astumian and Moss 1998).

This resonance can be studied from the points of view of spectral amplitude at the signal frequency, signal-to-noise ratio, residence-time histograms (i.e., interspike interval histograms), and others as well. In the first case, one computes for a given value of noise intensity $D$ the power spectrum averaged over many spike trains obtained with as many realizations of the noise process. The spectrum is usually in the form of a flat or curved background, on which the harmonics of the small stimulus signal are superimposed (see Figure 6.5D). A dip at low frequencies is often seen, which is due to phase jitter and to the refractory period. A signal-to-noise ratio can then be computed by dividing the height of the fundamental stimulus peak by the noise floor (i.e., the value of the noise background at the frequency of the stimulus). This signal-to-noise ratio can be plotted as a function of $D$, and the resulting curve will be unimodal, with the maximum corresponding to the stochastic resonance. Alternatively, one can measure the heights of the different peaks in the interspike interval histogram, and plot these heights as a function of $D$. The different peaks will go through a maximum at different values of $D$. While there is yet no direct analytical connection between stochastic resonance from these two points of view, it is usually the case that systems exhibiting stochastic resonance from one point of view will also exhibit it from the other.

The power spectrum measures the synchrony between firings and the stimulus. From the point of view of the interspike interval histogram, the measure of synchrony depends not only on the prevalence of intervals at integer multiples of a fundamental interval, but also on the width of the peaks of the interspike interval histogram. As noise increases past the resonance value, these widths increase, with the result that the phase locking is disrupted by the noise, even though there are many firings.

A simple theory of stochastic resonance for excitable systems is being developed. Wiesenfeld et al. (1994) have shown that stochastic resonance will occur in a periodically modulated point process. By redoing the calculation of the classic shot noise effect for the case of a periodic stimulus (the point process is then inhomogeneous, i.e., time-dependent), they have found an expression for the signal-to-noise ratio (SNR) (in decibels):

$$\text{SNR} = 10 \log_{10} \left[ \frac{4 I_1^2(z)}{I_o(z)} \exp(-U/D) \right], \tag{6.13}$$

where $I_n$ is the modified Bessel function of order $n$, $z \equiv rU/D$, and $U$ is a measure of the proximity of the fixed point to the firing threshold (i.e., some kind of activation barrier). For small $z$, this equation becomes

$$\text{SNR} = 10 \log_{10} \left[ \frac{U^2 r^2}{D^2} \exp(-U/D) \right], \tag{6.14}$$

which is almost identical to a well-known result for stochastic resonance in a bistable potential. More recently, analytical techniques have been devised to study stochastic resonance in two-variable systems such as the FitzHugh–Nagumo system (Lindner and Schimansky-Geier 2000).

The shape of the interspike interval histogram, and in particular, its rate of decay, is very sensitive to the stimulus characteristics. This is to be contrasted with the transition from sub- to suprathreshold dynamics in the absence of noise. There are no firings before the stimulus exceeds a threshold amplitude. Once the suprathreshold regime is reached, however, amplitude increases can bring on various $p : m$ locking patterns and even chaos. Noise allows the firing pattern to change smoothly and sensitively over a larger range of stimulus parameters.

The firing patterns of the neuron in the excitable regime are also interesting in the presence of noise only, i.e., without periodic forcing. In fact, such a noisy excitable system can be seen as a stochastic oscillator (Longtin 1993; Pikovsky and Kurths 1997; Longtin and Chialvo 1998; Lee and Kim 1999; Lindner and Schimansky-Geier 2000). The presence of a resonance in the deterministic dynamics will endow this oscillator with a well-defined preferred time between firings; this time scale is closely associated with the period of the limit cycle that arises when the system is biased into its autonomously firing regime. Recently, Pikovsky and Kurths (1997) showed that increasing the noise intensity from zero will lead to enhanced periodicity in the output firing pattern, followed by a decreased periodicity. This has been termed **coherence resonance**, and is related to the induction by noise of the limit cycle that exists in the vicinity of the excitable regime (Wiesenfeld 1985). The effect has also been predicted to occur in bursting neurons (Longtin 1997).

We close this section with a brief discussion of the origin of stochastic resonance in excitable neurons. Various aspects of this question have been discussed in Collins, Chow, and Imhoff 1995a; Collins, Chow, and Imhoff 1995b; Bulsara, Jacobs, Zhou, Moss, and Kiss 1991; Bulsara, Elston, Doering, Lowen, and Lindenberg 1996; Chialvo, Longtin, and Müller-Gerking 1997; Longtin and Chialvo 1998; Neiman, Silchenko, Anishchenko, and Schimansky-Geier 1998; Lee and Kim 1999; Shimokawa, Pakdaman, and Sato 1999; Lindner and Schimansky-Geier 2000. Here we focus on the distribution of the phases at which firings occur, the so-called "cycle histogram." Figure 6.10 shows cycle histograms for the FitzHugh–Nagumo model with subthreshold parameter settings similar to those used in previous figures, for high (left panels) and low frequency forcing (right panels). The lower panels (low noise) show that the cycle histogram is rectified, with firings occurring only in a restricted range of phases. The associated interspike interval histograms (not shown) are multimodal as a consequence of this phase preference. The rectification is due to the fact that the firing rate for zero forcing is low: When this rate is modulated downward by the signal, the rate goes to zero (and cannot go lower). The rectification for $T = 0.5$ is even stronger, because at higher frequencies, phase locking also occurs (Longtin and Chialvo 1998; Lee and Kim 1999): This is a consequence of the refractory period of the system, responsible for phase locking patterns in the suprathreshold regime, and increasingly important at higher frequencies. At higher noise, the rectification has disappeared: The noise has linearized the cycle histogram. The spectral power of the spike train at the signal frequency is maximal near the noise intensity that produces the "most sinusoidal" cycle histogram (as measured, for example, by a linear correlation coefficient).

This linearization is dependent on noise amplitude only for low frequencies (i.e., for $T > 2$ or so), such as those used in Collins, Chow, and Imhoff 1995a; Collins, Chow, and Imhoff 1995b; Chialvo, Longtin, and Müller-Gerking 1997: The neuron then essentially behaves as a static threshold device. As the frequency increases, linearization requires more noise, due to the increased importance of phase locking. This higher noise also produces an increased spontaneous rate of firing when the signal is turned off. Hence, this rate for the unmodulated system must increase in parallel with the frequency in order for the firings to be maximally synchronized with the stimulus. Also, secondary resonances at lower noise occur for higher frequencies (Longtin and Chialvo 1998) in both the spectra and in the peak heights of the interspike interval histogram, corresponding to the excitation of stochastic subharmonics of the driving force. The noise producing the maximal signal-to-noise ratio is itself minimal for frequencies near the best frequency (i.e., the resonant frequency) of the FitzHugh–Nagumo model (Lee and Kim 1999).

## 6.11   Noise May Alter the Shape of Tuning Curves

Tuning curves are an important characteristic of neurons and cardiac cells. They describe the sensitivity of these cells to the amplitude and frequency of periodic signals. For each sinusoidal forcing frequency, one determines the minimum amplitude needed to obtain a specific firing pattern, such as 1:1 firing. The frequency–amplitude pairs are then plotted to yield the

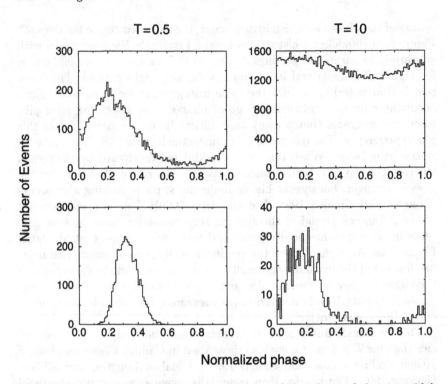

Figure 6.10. Probability of firing as a function of the phase of the sinusoidal forcing (left, $T = 0.5$; right, $T = 10$), obtained by averaging over 50 realizations of 100 cycles. The amplitude of the forcing is 0.01. For the upper panels, noise intensity $D = 8 \times 10^{-6}$, and for the lower ones, $D = 5 \times 10^{-7}$.

tuning curve. We have recently computed the behavior of the 1:1 and Arnold tongues of the excitable FitzHugh–Nagumo model with and without noise (Longtin 2000). Our work was motivated by recent findings (Ivey, Apkarian, and Chialvo 1998) that mechanoreceptor tuning curves can be significantly altered by externally added stimulus noise, and by an earlier numerical study that reported that noise could alter tuning curves (Hochmair-Desoyer, Hochmair, Motz, and Rattay 1984). It was also motivated by the tuning properties of electroreceptors (see, e.g., Scheich, Bullock, and Hamstra Jr 1973), and generally by ongoing research into the mechanisms underlying aperiodic phase locked firing in many excitable cells including cardiac cells.

Figure 6.11 shows the boundary (Arnold tongue) for 1:1 firing for noise intensity $D = 0$. It is V-shaped, highlighting again the resonant aspect of the neuronal dynamics. The minimum threshold occurs for the so-called best frequency which is close to the frequency of autonomous oscillations seen past the Hopf bifurcation in this system. For period $T > 1$, the region below these curves is the subthreshold 1:0 region. For $A > 0$, ratios as

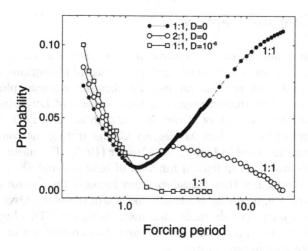

Figure 6.11. Effect of noise on the tuning curves of the FitzHugh–Nagumo model. Only the curves for 1:1 phase locking are shown; when noise intensity is greater than zero, the 1:1 pattern is obtained only on average.

parameters change (instead of the usual discontinuous Devil's staircases). We also compute the stochastic Arnold tongues for $D > 0$: For each $T$, the amplitude that produces a pattern with a temporal average of 1 spike per cycle is numerically determined. Such patterns are not periodic, but firings still exhibit phase preference. In contrast to the noiseless case, noise creates a continuum of locking ratios in the subthreshold region. For mid-to-long periods, noise "fans out" into this region all the tongues that are confined near the noiseless 1:1 tongue when $D = 0$. These curves can be interpreted as stochastic versions of the resonances that give rise to phase locking.

Increasing $D$ opens up the V-shaped 1:1 tongue at mid-to-long periods, while slightly increasing the threshold at low periods. Noise thus increases the bandwidth at mid-to-low frequency. The relatively invariant shape at low $T$ is due to the absolute refractory period, which cannot easily be overcome by noise. For larger $D$, such as for $D = 5 \times 10^{-6}$, the tongue reaches zero noise, namely, at $T = 2$ for the mean 1:1 pattern. This implies that for $T = 2$, noise alone (i.e., even for $A = 0$) can produce the desired mean ratio of one, while for $T > 2$, noise alone produces a larger than desired ratio. A more rigorous analysis of these noisy tuning curves, one that combines the noise-induced threshold crossing statistics with the resonance properties of the model, successfully accounts for the changes in shape for $T < 1$ (Longtin 2000). Our result opens the way for understanding the effect on tuning of changes in internal and external noise levels, especially in the presence of the filters associated with a given specialized transducer.

# 6.12   Thermoreceptors

We now turn our attention in these last two sections to physiological systems that exhibit multimodal interspike interval histograms *in the absence of any known periodic forcing*. The best-known examples are the mammalian cold thermoreceptors and the ampullae of Lorenzini (passive thermal and electroreceptors of certain fish such as sharks). The temperature fluctuates by less than 0.5 degrees Celsius during the course of the measurements modeled in Longtin and Hinzer (1996). The mean firing rate of these receptors is a unimodal function of temperature. Over the lower range of temperatures they transduce, they increase their mean firing rate with temperature, behaving essentially like warm receptors. Over the other half of their range, they decrease their mean firing rate. This higher range includes the normal body temperature, and thus an increase in firing rate signals a decrease in temperature.

This unimodal stimulus–response curve implies that a given firing rate can be associated with two constant temperatures. It has been suggested that the central nervous system resolves this ambiguity by responding to the pattern of the spike train. In fact, at lower temperatures, the firing is of bursting type with a long period between bursts and many spikes per burst (see Figure 6.12). As the temperature increases, the bursting period shortens, and the number of spikes per burst decreases, until there is on average only one spike per burst: This is then a regular repetitive firing, also known as a "beating" pattern. As the temperature increases further, a skipping pattern appears, as cycles of the beating pattern drop out randomly. The basic interval in the skipping pattern is close to the period of the beating pattern. This suggests that there is an intrinsic oscillation in these receptors that underlies all the patterns (see Longtin and Hinzer 1996 and references to Schafer and Braun therein).

Cold receptors are free nerve endings in the skin. The action potentials generated there propagate to the spinal cord and up to the thalamus. An ionic model for the firing activity of mammalian cold receptors has recently been proposed (Longtin and Hinzer 1996). The basic assumption of this model is that cold reception arises by virtue of the thermosensitivity of various ionic currents in a receptor neuron, rather than through a specialized mechanism. Other assumptions, also based on the anatomy and extracellular physiology of these receptors (intracellular recordings are not possible), include the following: (1) The bursting dynamics are of the slow-wave type, i.e., action potentials are not necessary to sustain the slow oscillation that underlies bursting; (2) the temperature modifies the rates of the Hodgkin–Huxley kinetics, with $Q_{10}$'s of 3; (3) the temperature increases the maximal sodium ($Q_{10} = 1.4$) and potassium ($Q_{10} = 1.1$) conductances; (4) the rate of calcium kinetics increases with temperature ($Q_{10} = 3$); (5) the activity of an electrogenic sodium–potassium pump increases linearly with temperature, producing a hyperpolarization; and (6) there is noise added to these deterministic dynamics, to account for skipping and for fluctuations in the number of spikes per burst and the interburst period.

Our model is modified from Plant's model (Plant 1981) of slow-wave bursting in the pacemaker cell of the mollusk *Aplysia*. For the sake of simplicity, we have assumed that the precise bursting mechanism is governed by an outward potassium current whose activation depends on the concentration of intracellular calcium. The model with stochastic forcing is governed by the equations

$$C_M \frac{dV}{dt} = G_I m_\infty^3(V)h(V_I - V) + G_x x(V_I - V)$$

$$+ G_K n^4(V_K - V) + G_{K-Ca}\frac{[Ca]}{0.5 + [Ca]}(V_K - V) \qquad (6.15)$$

$$+ G_L(V_L - V) + I_p + \eta(t),$$

$$\frac{dh}{dt} = \lambda \left[h_\infty(V) - h\right]/\tau_h(V), \qquad (6.16)$$

$$\frac{dn}{dt} = \lambda \left[n_\infty(V) - n\right]/\tau_n(V), \qquad (6.17)$$

$$\frac{dx}{dt} = \lambda \left[x_\infty(V) - x\right]/\tau_x, \qquad (6.18)$$

$$\frac{d[Ca]}{dt} = \rho \left[K_c x(V_{[Ca]} - V) - [Ca]\right], \qquad (6.19)$$

$$\frac{d\eta}{dt} = -\frac{\eta}{t_c} + \frac{\xi(t)}{t_c}. \qquad (6.20)$$

Here $h, n$, and $x$ are gating variables, and $[Ca]$ is the intracellular calcium concentration. *There is no periodic input:* The simulations are done for constant temperatures. The precise form of the voltage dependencies of the gating variables and time constants can be found in Plant (1981). The correlation time of the noise $\eta(t)$ (an Ornstein–Uhlenbeck process) was chosen as $\tau_c = 1.0$ msec, so that the noise has a larger bandwidth than the fastest events in the deterministic equations. Our choice of parameters yields action potential durations of 140 msec at $T = 17.8°C$ down to 20 msec at $T = 40°C$. The noise, which is intended to represent mainly the effect of conductance fluctuations in the ionic channels and ionic pumps, is made additive on the voltage variable for simplicity.

This model reproduces the basic firing patterns seen at different temperatures (Longtin and Hinzer 1996). The dynamics of our model can be separated into a fast subsystem and a slow subsystem. The slow subsystem oscillates autonomously with a period corresponding to the interval between the beginning of two bursts. When this oscillation sufficiently depolarizes the cell, the fast action potential dynamics become activated. These action potentials "ride" on top of the slow wave. The effect of the temperature is to decrease the period of the slow wave, to speed up the kinetics governing the action potentials, and to hyperpolarize the slow wave (due mostly to

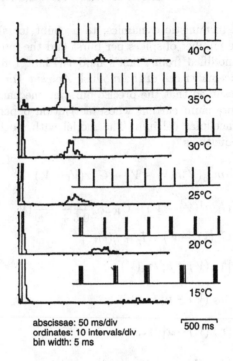

abscissae: 50 ms/div
ordinates: 10 intervals/div
bin width: 5 ms

500 ms

Figure 6.12. Interspike interval histograms and spike trains from cold receptors of the cat lingual nerve at different constant temperatures. From Braun, Schäfer, and Wissing 1990.

the Na–K pump). At higher temperatures, the slow wave is hyperpolarized so much that the threshold for the activation of the fast dynamics is not reached: The slow wave is subthreshold. From this point on, noise is important, since it can induce firings in synchrony with the slow wave. This is similar, then, to the situation seen above for neurons with external stimulation. The main difference is that the cold receptor has a stimulus built into its internal dynamics.

More recently, we have proposed a simplified phase model with periodic and stochastic forcing for the temperature dependence of such bursting activity (Roper, Bressloff, and Longtin 2000). The amplitude is proportional to the slow-wave amplitude, while the frequency is approximately $\sqrt{\lambda\rho}$ in our modified Plant model. This phase model can be analyzed for the boundaries between different solutions with different numbers of spikes per burst, and for the effect of noise on these boundaries.

## 6.13   Autonomous Stochastic Resonance

Near or in the skipping regime, i.e., at higher temperatures, the degree of phase locking between the spikes and the slow wave, as well as the characteristics of the interspike interval histogram, are again highly sensitive to variations in noise level, or in the period, amplitude, and bias of the slow wave. All these characteristics can change in response to changes in the physicochemical environment of the cold receptor. Thus, in this regime, the cell is sensitive to such changes, due to the presence of noise. In particular, noise appears to be essential for encoding higher temperatures, since without noise, there would be no firings (Braun, Wissing, Schäfer, and Hirsch 1994). The behavior of the interspike interval histogram as a function of the aforementioned parameters is similar to that seen in skipping neurons with external stimulation.

Noise helps express the characteristics of the slow wave in the output spike train. This sensitivity can be characterized by computing averaged power spectra and signal-to-noise ratios. For the slow-wave burster, the spectra are more complicated than in the cases studied above (Longtin 1997). In particular, the background is bumpy, i.e., there are more peaks than the usual harmonics, and the harmonics themselves are broader. This latter feature is expected because the phase of the autonomous oscillation fluctuates. Nevertheless, one finds that there is an increase in the signal-to-noise ratio with noise intensity $D$, from 0 up to large values. At the point when the signal-to-noise ratio starts to drop again, the numerical simulations, and indeed, the stochastic model itself become doubtful. In fact, at these high noise levels, the action potential waveforms themselves become noisy.

This autonomous stochastic resonance behavior can be contrasted to that studied in Gang, Ditzinger, Ning, and Haken (1994). In that paper, the first claim of autonomous stochastic resonance, the stochastic resonance behavior was characterized in a system right at a saddle-node bifurcation. There, the noise induces a limit cycle that has zero amplitude in the absence of noise. As the noise becomes too large, the coherence of this limit cycle decreases. Thus this behavior is similar to stochastic resonance. However, it requires a saddle-node bifurcation. Our study of the slow-wave burster shows that autonomous stochastic resonance does not directly require a saddle-node bifurcation. Neural dynamics are often sufficiently complex to generate their own autonomous oscillations (through a Hopf bifurcation in the case of our modified Plant model), and can be expected to exhibit stochastic resonance phenomena as the noise intensity varies. A related effect known as coherence resonance has been analyzed in the FitzHugh–Nagumo system (Pikovsky and Kurths 1997). In this latter case, the induction by noise of regularity in the firing pattern can be theoretically linked to the specific statistical dependencies of the escape time to threshold and of the refractory period on noise intensity.

## 6.14   Conclusions

We have given a brief overview of recent modeling efforts of noise in physiologically relevant dynamical systems, and studied in detail the response
of excitable cells to periodic input and noise. We have shown how noise
can interact with bifurcations, produce smooth transitions between firing
patterns as stimulus parameters are varied, and alter the frequency sensitivity of neurons. In most neural stochastic resonance studies, a neuron
was chosen in which the addition of noise to the periodic stimulus could
better transduce this stimulus. This approach of adding noise is warranted
because the noise level cannot be reduced in any simple way in the system
(and certainly not by cooling it down, as we suspect from our study of cold
receptors). It is interesting to pursue studies of how internal noise can be
changed.

There have been predictions of stochastic resonance in summing networks
of neurons (Pantazelou, Moss, and Chialvo 1993), in which the input was
aperiodic. There have also been theoretical studies of stochastic resonance
in a summing neuron network for slowly varying aperiodic signals, i.e., for
situations where the slowest time scale is that of the signal (Collins, Chow,
and Imhoff 1995a; Collins, Chow, and Imhoff 1995b; Chialvo, Longtin, and
Müller-Gerking 1997). There has also been a recent experimental study
of the enhancement by noise of the transduction of broadband aperiodic
signals (Levin and Miller 1996); in particular, that study investigated the
effect of the amount of overlap between the frequency contents of the signal
and of the noise added to the signal. In these latter two studies with aperiodic signals, the resonance is characterized by a maximum, as a function of
noise, in the cross-coherence between the input signal and the output spike
train, and by related information-theoretic measures such as the transin-
formation. Further work has been done, using simple neuron models driven
by periodic spike trains and Poisson noise, to address situations of periodic and noisy synaptic input (Chapeau-Blondeau, Godivier, and Chambet
1996). Since correlations in spike trains increase the variance of the associated input currents, stochastic resonance has been recently studied from
the point of view of increased coherence between presynaptic spike trains
(Rudolph and Destexhe 2001). There have been experimental verifications
of the stochastic resonance effect away from the sensory periphery, namely,
in a cortical neuron (Stacey and Durand 2001). In addition, there is an
increasing number of interesting applications of the concept of stochastic
resonance to human sensory systems (Collins et al. 1996; Cordo et al. 1996;
Morse and Evans 1996). All these studies emphasize the usefulness of the
combination of the subthreshold regime and noise, and in some cases such as
electroreceptors, of suprathreshold dynamics with noise (Chacron, Longtin,
St-Hilaire, and Maler 2000) to transduce biologically relevant signals.

To summarize our stochastic phase locking analysis, we have shown the following:

- Many neurons exhibit skipping in the presence of periodic stimulation.

- Modeling shows that skipping results readily from the combination of subthreshold dynamics and noise. In other words, no deterministic firings can occur; firings can occur only in the presence of noise. It can occur with suprathreshold dynamics, namely, with chaos and/or noise, with some deviations from the skipping picture presented here (such as tunability of the pattern).

- Noise helps in the detection of low-amplitude stimuli through an effect known as "stochastic resonance." This means that adding noise to these neurons allows them to transduce small stimuli that cannot by themselves make the cell fire. The effect relies on a linearization of the firing probability versus stimulus phase characteristic, which occurs in a stimulus-frequency-dependent fashion except when signals are slower than all neuron time scales. The simplest version of this effect in the presence of periodic forcing can be found in simple threshold-crossing systems when the signal is slow compared to all other time scales.

- Noise can extend the physical range of stimuli that can be encoded.

- Noise can alter the shape of tuning curves, and thus the frequency response characteristics of neurons.

- These results apply qualitatively to systems without external forcing, such as thermoreceptors.

Some important unresolved questions worthy of future investigations are the following:

- Can we identify more precise ionic mechanisms for the skipping patterns, based on data from intracellular recordings?

- To what extent are neurons in successive stages of sensory processing wired to benefit from stochastic resonance?

- What aspects of the skipping pattern determine the firing properties of the neurons they connect to, and why is this code, which combines aspects of a "random carrier" with "precisely timed firings," so ubiquitous?

- Another important issue to consider in modeling studies is that the axons connecting to receptor cells are driven synaptically, and the synaptic release is governed by the receptor generating potential. In the presence of periodic forcing on the receptor, such an axon can arguably be seen as driven by both a deterministic and a stochastic component. It will be useful to study how the synaptic noise can produce realistic skipping patterns, and possibly assist signal detection; such investigations will require using more detailed knowledge of synapses and connectivity between neurons.

- More generally, there is still plenty of work to be done, and, probably, effects to be discovered at the interface of stochastic processes and nonlinear physiological dynamics. Among the tasks ahead are the analysis of noise in systems with some form of memory, such as the pupil light reflex, or excitable systems that do not "recover" totally after a firing.

## 6.15   Computer Exercises: Langevin Equation

### Software

There is 1 **Matlab**\* program you will use for these exercises:

**langevin** A **Matlab** script that integrates the Langevin equation, equation (6.4), using the standard Euler–Maruyama method, equation (6.5). The solution is Ornstein–Uhlenbeck noise. The program generates the solution $x(t)$ starting from an initial condition $x(0)$. Time is represented by the vector $t$, and the solution by the vector $x$. The solution $x(t)$ can be seen using the command **plot(t,x)**. The histogram of solution can be seen using the command **plot(rhox,rhoy)**.

The following exercises are to be carried out using the **Matlab** program **langevin.m**. The parameters of the equation, i.e., the decay rate $\alpha$ **alpha**, the intensity of the Gaussian white noise $D$ **dnz**, and the initial condition for the state variable **xinit**, are in **langevin.m**. The integration parameters are also in that program. They are the integration time step **dt**, the total number of integration time steps **tot**, the number of times steps to discard as transients **trans**, and the number of sweeps (or realizations) **navgs**. The description of the program parameters is given in comment in the **Matlab** file. These parameters can thus be changed by editing **langevin.m**, and then the numerical integration can be launched from the **Matlab** command window. The program generates the solution $x(t)$ starting from an initial condition $x(0)$. Time is represented by the vector $t$, and the solution by the vector $x$. A solution $x(t)$ can thus be plotted at the end of the simulation using the command **plot(t,x)**.

---

\*See Introduction to **Matlab** in Appendix B.

A simulation involving a given set of random numbers is called a "realization." These random numbers provide the source of noise for the code, and thus mimic the noise in the stochastic differential equation (which, as we have seen, models the noise in the system under study). For our problems, Gaussian-distributed random numbers are needed; they are generated internally by Matlab using the function **randn**. A random number generator such as the one built into Matlab needs a "seed" value from which it will generate a sequence of independent and identically distributed random numbers. Matlab automatically handles this seeding. Since random number generation is done on a computer using a deterministic algorithm, the independence is not perfect, but good enough for our purposes; nevertheless, one should keep this in mind if one uses too many random numbers, since such "pseudorandom" number generators will repeat after a (usually very large) number of iterations.

From a time series point of view, each realization will differ from the other, since it uses different sets of random numbers. However, each realization has the same statistical properties (such as, e.g., moments of probability densities of the state variables and correlation functions). In order to get a good idea of these properties, one typically has to average over many realizations. Each realization can start from the same initial condition for the state variables, or not, depending on which experimental protocol you are trying to simulate, or what kind of theory you are comparing your results to. In some cases, it is also possible to estimate these properties from one very long simulation. But generally, it is best to average over many shorter realizations, each one using a different initial value for the noise variable. This avoids the problem of finite periods for the random number generator, and allows good averaging over the distribution of the noise process. Averaging over multiple realizations also has the advantage of reducing the estimation error of various quantities of interest, such as the amplitude of a peak in the power spectrum. The codes provided here can easily be modified to perform one or many realizations.

The numerical integration scheme used here is a stochastic version of the Euler algorithm for ordinary differential equations, known as the Euler–Maruyama algorithm (6.5).

## Exercises

The purpose of these exercises is to study the effect of the noise intensity and of the parameter $\alpha$ on the dynamics of the Langevin equation

$$\frac{dx}{dt} = -\alpha x + \xi(t).$$ 
(6.21)

Note that this equation has, in the absence of noise, only one fixed point, at $x = 0$.

Ex. 6.15-1. **Effect of the parameter $\alpha$ on the dynamics.**

Run simulations for various values of $\alpha$ and Gaussian white noise intensity $D$. You should find that the solution looks smoother when $\alpha$ is larger. Also, you should find Gaussian densities for the state variable $x$; note that the program actually estimates these densities by unnormalized histograms of the numerically generated solution. Increasing $\alpha$ for a given noise intensity will reduce the variance of the histogram of the $x$-solution. You should also find that the variance is given by $D/\alpha$; you can try to show this by calculating the stationary solution of the associated Fokker–Planck equation.

**Ex. 6.15-2. Effect of noise intensity on the dynamics.** Also, increasing the noise intensity while keeping $\alpha$ constant has the opposite effect. The solution of this equation can be used as a source of "Ornstein–Uhlenbeck" colored noise.

## 6.16   Computer Exercises: Stochastic Resonance

These exercises use the **Matlab** file **fhnnoise.m** to simulate the FitzHugh–Nagumo excitable system with sinusoidal forcing and colored (Ornstein–Uhlenbeck) additive noise on the voltage equation:

$$\epsilon \frac{dv}{dt} = v(v - a)(1 - v) - w + I + r\sin\beta t + \eta(t),  \tag{6.22}$$

$$\frac{dw}{dt} = v - dw - b,  \tag{6.23}$$

$$\frac{d\eta}{dt} = -\lambda\eta + \lambda\xi(t).  \tag{6.24}$$

Here $\lambda = 1/t_{\text{cor}}$; i.e., $\lambda$ is the inverse of the correlation time of the Ornstein–Uhlenbeck process; we have also chosen a commonly used scaling of the noise term by $\lambda$. You can explore the dynamical behaviors for different system, stimulus, and noise parameters. Note that in this program, the periodic forcing is added to the voltage equation. The simulations can be lengthy if smooth interspike interval histograms are desired, so you will have to decide, after a few tries, how long your simulations should be to answer the questions below. An integral (order-1) stochastic Euler–Maruyama algorithm is used here to integrate this system of stochastic differential equations. The time step has to be chosen very small, which limits the integration speed.

### Software

There is one **Matlab**[†] program you will use for these exercises:

---

[†]See Introduction to **Matlab** in Appendix B.

**fhnnoise** A `Matlab` script (operates in the same way as **langevin.m**) integrates the FitzHugh–Nagumo system equations (6.12) driven by sinusoidal forcing and colored noise. The colored noise is the Ornstein–Uhlenbeck process $\eta(t)$. The program uses the integral Euler–Maruyama algorithm proposed in Fox, Gatland, Roy, and Vemuri (1988). The outputs of the program are the solution $x(t)$ and the interspike interval histogram $\rho(I)$. The solution can be seen using the command **plot(t,v)**. The interspike interval histogram can be seen using the command **plot(rhox,rhoy)**. The sequence of intervals can be plotted using **plot(interv)**.

The description of the program parameters is given in comment in the `Matlab` file. These parameters can thus be changed by editing **fhnnoise.m**, and then the numerical integration can be launched from the `Matlab` command window.

The program **fhnnoise.m** operates in the same way as **langevin.m**. The main equation parameters you may be interested in changing are the amplitude **amp** and frequency **f** of the sinusoidal forcing; the bias current **ibias**, which brings the system closer to threshold as it increases; and the noise intensity **dnz**. The noise intensity specified in the file refers to the intensity of the Gaussian white noise $D$, which, when lowpass filtered, gives the colored Ornstein–Uhlenbeck noise. This latter noise is also Gaussian, with an exponentially decaying autocorrelation function, and its variance is given by $D/\text{tcor}$, where `tcor` is the noise correlation time.

The simulation parameters are controlled by the total number of integration time steps **tot**, the number of time steps considered as transients **trans**, and the number of sweeps (or realizations) **navg**. The integration time step is controlled by **dt**, and may have to be made smaller than its reference value 0.0025 for higher sinusoidal forcing frequencies, i.e., for $f > 1$ or so. You can always check that results are accurate with a given time step by checking that they are statistically the same as for a new simulation with smaller time step.

## Exercises

**Ex. 6.16-1. Hopf bifurcation in the absence of noise.** In the absence of noise and periodic forcing, find the Hopf bifurcation point in the FitzHugh–Nagumo model by varying the bias current parameter $I$. You can also try to calculate this value analytically. This parameter controls the distance between the resting potential and the threshold. Below the Hopf bifurcation, the system is said to be in the subthreshold or excitable regime. Above this bifurcation, it is said to be in the repetitive firing regime.

Ex. 6.16-2. **Effect of $I$ on firing frequency.** For the same conditions as above, how does the frequency of firing depend on $I$? (It changes abruptly near the bifurcation point, but varies little thereafter.)

Ex. 6.16-3. **Effect of noise intensity on interspike interval histogram (ISIH) with noise.** Compute the interspike interval histogram with noise only (amplitude of sinusoidal forcing is set to zero), and study the behavior of this interspike interval histogram as a function of noise intensity $D$. (You should find that the maximum of the distribution shifts slightly to smaller interval values, but that the mean shifts over more significantly.)

Ex. 6.16-4. **Effect of stimulus amplitude on the interspike interval histogram envelope.** Study the behavior of the interspike interval histogram envelope as a function of stimulus amplitude, in the subthreshold regime. Can the second peak be the highest? (The envelope decays more rapidly the higher the amplitude of the sinusoidal forcing. Yes, the second peak can be the highest, especially at higher frequencies or low noise.)

Ex. 6.16-5. **Effect of stimulus frequency on interspike interval histogram envelope.** Study the behavior of the interspike interval histogram envelope as a function of stimulus frequency $\beta/(2\pi)$, in the subthreshold regime. Is the first peak always the highest? (No; see previous question.)

Ex. 6.16-6. **Effect of noise intensity on interspike interval histogram envelope.** Study the behavior of the interspike interval histogram as a function of the noise intensity $D$, in the subthreshold regime. (The noise plays a similar role to the amplitude of the sinusoidal forcing: The higher the noise, the faster the decay of the histogram envelope. However, increasing noise also broadens the peaks in the interspike interval histogram.)

Ex. 6.16-7. **Subthreshold and suprathreshold regimes.** Compare the behavior of the interspike interval histogram in the subthreshold and suprathreshold regimes. In the subthreshold regime, increasing noise always increases the probability of shorter intervals. In the suprathreshold regime, noise can perturb the limit cycle, producing longer intervals than the cycle period. Hence, in this case, increasing noise does not necessarily decrease the intervals.

Ex. 6.16-8. **Stochastic Resonance**
Plot the maximum of the peaks (or some area in the interspike interval histogram around this peak, to average out fluctuations) as a function of the noise intensity in the subthreshold regime, and in the suprathreshold regime. You should find that the value of the maximum for the first peak goes through a maximum as a function of $D$.

Note that, in this regime, no spikes can be generated without noise, and too much noise leads to solutions dominated by noise. Hence, a moderate value of noise causes a predominance of firing intervals around the stimulus period, which is a manifestation of stochastic resonance. You should find that the other peaks also go through maxima. You can also modify the code to compute power spectral densities for the spike trains generated by this stochastic system. For example, you can generate a vector of zeros and ones that resamples the solution $v(t)$ at a lower frequency; a zero represents no spike in the corresponding time bin of this vector, while a one represents one (or more) firing events in that time bin. You can then call the spectral functions (such as the fast Fourier transform) in `Matlab`, and average the results over many realizations to reduce the fluctuations in the spectra. This computation requires more background material and is not pursued here.

# 7

# Reentry in Excitable Media

## Marc Courtemanche
## Alain Vinet

## 7.1 Introduction

A normal coordinated and effective ventricular contraction is initiated and controlled by the cell-to-cell spread of electrical excitation through the myocardium. The normal excitation sequence starts from the sinoatrial node, a group of spontaneous active cells lying in the wall of the right atrium that acts as the pacemaker of the heart. The activation wave then spreads through the atria and converges to the atrioventricular node. The atrioventricular node is the only electrical pathway connecting the atria to the ventricles. From there, the activation follows the Purkinje fibers, a bundle of fast-conducting fibers that distributes the excitation across the inner surface of both ventricles. The activation finally travels through the ventricular muscle toward the endocardium. The spread of the electrical activation wave throughout the cardiac excitable tissue is the first step of a cascade of events leading to the contraction of the muscle.

If, for some reason, the normal excitation sequence is disrupted, the effectiveness of the cardiac contraction may be substantially reduced. The various types of disruption of the normal excitation and repolarization patterns occurring during propagation through the myocardium are known under the generic term of cardiac arrhythmia. There are several types of cardiac arrhythmia, ranging from the relatively benign supraventricular flutter to life-threatening ones such as ventricular tachycardia and fibrillation. The rhythm of excitation might become too fast for a complete refilling of the ventricles between the contractions, thus limiting the amount of blood released in the circulation. The propagation of the activation might also become disorganized, as in fibrillation, leading to uncoordinated contraction of the tissue and to a fatal drop of the blood pressure.

Reentry is a major mechanism responsible for the initiation and maintenance of tachycardia and fibrillation (Wit and Cranefield 1978; Allessie, Bonke, and Shopman 1977; Bernstein and Frame 1990; Davidenko, Pertsov,

Salomonsz, Baxter, and Jalife 1992). Reentry refers to an electrophysiological disorder in which single or multiple activation fronts continuously find
a propagation pathway through the myocardium, leading to high-frequency
repeated stimulations that supersede the normal pacemaker activity of
the sinus node. The reentry pathway might be around an anatomical obstacle, such as the orifice of a major vessel (Frame, Page, and Hoffman
1986; Bernstein and Frame 1990), or around an island of infarcted tissue
in which excitability might be depressed or abolished (De Bakker et al.
1988). It might also form around functionally refractory, but otherwise
normal, regions of the myocardium (Allessie, Bonke, and Shopman 1973;
Allessie, Bonke, and Shopman 1976; Allessie, Bonke, and Shopman 1977;
Frazier, Wolf, Wharton, Tang, Smith, and Ideker 1989; Davidenko, Pertsov,
Salomonsz, Baxter, and Jalife 1992).

Any type of mathematical model dealing with reentry must incorporate
in one way or the other a representation of the active electrical properties of
the cells and of the intercellular connections. The following sections present
a brief review of the basic physiological properties of the cardiac excitable
tissue that are known to influence reentry. Then, three types of modeling
approaches to the reentry problem are examined: cellular automata, iterative models, and partial differential equations. The premises of each type of
model and their relation to the physiological data are presented, the main
results are recalled, and their relevance is discussed.

## 7.2   Excitable Cardiac Cell

### 7.2.1   Threshold

Normal atrial and ventricular myocardial excitable cells have a unique stable resting state to which they return after a perturbation. However, when
the resting potential is reduced experimentally from its normal value of
$\approx -90$ mV to less than $\approx -60$ mV, automatic oscillatory activity may
occur (Katzung, Hondeghem, Craig, and Matsubura 1985). This may be
done either by injecting a depolarizing current into the cell or by modify-
ing the extracellular concentrations of different ionic species. This type of
bifurcation from quiescent to oscillatory state has been studied in different
models of the cardiac myocyte (Chay and Lee 1985; Vinet and Roberge
1990; Varghese and Winlow 1994) and has been suggested to play a role in
atrial flutter and fibrillation (Saito, Otoguto, and Matsubara 1978; Wit and
Cranefield 1977; Singe, Baumgarten, and Ten Eick 1987). These properties
will not be discussed further because we focus on reentry taking place in
purely excitable tissues.

Excitable cells with a unique resting state display threshold-like (quasi-
threshold) behavior for action potential generation (FitzHugh 1960; Rinzel
and Ermentrout 1989; Clay 1977). When a square-pulse stimulus of current

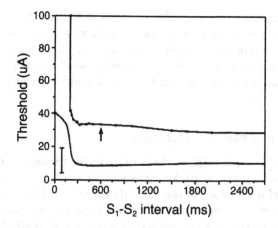

Figure 7.1. Threshold as function of S1-S2, the time between the last pacing stimulus (S1) and the premature stimulation (S2) observed in Purkinje fibers for 20 ms square-pulse stimulations applied by a suction pipette. Pacing was done with an S1-S1 interval of 3 sec; activity was recorded by a glass microelectrode positioned 1 to 2 mm from the stimulation site. Preparation was superfused with a 7 mM KCl concentration. This same the protocol was also used to produce Figure 7.2. From Chialvo, Michaels, and Jalife (1990).

is applied to an isolated cell or to a space-clamped model of the membrane, there is an abrupt change from short electrotonic responses to prolonged action potentials whenever the amplitude of the stimulus is raised above a specific fixed level. The value of the current threshold varies with the duration of the stimulus and with the timing of its application following a preceding action potential (see Figure 7.1). After an action potential, there is a time interval, the absolute refractory period, during which it is impossible to elicit a new action potential. A similar type of behavior is also observed in extended tissue.

In an isolated cell or a space-clamped model, the threshold and the absolute refractory period depend mainly on the availability of the fast sodium current and can be related to the state of the membrane. However, there is always some arbitrariness in the criteria chosen to discriminate between active and passive responses, particularly near the end of the absolute refractory period (Rinzel and Ermentrout 1989; Vinet, Chialvo, Michaels, and Jalife 1990; Vinet and Roberge 1994c).

In extended tissue, this arbitrariness is much reduced by defining superthreshold stimuli as those resulting in an action potential propagating at some distance from the stimulation site (Kao and Hoffman 1958; Noble and Hall 1963; Chialvo, Michaels, and Jalife 1990; Lewis 1974). The threshold is then a function of the state of the tissue around the stimulation site and of the spatial and temporal characteristics of the stimulus (e.g., Rushton 1937; Noble and Hall 1963; Quan and Rudy 1991; Starmer,

Biktahev, Romashko, Stepanov, Makarova, and Krinsky 1993). For a simple FitzHugh–Nagumo-type model on an infinite cable, it has been shown that an unstable propagated solution exists, acting as a separatrix between the quiescent state and the stable propagating solution, such that there is a formal threshold in this case (see Chapter 3, and for review, Zykov 1988; Britton 1986).

## 7.2.2  Action Potential Duration

The normal action potential of the cardiac myocytes can be divided into three phases: a very fast upstroke; a plateau phase during which the membrane potential is sustained or repolarized slowly; a final fast repolarization phase. During each phase of the action potential, specific ionic processes, acting on distinct time scales, become alternately dominant (Noble 1995). The simplest models of the cardiac myocyte usually involve at least three separated times scales, associated with each phase of the action potential (Beeler and Reuter 1977). At slow rhythm, the action potential may last from $\approx 100$ to $\approx 300$ ms, depending on the species and on the type of cells, related to their location in the tissue. The plateau phase is characteristic of the cardiac action potential. It controls the duration of the action potential and of the absolute refractory period and makes them much longer than in nerve cells.

The action potential duration can be defined either as the time to reach a reference level of membrane potential during repolarization or to repolarize at some fraction of the peak depolarization. In experimental preparations, it can be measured with intracellular microelectrodes, with suction pipettes giving monophasic action potential recordings (Franz 1991), or through voltage-sensitive fluorescence (Davidenko, Pertsov, Salomonsz, Baxter, and Jalife 1992).

Action potential duration can be measured during pacing at a basic cycle length, after a step change in driving frequency, or after a single premature stimulus (S2) following entrainment at a fixed rate (S1-S1). In most instances, the action potential duration was found to increase monotonically with the prolongation of the interstimulus interval (Boyett and Jewell 1978; Elharar and Surawicz 1983; Robinson, Boyden, Hoffman, and Hewlet 1987) (see Figure 7.2). But some cases were reported in which there was an increase of the action potential duration at short coupling intervals (Watanabe, Otani, and Gilmour 1995) or passive response in an intermediate range of S1-S2 coupling intervals because of supernormality in the threshold (Chialvo, Michaels, and Jalife 1990).

The restitution curve describes the relation between the action potential duration and the interstimulus interval. The recovery time, or diastolic interval defined as the time from the end of the action potential duration to the onset of the next stimulus, has also been used as an independent variable. The dependency between the action potential duration and the

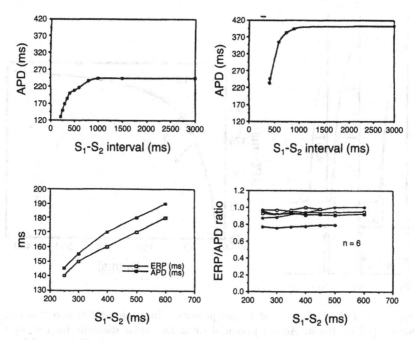

Figure 7.2. Cardiac action potential duration (APD). Top: As a function of inter-stimulus interval S1-S2 (see Figure 7.1 for protocol). Signals were recorded with a glass microelectrode in Purkinje fibers superfused with 7 (A) or 4 (B) mM KCl from Chialvo, Michaels, and Jalife (1990). Bottom: The action potential duration for stable entrainment at different S1-S1 intervals. Signals were recorded on ca-nine epicardium with suction electrode, from Franz (1991). In both cases, action potential duration was measured from the onset of stimulus to 90% repolarization. In the bottom figure, ERP is the effective refractory period.

diastolic interval has been most often fitted using a single or a double ex-ponential function (see Figure 7.3), with a steep portion (i.e., slope greater than one) at low values of the diastolic interval. However, curves of action potential duration versus diastolic interval with a steep portion surrounded by two regions with slope less than one or with a biphasic shape have also been reported (Fox, McHarg, and Gilmour 2002; Watanabe, Otani, and Gilmour 1995; Garfinkel et al. 1997).

Although the action potential duration depends primarily on the diastolic interval, it is also modulated by long-term memory effects that depend on the previous pacing history of the system. Slow processes, like the dynamics of the slow gate variables or the change of ionic concentrations within and outside of the cell, tend to produce a gradual decrease in the action po-tential duration during fast pacing. In extended tissue, the action potential duration does not depend only on the local state of the membrane, but also on the curvature of the wavefront (Comtois and Vinet 1999), the state of the neighbors (Vinet and Roberge 1994b; Vinet 1995; Vinet 1999; Watan-

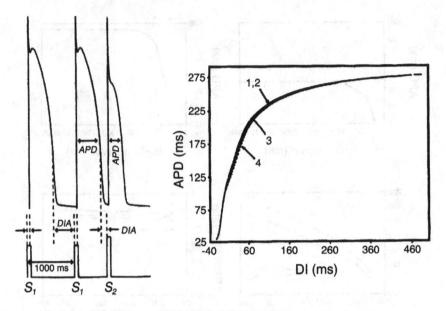

Figure 7.3. Left: Definition of action potential duration (APD) and diastolic interval (DIA), Right: Action potential duration versus diastolic interval in a model of myocyte. Left panel from Vinet and Roberge (1994c). Right panel from Vinet, Chialvo, Michaels, and Jalife (1990).

abe, Fenton, Evans, Hastings, and Karma 2001), and on the nature of the intercellular coupling (Joyner 1986; Tan and Joyner 1990; Jamaleddine, Vinet, and Roberge 1994; Jamaleddine and Vinet 1999).

## 7.2.3 Propagation of Excitation

The conduction velocity is an increasing function of the diastolic interval (Frame and Simson 1988) (Figure 7.4). This is related to the fact that the fast sodium current is the main current responsible for the upstroke formation and the propagation. Consequently, any factor interfering with the postupstroke repolarization, increasing the resting potential or reducing the rate of depolarization may delay or reduce the sodium reactivation and decrease the speed of propagation (Zykov 1984; Cabo et al. 1994; Rudy 2000; Kucera and Rudy 2001).

## 7.2.4 Structure of the Tissue

The current flowing from one cell to another is responsible for the propagation (Joyner 1982). As a first approximation, the myocardium substratum may be viewed as a network of parallel interconnected rod-shaped myocytes having an average diameter of 12–20 $\mu$m and a length of about 100 $\mu$m.

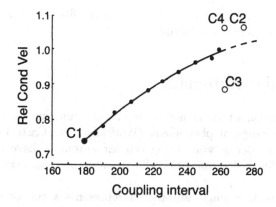

Figure 7.4. Relations between conduction velocity and coupling interval (i.e., S1-S2 interval), measured in an in vitro preparation of tissue surrounding the tricuspid valve. From Frame and Simson (1988).

The myocytes are physically and electrically interconnected through resistive channels called nexuses or gap junctions. Despite their often postulated "low resistance," these junctions still present resistive hindrance to intercellular current and thus create discontinuities in the cell-to-cell transmission of excitation (Spach, Miller, Geselowitz, Barr, Kootsey, and Johnson 1981). When myocytes are abutted end to end to form elongated cardiac fibers, two consecutive cells are generally linked by a relatively high number of gap junctions that offer a low resistance to current flow such that propagation can be considered reasonably continuous (Spach and Kootsey 1983). On the other hand, in the direction normal to the long cell axis, two neighboring fibers are loosely coupled via a low number of gap junctions, resulting in a substantially larger equivalent junction resistance that makes transverse propagation essentially discontinuous.

During propagation in the anisotropic cardiac muscle, unidirectional block may very well arise in the longitudinal direction, while transverse propagation may be more secure (Spach, Miller, Geselowitz, Barr, Kootsey, and Johnson 1981; Spach and Kootsey 1983; Spach and Dolber 1985; Spach 1990; Quan and Rudy 1990b; Keener 1988a; Keener and Phelps 1989; Keener 1987; Keener 1988b). In addition, the gap junction resistance is known to vary substantially during repetitive activity through changes in factors such as the intracellular $[Ca_i]$ concentration and the intracellular pH (De Mello 1989; Jamaleddine and Vinet 1999). The current flow is not restricted to the intracellular media, but occurs also through the external medium whose resistivity and organization can deeply influence the propagation (Wikswo 1995; Geselowitz and Miller 1983; Keener 1996; Pumir and Krinsky 1996).

In reality, the structure of the myocardium is much more complex, since the fibers form bundles separated by connective tissue and extracellular space, while there is a continuous rotation of the fiber's main axis between the endocardium and the epicardium.

# 7.3   Cellular Automata

Cellular automata, a term coined by John von Neumann, has been used to model a wide range of phenomena (Wolfram 1986; Casti 1989). In the cardiac field, modeling work using cellular automata have been mostly restricted to finite deterministic cellular automata (Saxberg and Cohen 1991).

In this context, a single automaton represents a cell or some space-clamped portion of the tissue. Time is sliced in discrete and equal time intervals, during which the automaton occupies one of a finite collection of distinct states. At each time step, all automata are synchronously updated through a set of transition rules. These are functions of the actual state of the automaton and of its inputs. Although deterministic cellular automata have often been presented as no more than a necessary compromise to simulate large pieces of cardiac tissue (Okajima, Fujimo, Kobayshi, and Yamada 1968), they have played an important part in the development of concepts about tachycardia and fibrillation.

## 7.3.1   Wiener and Rosenblueth Model

The Wiener and Rosenblueth model (Wiener and Rosenblueth 1946), also often referred to as the Greenberg and Hastings model, since it has been formalized and extended by these authors (Greenberg and Hastings 1978; Greenberg, Hassard, and Hastings 1978), has become the classical deterministic cellular automata representation of excitable tissues. Each cell is represented as a point on a regular lattice. The state space includes a resting state (R), an excited state (E), and $N$ absolute refractory states (AR$_i$, $i = 1, \ldots, N$) with the local transition rules

$$E \mapsto AR_1,$$
$$AR_i \mapsto AR_{i+1} \quad , i = 1, \ldots, N - 1,$$
$$AR_N \mapsto R.$$

The transition from state R to E, the propagation rule, takes place whenever there is at least one neighbor in state E in a circle of radius $L$ centered on each automaton. It is also possible to induce the transition by applying an external input. Assuming that the automata are on the nodes of a square lattice of side $\Delta x$ and that each time step represents a time $\Delta t$, the model represents a medium of fixed absolute refractory period $(N + 1)\Delta t$. The speed of propagation along the main axes is $K\Delta x/\Delta t$, where $K$ is the integer part of $L/\Delta x$.

## Scaling

The correspondence with cardiac tissue can be made by assigning numerical values to the parameters $\Delta t$, $\Delta x$, the extent of the neighborhood $L$, and the number of refractory states $N$. The speed of the propagation ($\approx 30$ to 60 cm/sec) and the duration of the absolute refractory period ($\approx 100$ ms) can be scaled independently thanks to the parameters $L$ and $N$. The parameters $\Delta t$ and $\Delta x$ control the temporal and spatial resolution of the system, while $L$ gives the distance over which the cells are assumed to influence one another. For example, if $L$ is chosen to be in the range of the space constant of diffusion (typically $\approx 1$ mm or less), $\Delta t$ should be fixed around 2 ms to get a realistic value of speed ($\approx 1$ m/s), and $N$ should be set around 50 to get a realistic value of the absolute refractory period.

## Nonuniformity

One weakness of this formulation is that the speed of propagation is not uniform, but depends on the direction. The continuity of the activation (sites in state E) and repolarization (sites at state $AR_N$) fronts is disrupted, such that curvature effects that are known to be important (Cabo et al. 1994; Zykov 1988) in excitable media cannot be reproduced. In Section 7.3.2, we described various modifications that can be introduced in the model to correct these shortcomings.

## Circulation on a Loop

We consider first a set of $M$ Wiener and Rosenblueth automata connected to form a one-dimensional ring. An automaton that has been excited (i.e., transition from R $\mapsto$ E) can be reactivated when it returns to state R after $N + 1$ time steps. Accordingly, an excitation front traveling in one direction around the ring will persist if $M > N + 1$. In this simple model, reentry is sustained and periodic as long as there is a minimum excitable gap (i.e., $M - (N + 1) > 0$) between the head (automaton in state E) and the tail (automaton in state $AR_N$) of the moving wavefront. When each automaton is connected only to its adjacent neighbors, the period of the reentry is proportional to M. We will examine in Section 7.4 whether these predictions still hold for more complex models of propagation. The model also provides a basic framework to investigate different questions. One might ask, for example, how the effect of external stimulation might be introduced in the transition rules to allow the initiation of reentry by a sequence of stimuli.

## Vortices in 2-D

Greenberg et al. (Greenberg, Hassard, and Hastings 1978; Greenberg and Hastings 1978), using a slightly different version of the Wiener and Rosenblueth model, have provided conditions to obtain persistent patterns of propagation in a two-dimensional lattice. Each automaton is coupled only to its first neighbors along the two main axes, a connection pattern referred to as a Moore neighborhood. The state space is characterized by two parameters: $C$, the number of excited states, and $N$, the number of refractory states. When it is excited, the automaton steps through a sequence of $C$ states (from $E_1$ to $E_C$) from which it can excite its neighbors, and then through the $N$ refractory states.

Greenberg et al. have provided a condition for the system to return uniformly to the resting state (i.e., all automata in the state R). Assigning the integers from 0 to $C + N$ to the sequence of states [R, $E_1, \ldots, E_C, AR_1, \ldots, AR_N$], the distance between two states $m$ and $n$ is defined as

$$d(m,n) = \min([|\, m - n \,|, C + N + 1 - |\, m - n \,|]).  \qquad (7.1)$$

The distance thus corresponds to the minimum number of time steps for an automaton to evolve either from state $m$ to state $n$, or vice versa. Denoting the state at a node $(i,j)$ by $u_{i,j}$, Greenberg et al. prove that all the automata reach the state 0 in a finite time if the initial pattern contains only a finite number of nonzero nodes and if it satisfies the condition

$$d(u_{i,j}, u_{i',j'}) < \min\left(C + 1, \frac{C + N + 1}{4}\right),  \qquad (7.2)$$

for all connected nodes $(i,j)$ and $(i',j')$. This means that the initial pattern must contain discontinuities in order to lead to sustained activity.

A cycle is a sequence of states $[S_1, \ldots, S_k]$ over a closed path of connected nodes. Its winding number is

$$W = \frac{1}{C + N + 1} \sum_{l=1}^{k} \sigma(u_{l+1}, u_l),  \qquad (7.3)$$

where $\sigma(u_{l+1}, u_l)$ is the oriented distance between two states ($+$ for clockwise and $-$ for counterclockwise progression).

If the initial pattern contains a cycle P of length $k$ such that

$$d(u_{l+1}, u_l) \leq C, \quad l = 1, k,  \qquad (7.4)$$

and if $W(P) \neq 0$, then the solution is persistent and gives self-sustained propagation.

Figure 7.5 shows an example of an initial condition leading to periodic self-sustained activity. As in the case of the ring, it might be asked how such initial conditions can be induced by external stimulation. Investigating the periodicity and spatial complexity of the patterns that can be generated by this model is part of this chapter's computer exercises dealing with cellular automata.

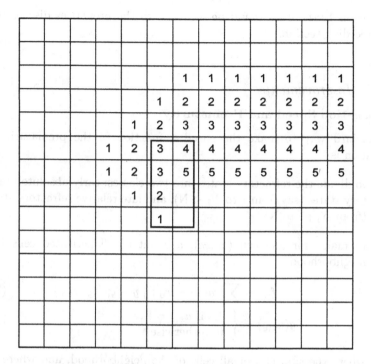

Figure 7.5. Examples of initial conditions leading to a vortex. Cellular automata with $C = 2$, $N = 3$. Blank cells are in state 0. From Greenberg, Hassard, and Hastings (1978).

## Fibrillation

Spatial dispersion of refractoriness has been invoked as a potential source of fibrillation. This has been investigated using the Wiener and Rosenblueth model with a random spatial distribution in the duration of the refractory period ($N$ drawn from normal distribution (Smith and Cohen 1984; Kaplan, Smith, Saxberg, and Cohen 1988) or half-normal distribution (Auger and Bardou 1988), for review (Saxberg and Cohen 1991)). Similar results were obtained by an early work of Moe et al. (1964) in a quite different automaton (see below).

Repetitive disorganized activity has been reported on square (Auger and Bardou 1988) or hexagonal lattices (Kaplan, Smith, Saxberg, and Cohen 1988) and on a cylinder (Smith and Cohen 1984). There is some doubt that the activity was really persistent. It was induced either by fast pacing, or by pacing followed by premature stimulations.

The results indicated that the likelihood of fibrillatory-like behavior was increased by enlarging the standard deviation in the distribution of the absolute refractory period, but was diminished by increasing its mean (Smith and Cohen 1984). This would be in line with the properties discussed in the preceding section.

## 7.3.2  Improvements

### Propagation Speed and Prematurity

A standard way to make the speed a function of the prematurity of activation is

- to divide the refractory states into two classes: early absolute refractory states (say from 1 to $N_1 < N$), and late relative refractory states (from $N_1+1$ to N).

- to make, for any cell $(x_0, y_0)$, a count of all activated cells in a neighborhood,

$$f_{\text{act}} = \sum a(|\, x - x_0\,|, |\, y - y_0\,|) g(u_{x,y}), \qquad (7.5)$$

$$g(u_{x,y}) = \begin{cases} 1 & \text{if } u_{x,y} = \text{E}, \\ 0 & \text{otherwise,} \end{cases} \qquad (7.6)$$

where the sum is over all cells of the neighborhood, and where the $a(|\, x - x_0\,|, |\, y - y_0\,|)$ are weighting factors that may depend of the relative position of each cell in the neighborhood. The set of weights can be used to approximate a Laplacian (Fast, Efimov, and Krinsky 1991; Weimar, Tyson, and Watson 1992), or to introduce anisotropy or other connective features in the medium.

- to redefine the transition rules as

$$\text{R} \mapsto \text{E} \quad \text{if } f_{\text{act}} \geq f_0, \qquad (7.7)$$
$$\text{AR}_i \mapsto \text{E} \quad \text{if } f_{\text{act}} \geq f_0 + f(\text{AR}_i), N \geq i > N_1, \qquad (7.8)$$

where $f(\text{AR}_i)$ is a decreasing function of $\text{AR}_i$. This is equivalent to introducing a threshold that varies with prematurity.

Examining how these changes modify the properties of reentry in a one-dimensional ring and of vortices in a two-dimensional lattice is among the exercises at the end of the chapter.

Moe et al. (1964) have introduced a relationship between prematurity and speed by allocating to each relative refractory state a temporal delay between the activation and the effective transition to the excited state. In alternative formulations (Rosenblueth 1958; Ito and Glass 1991; Courtemanche 1995), each excited cell sends an excitatory pulse traveling in a neighborhood with a speed that can be a function of prematurity.

## Uniformity

The nonuniformity of the propagation speed disrupts the continuity of the activation and repolarization fronts. On a square lattice, when the neighborhood corresponds to all cells falling within a radius $R$, one way to increase the spatial smoothness of the activation and repolarization fronts is to consider each cell as a point whose coordinates are allocated randomly within its corresponding square of the lattice. The effective neighbors of a cell are those whose representative point falls within a distance $R$ (Markus and Hess 1990; Courtemanche 1995).

## Action Potential Duration and Prematurity

The variation of the action potential duration or of the absolute refractory period as a function of prematurity is more difficult to include in a classical deterministic cellular automaton. One way would be to introduce multiple excited states and partially relative refractory states, with transitions to different excited states depending on the partially relative state occupied by the automaton at the moment of the excitation. In the model of Moe and al. (1964), the length of the absolute refractory period was set proportional to the square root of the time between the successive activations. In Ito and Glass (1992), the action potential duration was set as a function of the diastolic interval.

## Action Potential Duration and Coupling

In a diffusive medium, the coupling should tend to flatten the spatial action potential duration profile, putting a limit to the differences in the action potential durations that can exist in a region. Smoothing the spatial profile of action potential duration modifies the form of the repolarization front. Gerhardt et al. (1990) have proposed a version in which the state of each node is evaluated in two steps. A first evaluation is done using standard rules of deterministic cellular automata. Then, if this first evaluation gives a state in the refractory period, the final state is an average of this first evaluation over a neighborhood.

# 7.4   Iterative and Delay Models

Let us start by considering the case of a stable propagation around a ring of radius $R$, with an action potential of fixed duration traveling at a constant speed $(\theta)$.

### 7.4.1  Zykov Model on a Ring

Assume that the speed $\theta$ is an increasing function of $P$, the time elapsed between two returns that corresponds to the period of the rotation. Since we assume that propagation is at a constant speed during one turn, $P$ must be the same at all sites. Then $P_{n+1}$, the period of the $(n+1)$st turn, is given by

$$P_{n+1} = \frac{2\pi R}{\theta(P_n)} = 2\pi R f(P_n), \qquad (7.9)$$

where $R$ is the radius of the ring.

This system corresponds to a finite difference equation, a topic that has been already discussed in Chapter 2. It can be solved graphically using the cobweb method. The fixed point $P_s(R)$ is stable as long as

$$\left| 2\pi R \frac{df(P_s)}{dP} \right| = 2\pi |\alpha| < 1. \qquad (7.10)$$

Otherwise, the stability is lost through a sequence of growing alternations in $P$, since $\alpha = d(1/\theta)/dP < 0$.

### 7.4.2  Delay Equation

Equation (7.9) was used to find the fixed point. To study the stability of the period, we follow the evolution of the system starting from any initial distribution of the period around the ring. Each point on the ring can be designated by an angular coordinate $\phi$ between 0 and $2\pi$. The time between two successive returns is then

$$P(\phi) = R \int_{\phi-2\pi}^{\phi} f(P(\psi))d\psi. \qquad (7.11)$$

To describe the evolution of a perturbation $\xi(\phi)$, we linearize around $P(\phi) = P_s(R)$ and obtain the neutral delay equation (MacDonald 1989)

$$\frac{d\xi(\phi)}{d\phi} = R\frac{df(P_s)}{dP}(\xi(\phi) - \xi(\phi - 2\pi)) \qquad (7.12)$$

$$= \alpha(\xi(\phi) - \xi(\phi - 2\pi)) \qquad (7.13)$$

Assuming that $\xi(\phi) = A\,e^{\lambda\phi}$, this becomes

$$\lambda = \alpha(1 - e^{-\lambda 2\pi}), \qquad (7.14)$$

which, for $\alpha < 0$, has no real nontrivial solution. Assuming $\lambda = \lambda_r + i\lambda_I$, we have

$$\lambda_r = \alpha - \alpha e^{-\lambda_r 2\pi}\cos(\lambda_I 2\pi) \qquad (7.15)$$

$$\lambda_I = \alpha e^{-\lambda_r 2\pi}\sin(\lambda_I 2\pi) \qquad (7.16)$$

Hence

$$(\lambda_r - \alpha)^2 + \lambda_{rmI}^2 = \alpha^2 e^{-\lambda_r 2\pi}, \tag{7.17}$$

and $\lambda_r \leq 0$ for all $\alpha < 0$.

This means that reentry is stable if the speed of propagation increases as a function of the local period between activations. This is completely different from the result established in Section 7.4.2, in which reentry was unstable if $2\pi\alpha < -1$. However, this result was established by assuming a constant speed of propagation for each turn around the loop, with a discontinuous turn-to-turn variation of the period measured at an arbitrary point of observation. This is an obviously unrealistic approximation that can lead to a wrong conclusion. The continuous model shows that the local fluctuation of the speed acts as a self-correcting process: Acceleration of the propagation induces a local reduction of the period, which decreases the speed and stabilizes the reentry.

### 7.4.3  Circulation on the Ring with Variation of the Action Potential Duration

If we now postulate that the action potential duration (APD) is a function of diastolic interval (DIA) while the propagation speed $\theta$ remains constant, we obtain:

$$\text{DIA}(\phi) = \text{BCL} - \text{APD}(\phi - 2\pi), \tag{7.18}$$

$$\text{BCL} = \frac{2\pi R}{\theta}, \tag{7.19}$$

where BCL is basic cycle length. This means that each point on the ring behaves as an isolated cell and that the sequence of diastolic intervals at each site follows the equation

$$\text{DIA}_{n+1} = \text{BCL} - \text{APD}(\text{DIA}_n) \quad \text{if } \text{DIA}_n \geq \text{ARF}, \tag{7.20}$$

where ARF is the absolute refractory period and $(\text{DIA})_n$ is the $n$th diastolic interval at any site. This finite difference equation has been much studied for cardiac excitable cells (Guevara et al. 1984; Guevara et al. 1989; Guevara et al. 1990; Lewis 1974; Chialvo et al. 1990; Vinet et al. 1990; Vinet and Roberge 1994a).

Periodic reentry corresponds to $\text{DIA}_i = \text{DIA}_{i+1}$, a fixed point that is stable if $|d\,\text{APD}/d\,\text{DIA}| < 1$. Shortening the circumference of the ring reduces the basic cycle length BCL. If the action potential duration as a function of diastolic interval, APD(DIA), is a monotonically increasing function, this may lead to 1:1 reentry becoming unstable, with the appearance of stable 2:2 solutions. However, this representation has the same type of weakness as that of the system of Section 7.4.2. Stable 2:2 solutions would involve abrupt transitions between two values of action potential duration at neighboring locations around the ring, a situation that is unrealistic in a coupled system with diffusion.

## 7.4.4  Delay Equation with Dispersion and Restitution

The analyses of the previous sections show that, for the case in which we consider only the dispersion in the speed of propagation, we observe stable dynamics whenever the slope of the dispersion relation is negative. On the other hand, assuming constant speed and considering only a monotonically increasing restitution curve in action potential duration, we obtain a change to a periodic sequence of action potential durations when the ring is made small enough. The same period-two sequence of APD should be observed at all locations, but there is no constraint on the relative phase of the alternation along the ring. This is *not* what has been observed in experiments (Frame and Simson 1988).

What can be the effect of adding *both* dispersion and restitution? Spatial fluctuations of the recovery time impact upon the distribution of both the action potential duration and the speed of propagation. As seen in Section 7.4.2 an acceleration of the propagation reduces the local recovery time, such that the propagation should slow down. But the reduction of the recovery time also abbreviates the next action potential duration, such that the recovery time might be prolonged at the next passage of the front in the same region. This is a potential scenario to destabilize the propagation through a process of growing alternation. To examine the dynamics of the system, we will construct a delay equation model that includes these two features. We begin by introducing new coordinates to describe the dynamics of the propagating action potential on the ring.

We consider the changes in $A$ (action potential duration or APD), $C$ (propagation speed), $t_r$ (recovery time or diastolic interval), and $T$ (the period or $P$) as a function of the location $x$ of the wavefront along the ring. The space-time diagram of Figure 7.6, which displays a continuous solution observed at a ring length $L = 13.15$ cm in a system of partial differential equations (these solutions are examined in much more detail in the following section), illustrates how we will think about these quantities. Each trace in the figure gives the spatial profile of the potential $V$ as a function of distance along the ring at a fixed instant. Each trace is repeated once horizontally to highlight continuity at the boundaries. Traces are stacked vertically at the rate of one every 10 milliseconds. We define a time-dependent coordinate $x(t)$ corresponding to the location of the wavefront along the ring. This coordinate increases continuously as the front circulates around the ring. Two values $x$ and $x'$ are identified with the same location on the ring if $x = x'(\mathrm{mod}\, L)$. In Figure 7.6, the bottom trace is taken at $t = 0$ and the trace whose excitation front is labeled with an asterisk corresponds to $t = 400$ msec. Let $x^*$ be the location of the labeled wavefront as described above. We define four quantities: $C(x^*)$, the speed of the excitation front at $x^*$; $A(x^*)$, the duration of the action potential that is produced at $x^*$ by the incoming activation front; $t_r(x^*)$, the recovery time from the end of the previous action potential to the onset of the new one that begins when the activation front reaches $x^*$; $T(x^*)$, the circulation time from $x^* - L$ to $x^*$ that corresponds to the time taken by the activation front to complete its last rotation.

Using the quantities $A(x)$, $t_r(x)$, and $C(x)$, we write an equation stating that for any point $x$ along the ring, the recovery time is the difference between the circulation time and the previous action potential duration. From Figure 7.6, we see that

$$T(x) = t_r(x) + A(x - L),\tag{7.21}$$

an equation analogous to equation (7.18). Substituting $\int_{x-L}^{x} ds/C(s)$ for the circulation time $T(x)$, as in equation (7.11), yields the equation

$$t_r(x) = \int_{x-L}^{x} \frac{ds}{C(s)} - A(x - L).\tag{7.22}$$

The next theoretical step is based on the assumption that both $A$ and $C$ are functions of the recovery time $t_r$ (Courtemanche, Glass, and Keener 1993; Courtemanche, Glass, and Keener 1996). Assuming that the time course of recovery (and its effect on the response of the system to excitation) does not depend on past history of the system is an approximation that has been often used in the analysis of cardiac propagation (Vinet, Chialvo, Michaels, and Jalife 1990; Quan and Rudy 1990a; Lewis 1974; Ito and Glass 1992; Ito and Glass 1991). In other words, we assume that recovery follows an invariant trajectory, a statement that can be formally demonstrated for simple two-variable models of excitable media such as the FitzHugh–Nagumo model presented in Chapter 4 and the computer exercises in Section 4.8. In addition, we neglect the effect of intercellular coupling on action potential duration.

Using the restitution and dispersion curves, equation (7.22) may be rewritten as

$$t_r(x) = \int_{x-L}^{x} \frac{ds}{c(t_r(s))} - a(t_r(x - L)),\tag{7.23}$$

which is an integral delay equation for the recovery time $t_r$. It describes the dynamics of pulse circulation on the ring.

Equation (7.23) can be reduced to a neutral delay differential equation (Hale and Lunel 1993) by taking derivatives with respect to $x$, which yields

$$\frac{d}{dx}\left(t_r(x) + a\left(t_r(x - L)\right)\right) = \frac{1}{c\left(t_r(x)\right)} - \frac{1}{c\left(t_r(x - L)\right)}.\tag{7.24}$$

We shall use the integral delay equation and analyze the dynamics and stability of equation (7.24).

Before pursuing the analysis, it is convenient to change variables in equation (7.23). Setting $y = Qx/L$ and $t_r(Ly/Q) = z(y)$ into equation (7.23)

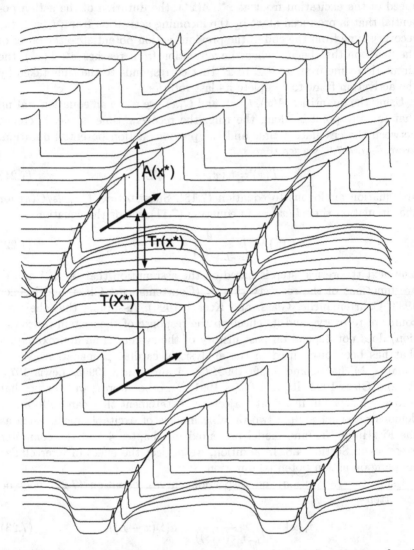

Figure 7.6. Dynamics of a circulating pulse on a ring of length $L = 13.15$ cm found
from numerical integration of the Beeler–Reuter equations. Each horizontal trace
represents the profile of voltage along the ring, repeated once to illustrate continuity. Traces are then stacked vertically once every 10 msec (time increases from
bottom to top). Note the large changes in waveform as the excitation propagates
from left to right along the ring. At a given location of the excitation front along
the ring, labeled with an asterisk ($x = x^*$), we illustrate the quantities used
to characterize the dynamics of the circulating pulse, namely, the pulse speed
$C(x^*)$, the pulse duration $A(x^*)$, the recovery time $t_r(x^*)$, and the circulation
time $T(x^*)$. From Courtemanche, Glass, and Keener (1996).

yields

$$z(y) = \frac{1}{Q} \int_{y-Q}^{y} d(z(s))ds - a(z(y-Q)), \tag{7.25}$$

where $d(z) = L/c(z)$. Equation (7.25) has steady-state solutions $z = z^*$ satisfying

$$z^* = d(z^*) - a(z^*). \tag{7.26}$$

Let $a' = da/dz(z^*)$ and $d' = dd/dz(z^*)$. Linearizing equation (7.25) near $\tilde{z}(y) = z(y) - z^*$ yields

$$\tilde{z}(y) = \frac{1}{Q} \int_{y-Q}^{y} d'\tilde{z}(s)ds - a'\tilde{z}(y-Q). \tag{7.27}$$

We look for solutions to equation (7.27) of the form $\tilde{z} = Ce^y$. In the original coordinates, these solutions are of the form $t_r = t_r^* + be^{Qx/L}$, so that $Q/L$ is an eigenvalue of the linearized integral equation (7.23). Substituting for $\tilde{z}$ in equation (7.27) gives

$$Q(1 + a'e^{-Q}) = (1 - e^{-Q})d'. \tag{7.28}$$

The stability of the steady state is determined by the roots of equation (7.28), which is the characteristic equation for the integral delay equation. Stability requires $\Re(Q) < 0$, where $\Re(Q)$ and $\Im(Q)$ represent the real and imaginary parts of $Q$, respectively. The steady state loses its stability when the roots cross from the left to the right complex plane, indicated by $\Re(Q) = 0$. Given that both the restitution and dispersion curves are monotonically increasing functions of the recovery time, we limit ourselves to the case in which $a' \geq 0$ and $d' < 0$. The analysis of this case leads to the following property: Let $Q \neq 0$ be a non trivial root of equation (7.28) for $a' \geq 0$, $d' < 0$. Then $\Re(Q) = 0$ if and only if $a' = 1$, $\Re(Q) < 0$ if and only if $0 \leq a' < 1$, while $\Re(Q) > 0$ if and only if $a' > 1$.

This property gives the stability region of the steady-state for $a' \geq 0$ and $d' < 0$. The steady state is stable for $0 \leq a' < 1$, and stability is lost at $a' = 1$ through an infinite-dimensional Hopf bifurcation. Keeping $d' < 0$ fixed and starting from $a' = -1$, the root crosses the imaginary axis into the left-hand plane near $2\pi i$. It then curves around and crosses the imaginary axis into the right-hand plane near $\pi i$. This behavior appears typical of all roots $Q^{(k)}$, and occurs symmetrically across the real axis. For $d' < 0$ and $a' < 0$, there is an additional real root that crosses from the left to the right complex plane when $a' = -1 + d'$. We do not consider here the more complicated behavior of the roots in the case $d' > 0$.

Our application to excitable media sets $a' \geq 0$ and $d' < 0$. We are particularly interested in the bifurcation at $a' = 1$, where we find that the imaginary part $q_0$ of the roots satisfies

$$\tan\left(\frac{q_0}{2}\right) = \frac{q_0}{d'}. \tag{7.29}$$

The solutions of equation (7.29) can be viewed graphically as the intersection between the straight line of slope $1/d'$ and the tangent function. There are an infinite number of solutions (this is true in general for equation (7.28)) associated with an infinite number of roots $Q^{(k)} = iq_0^{(k)}$ crossing the imaginary axis. If we assume that the dispersion relation is nearly flat at the bifurcation point, then $|d'| \ll 1$, and the imaginary parts $q_0^{(k)}$ of the roots can be approximated as

$$q_0^{(k)} = (2k+1)\pi - \frac{2d'}{(2k+1)\pi} + \mathcal{O}(d'^2), \quad k = 0, 1, 2, \ldots. \tag{7.30}$$

For the partial differential equation solution given in Figure 7.6, we have $d' \approx -0.088$ at the bifurcation.

Given that an infinite number of unstable modes exist for $a' > 1$, we carry out an expansion of the roots of equation (7.28) for $a'$ close to 1. This yields an estimate for the wavelength of the unstable modes beyond the bifurcation, for $|d'| \ll 1$,

$$\Lambda^{(k)} = \frac{2\pi L}{\Im(Q)} = \frac{2L}{2k+1} + \frac{2d'L(2-\epsilon)}{(2k+1)^3\pi^2} + \mathcal{O}(d'^2). \tag{7.31}$$

The mode of lowest frequency, $\Lambda^{(0)}$, has wavelength slightly less than twice the ring length. The linear theory does not predict a dominance of this mode over the others based on its initial growth rate. This analysis thus suggests that the loss of stability of the periodic reentry may occur through the appearance of multiple modes of propagation corresponding to continuous oscillations of action potential duration and diastolic interval as propagation proceeds around the ring. However, it does not indicate which of these modes are stable, and if so, which set of initial conditions can lead to a specific one.

Equation (7.23) can be integrated using a simple forward method, subject to the constraints imposed by the ring geometry. Initial conditions must be specified by giving the value of $t_r(x)$ over the interval $[-L, 0)$. The finite difference equation for $t_r$ is given by

$$t_r^{n+1} = \sum_{i=n-N+1}^{n} \frac{\Delta x}{c(t_{ri})} - a(t_r^{n-N+1}), \tag{7.32}$$

where the integral is evaluated using a simple trapezoid rule, using $\Delta x = L/N$, $N = 200$. This discretized version of the delay equation is identical to the iterative model of propagation on a ring developed by Ito and Glass (1992). Below are two different solutions of the discretized integral delay equation, showing two different wavelengths for the quasi-periodic solutions. These were obtained using two different initial conditions, corresponding to sinusoidal profiles of diastolic interval with wavelengths given by equation (7.31).

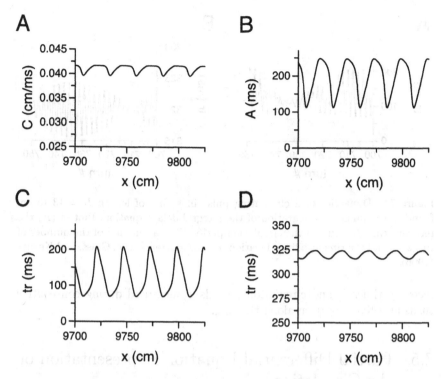

Figure 7.7. Dynamics of a circulating pulse on a ring of length $L = 13.15$ cm found from numerical integration of the integral delay equation. (A) Speed $C(x)$, (B) pulse duration $A(x)$, (C) recovery time $t_r(x)$, (D) circulation period $T(x)$. Transients have dissipated. Wavelength is $\Lambda = 25.7$ cm. From Courtemanche, Glass, and Keener (1996).

Figure 7.7 shows a trace of $C(x)$, $A(x)$, $t_r(x)$ and $T(x)$ after stabilization. The wavelength of the oscillation is $\Lambda \approx 25.7$ cm. Note how the sequence of measurements ($T^n$ and $t_r^n$ shown here) are quasi-periodic, as opposed to the periodic sequences we deduced in the case in which only restitution was assumed in the model. Figure 7.8 shows the sequence of recovery times $t_r$ and circulation periods $T$ measured during consecutive turns at a fixed location along the ring.

Based on the analysis of the integral delay equation, it might be possible to observe oscillating solutions at one of the other frequencies that are known to become unstable at the bifurcation. For example, the second lowest frequency corresponds to a wavelength $\Lambda_1$ slightly less than two-thirds the ring length. We have obtained such a solution using initial conditions in the form of a low-amplitude sine wave of the correct frequency (as predicted by equation (7.31)) over the initial interval $[0, L]$. Figure 7.9 shows a plot of $C(x)$, $A(x)$, $t_r(x)$, and $T(x)$, illustrating the solution with frequency close to $2L/3$ after transients have dissipated. Figure 7.10 shows the sequence of

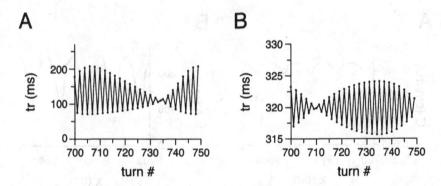

Figure 7.8. Dynamics of a circulating pulse in a ring of length $L = 13.15$ cm found from numerical integration of the integral delay equation. Plot of (A) the recovery time $t_r$ and (B) the circulation period $T$ as a function of the number of turns around the ring at a fixed location. From Courtemanche, Glass, and Keener (1996).

recovery times $t_r$ and circulation periods $T$ measured during consecutive turns at a fixed location along the ring.

## 7.5    Partial Differential Equation Representation of the Circulation

### 7.5.1    Ionic Model

A partial differential equation model of the electrical behavior of the cardiac excitable tissue must include a representation of the membrane ionic properties. It is mandatory, since without the nonlinear behavior of the membrane, the system would be diffusive and there would be no sustained propagation.

Two models of excitable membrane have been presented in a preceding chapter: the FitzHugh–Nagumo equations and Hodgkin–Huxley model (Chapter 4). In these models, the temporal evolution of membrane potential $(V = V_{\text{int}} - V_{\text{ext}})$ of the space-clamped membrane and of the gate variables $(Y_i)$ is expressed by a system of ordinary differential equations

$$C_m \frac{dV}{dt} = -I_{\text{ion}}(V, Y_i), \tag{7.33}$$

$$\frac{Y_i}{dt} = f_i(V, Y_i), \tag{7.34}$$

where $I_{\text{ion}}$ is the ionic current, which is a function of $V$ and of a set of dynamic variables $Y_i$ controlling the different ionic channels. Each model differs in the number of gate variables and in the definition of the $Y_i$ and $f_i$.

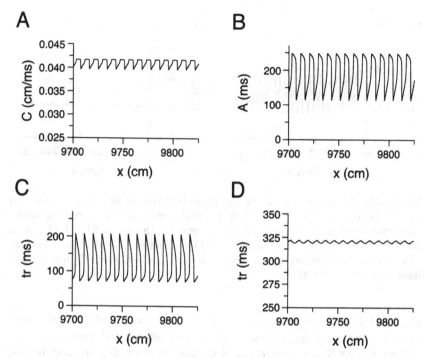

Figure 7.9. Dynamics of a circulating pulse in a ring of length $L = 13.15$ cm found from numerical integration of the integral delay equation. Initial conditions were chosen so that an alternate solution of wavelength close to $2L/3$ is observed. (a) Speed $C(x)$, (b) pulse duration $A(x)$, (c) recovery time $t_r(x)$, (d) circulation period $T(x)$. Transients have dissipated. Wavelength is $\Lambda = 8.74$ cm. From Courtemanche, Glass, and Keener (1996).

The Hodgkin–Huxley model was developed to represent the dynamics of nerve cells. Its original formulation includes two gate-controlled currents (sodium, $I_{Na}$, and potassium, $I_K$), and two characteristic time scales: the fast activation of the sodium current responsible for the upstroke ($\tau \approx 0.5$ ms), and the slower closure of the sodium current and opening of the potassium current responsible for the repolarization ($\tau \approx 5$ ms). The model has a sharp threshold behavior. Its action potential duration and its absolute refractory period are on the order of a few milliseconds. Because of this division between two distinct time scales, its dynamics have been shown to be largely equivalent to the much simpler FitzHugh–Nagumo model (FitzHugh 1960).

The Beeler–Reuter model (Beeler and Reuter 1977) was one of the first models of cardiac myocytes. It includes three gate-regulated currents: the fast-activating $I_{Na}$ still controlling the upstroke; an intermediate inward current $I_{Si}$ responsible for the plateau; the slow $I_K$ current involved in the late repolarization. There are three distinct time scales: fast (activation of

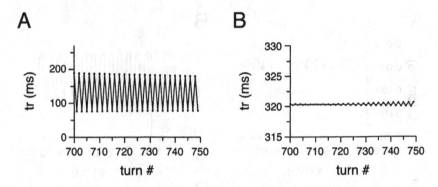

Figure 7.10. Dynamics of a circulating pulse in a ring of length $L = 13.15$ cm found from numerical integration of the integral delay equation. Initial conditions were chosen so that an alternate solution of wavelength close to $2L/3$ is observed. Plot of (A) the recovery time $t_r$ and (B) the circulation period $T$ as a function of the number of turns around the ring at a fixed location. From Courtemanche, Glass, and Keener (1996).

$I_{Na}$, $\tau \approx 0.5$ ms), intermediate (closure of $I_{Na}$, opening of $I_{Si}$, $\tau \approx 5$ ms) and slow (closure of $I_{Si}$, activation of $I_K$, $\tau \approx 100$ ms and more).

There is now a new generation of cardiac action potential models that include mechanisms describing the cytoplasmic ionic concentrations of sodium, calcium, and potassium (e.g., Luo and Rudy 1994; Nordin and Ming 1995). In cardiac myocytes, as in all muscle cells, the internal calcium concentration controls the contraction. The calcium is stored in an inner reservoir called the sarcoplasmic reticulum. The entry of calcium through the membrane gate-controlled calcium ionic channels does not increase much the internal calcium concentration, but rather triggers its release from the sarcoplasmic reticulum. The calcium is recaptured by the sarcoplasmic reticulum or is returned to the external medium by active pumping and the action of the sodium–calcium exchanger. The plateau phase, instead of depending on a single current ($I_{Si}$) as in the Beeler–Reuter model, results from the interplay of different mechanisms, mainly the gate-controlled calcium current and the sodium–calcium exchanger. Since repetitive stimulations may induce long–term depletion or accumulation of ions, the restitution curve has, in these models, a strong memory component (i.e., dependence on previous pacing history), that may induce hysteresis and bistability in the response patterns (Friedman, Vinet, and Roberge 1996). However, these mathematical models are high-dimensional (10 variables or more), and imply a high computational load even in their space-clamped version.

Regarding one-dimensional and two-dimensional models of propagation, it is often claimed that the FitzHugh–Nagumo model provides a representation of cardiac excitable tissues. But since both the cellular automata models and the iterative models on loops have shown that action potential duration restitution is a main determinant of the dynamic in extended medium, there are still questions whether ionic models with such important differences regarding the action potential duration rhythm dependence really give dynamics similar to the FitzHugh–Nagumo model in extended tissues.

## 7.5.2  One-Dimensional Ring

In this section, we discuss the results obtained on one–dimensional rings of Beeler–Reuter cells. We consider that the cells are well connected such that the intracellular medium and the membrane can be considered as a continuous and isotropic medium. The external medium is shunted, or has a fixed resistivity. The meaning of these assumptions is discussed in the section on two-dimensional reentry. The equations for a ring of radius $R$ are

$$\frac{1}{\rho S} \frac{\delta V^2}{\delta^2 t} = C_{\mathrm{m}} \frac{\delta V}{\delta t} + I_{\mathrm{ion}}(V, Y_i), \tag{7.35}$$

$$V(0) = V(2\pi R), \tag{7.36}$$

$$\frac{\delta V(0)}{\delta x} = \frac{\delta V(2\pi R))}{\delta x}, \tag{7.37}$$

$$\frac{Y_i}{dt} = f_i(V(x), Y_i(x)), \tag{7.38}$$

$$Y_i(0) = Y_i(2\pi R), \tag{7.39}$$

where $\rho$ is the lumped resistivity, $S$ is the surface-to-volume ratio of the internal medium, and $C_{\mathrm{m}}$, the membrane capacitance. Reentry has been studied for different versions of the Beeler–Reuter model. For the original Beeler–Reuter membrane model with nominal values of the parameters, the evolution of the sustained propagated solution fits well with the prediction of the integral delay model presented in Section 7.4.4 (Courtemanche, Glass, and Keener 1996; Courtemanche, Glass, and Keener 1993; Vinet and Roberge 1994b; Vinet 1995; Karma 1994; Karma 1993; Karma, Levine, and Zou 1994). The main characteristics of the solutions obtained by numerical simulations were as follows:

- The propagation becomes quasi-periodic at a critical value of $R_{\mathrm{c}}$ of the radius.

- The transition to quasi-periodicity occurs when the action potential duration of the stable 1:1 solution approaches the region where the slope of the relation between action potential duration and diastolic interval relation increases to 1.

- Near $R_c$, the system has multiple modes of rotation similar, at least for the first few ones, to those predicted by the integral delay equation model.

- At lower radius, only the first mode persists.

- At lower radius, there are moments when the propagation becomes almost electrotonic. This speeds up the repolarization behind the front, such that for a while, two disjoint excitable gaps exist.

The discrepancies between the integral delay equation and Beeler–Reuter-type models (infinite versus finite number of quasi-periodic modes, fine structure in the organization of the solutions near the bifurcation point) were resolved by a modification of the calculation of the local action potential duration in the integral delay model, expressing it as a weighted average of the values of action potential duration as a function of diastolic interval, APD(DIA), over a neighborhood (Vinet 1999; Vinet 2000; Watanabe, Fenton, Evans, Hastings, and Karma 2001). The effect of speeding the $I_{Si}$ dynamics has also been studied (Vinet 1999). The value of $R_c$ is decreased, the interval of aperiodic regime is shortened, and there is multistability between the 1:1 regime and different modes of aperiodic rotation. It remains to be seen whether these aperiodic regimes are still quasi-periodic and whether the memory effect in action potential duration and propagation speed coming with the speeding of $I_{Si}$ may induce more complex dynamics.

## 7.6    Reentry in Two Dimensions

### 7.6.1    Reentry Around an Obstacle

We have examined one-dimensional ring models of reentry of increasing complexity, from cellular automata, to iterative models and delay equations, and finally to partial differential equations. Clearly, reentry on a one-dimensional ring is a caricature of reentry in living hearts that may be appropriate when cardiac excitation circulates around an inexcitable obstacle. The wave extending out from excitation anchored to a sizable obstacle may have little influence in determining the details of propagation around the obstacle, and the results of the ring models may be applicable. However, it is dubious that this approximation can hold in all circumstances. Consider the case illustrated in Figure 7.11 of reentry around a two-dimensional annulus. The radius of the central obstacle can be reduced to reach the circumference at which periodic reentry becomes unstable in a one-dimensional ring. Does reentry still remain periodic in the annulus, meaning that the dynamics at the tip of a wave anchored on the obstacle cannot be separated to what happens to the extended wavefront? Different

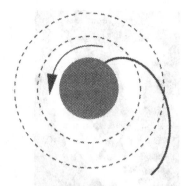 

Figure 7.11. What happens to an excitation wave anchored on an obstacle if one shrinks and removes the obstacle?

results rather show that the dynamics of excitation and repolarization are dependent on both the inner and outer radii of the annulus (Xie, Qu, and Garfinkel 1998; Comtois and Vinet 1999)

On the other hand, if the central obstacle is reduced below the minimum circumference allowing self-sustained propagation in the ring, will sustained propagation also be abolished in the two-dimensional medium? Results shown below rather demonstrate that complex sustained dynamics may exist in a continuous two-dimensional medium.

Can we use simple principles to derive the shape of the rotating wavefront emanating from the obstacle? As a first approximation, let us assume that the propagation speed normal to the wavefront is fixed. This assumption neglects the effect of wavefront curvature on the speed of propagation, although it is known to be important in excitable media. Assume also that the wavefront is rotating rigidly around the obstacle, without changing shape, and that the wavefront tip always forms a right angle with the edge of the circular obstacle, a consequence of the null-flux boundary condition. As a result, the "circulation" velocity of wavefront points located along concentric circles of increasing radii must increase to maintain a rigid shape and a fixed angular velocity. To meet these conditions, the activation front must have the shape of a spiral, converging toward an *Archimedean spiral* (i.e., in polar coordinates $\theta \propto r$) as the inner radius of the obstacle is reduced. The existence of spiral waves as the stable mode of propagation of a fixed-shape activation front around an obstacle was also predicted in the case where the relation between propagation speed and the curvature of the front as well as the dispersion of the conduction speed as a function of the frequency of the rotation wave were both taken into account (Zykov 1988; Tyson and Keener 1987; Keener 1991; Meron 1991; Meron 1991; Mikhailov, Davydov, and Zykov 1994).

It turns out that in numerical simulations, at least in the cases in which it is possible to carry out this "obstacle removal" by decreasing gradually the inner radius of the ring, the rotating wave becomes a spiral wave that

Figure 7.12. Experimental spiral waves in a canine epicardial preparation. From Pertsov et al. (1993).

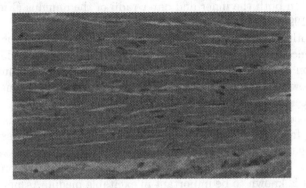

Figure 7.13. A microscopic view of cardiac muscle.

is a self-sustained solution in excitable medium. Hence spiral waves may be thought of as a mathematical representation of *functional* reentry in the absence of an obstacle if complex cardiac tissue may be realistically represented as an excitable media. This is partly the subject of the next section. As mentioned in the introduction and illustrated in Figure 7.12, experimental evidence supports the existence of spiral waves as a mechanism of reentry.

## 7.6.2 Simplifying Complex Tissue Structure

In going from the ring to a sheet of tissue (or a three-dimensional volume!), a number of complicating factors related to the structure of the tissue become important. The preferential direction of conduction along cardiac fibers (anisotropy, see Figure 7.13), the discreteness of intercellular connections and their dependence on tissue parameters, the changing direction of cardiac fibers, the presence of connective tissue barriers, the conduction properties of the extracellular medium, all may become crucial to electrical propagation. We will briefly address those issues at the end of this section. We begin, however, by assuming the simplest isotropic and continuous-sheet model, i.e., a straightforward extension of the one-dimensional cable equation to two dimensions of the form

$$\frac{\partial V}{\partial t} + \frac{I_{BR}}{C_m} = D(\frac{\partial^2 V}{\partial x^2} + \frac{\partial^2 V}{\partial y^2}),  \qquad (7.40)$$

where $I_{BR}$ is the total current flowing through the cardiac cell membrane, obtained from the Beeler–Reuter equations (Beeler and Reuter 1977). This is an excitable medium. The major assumption made here is that the tissue acts as a *syncytium*, i.e., that adjacent intracellular domains are fused into a unique uniform interior, neglecting discrete intercellular connections. Also, the extracellular space is assumed to be a uniform, zero-resistance conductor. This is the *monodomain* formulation, as opposed to a *bidomain* formulation where both the intracellular and extracellular potentials are explicitly considered (Henriquez 1993). Have we removed all the interesting complexities of two-dimensional cardiac tissue such that studying equation (7.40) has become worthless? In Section 7.5.2 on the dynamics of the Beeler–Reuter model on a one-dimensional ring, it was shown that there is a transition from periodic to quasi-periodic reentry when the ring becomes shorter than a critical radius. In a two-dimensional medium, transition from stable to meandering vortex, in which the tip of the spiral wave follows a periodic or quasi-periodic trajectory, has been observed for a combination of parameters both in the FitzHugh–Nagumo (for review, Winfree 1991) and Beeler–Reuter models (Efimov, Krinsky, and Jalife 1995). Transition to meandering has been observed in a host of different models and is a very common scenario by which rigidly rotating spiral waves lose stability. However, there is not yet a general analytical theory for spiral wave motion, except in the weakly excitable limit (Hakim and Karma 1999). This context seems to be far removed form the dynamics in cardiac tissue, in which the action potential duration restitution curve (Gerhardt et al. 1990; Ito and Glass 1991; Winfree 1989; Panfilov and Holden 1991; Courtemanche and Winfree 1991; Baer and Eiswirth 1993; Panfilov and Hogeweg 1993; Leon et al. 1994; Karma 1994; Karma 1993; Qu et al. 2000; Hall et al. 1999; Fox et al. 2002), the speed dispersion curve (Giaquinta et al. 1996; Watanabe et al. 2001; Qu et al. 2000), as well as the effect of curvature on the speed of propagation and the action potential duration (Cabo et al. 1994; Cabo et al. 1998; Comtois and Vinet 1999) have all been shown to have an impact on the stability of spiral waves.

## 7.6.3  Spiral Breakup

Spiral breakup refers to the situation in which a spiral wave dislocates and gives rises to disconnected propagation fronts. We have introduced in the

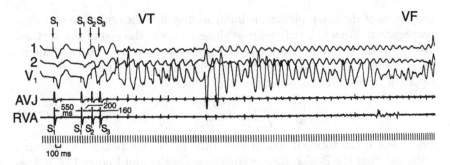

Figure 7.14. Transition from organized ventricular tachycardia (VT) to disorganized fatal ventricular fibrillation (VF) in a patient as observed on the electrocardiogram. From Ruskin, DiMarco, and Garan (1980).

Beeler–Reuter model a parameter $\sigma$ to control the maximum action potential duration. This parameter $\sigma$ multiplies the time constants controlling the activation and inactivation of the gate variables of the calcium current $I_{Si}$ that is responsible for the plateau of the action potential. Using $\sigma \leq 1$, ($\sigma = 1$ in the original Beeler–Reuter model) speeds up the dynamics of the calcium current, abbreviates the plateau of the action potential, and produces briefer action potential.

Stable spirals are observed at $\sigma = 0.5$, while for higher values of $\sigma$, **spiral breaks** occur, meaning that there are repeated localized propagation failures along the reentrant wavefront resulting in overall more or less disorganized activity within the simulated sheet of tissue. Could this be how stable ventricular tachycardia degenerates into fatal ventricular fibrillation, a transition that has been observed both in patients and animal preparations (see Figure 7.14)?

What is the nature of the instability responsible for spiral breakup? One possible explanation may be that when the period of the spiral drops into the region of action potential duration instability, large changes in action potential duration, as those observed in a ring, may lead to spiral wave breaks. A large rapid change in action potential duration during action potential propagation results in slow-moving repolarization fronts that act like "arcs of conduction block" (Figure 7.15 and Figure 7.16).

The spiral front may also be shown to propagate into tissue with alternating recovery properties (as measured by the recovery time of the tissue or diastolic interval) with larger alternations for larger values of $\sigma$ (Figure 7.17).

Among the problems in applying this hypothesis to obtain a definite criterion for spiral stability is the difficulty in measuring spiral period in presence of pronounced meander and pulse duration instability. This is true even in the absence of spiral wave breaks, e.g., at $\sigma = 0.5$ (Figure 7.18 and Figure 7.19). The same problem also appears for the measure of the action potential duration (Koller, Riccio, and Gilmour 1998).

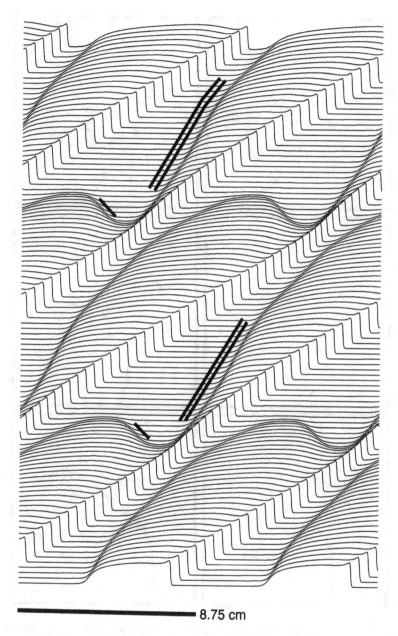

**8.75 cm**

Figure 7.15. Illustrating slow-moving recovery fronts during action potential duration instability on the ring ($\sigma = 0.7$).

Figure 7.16. Illustrating slow recovery fronts during spiral motion. The length of the arrows is proportional to the recovery front speed, which is slower than the incoming excitation front speed at the point of the wave break ($\sigma = 0.7$).

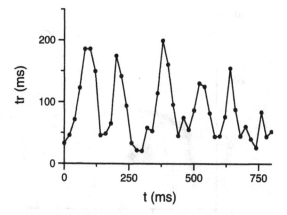

Figure 7.17. Alternations in diastolic interval seen in the moving frame of the spiral front ($\sigma = 0.7$).

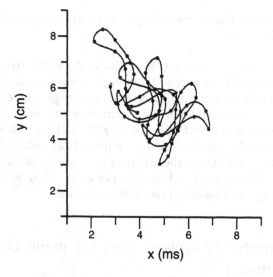

Figure 7.18. Meander path at $\sigma = 0.5$.

## 7.7  Conclusions

For over half a century, since the publication of the seminal work of Wiener and Rosenblueth, mathematical modeling has complemented experimental work and has contributed significantly to the understanding of cardiac arrhythmia. However, much work is still needed, since even in the simplified setting of an isotropic and continuous two-dimensional representation of the cardiac tissue, we remain far from a global understanding of the possible regimes of reentry and of the factors governing their dynamics. The consequences of this lack of understanding have been dramatically illustrated

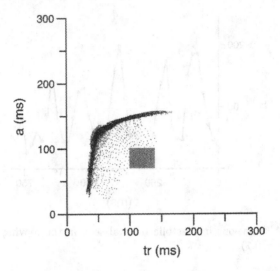

Figure 7.19. Restitution measured during spiral reentry at $\sigma = 0.5$. Note the wide range of periods observed.

by several antiarrhythmic drug trials (Naccarelli, Wolbrette, Dell'Orfano, Patel, and Luck 1998) in which the mortality was actually increased by the administration of different antiarrhythmic drugs. There is hope that mathematical models might not only contribute to elucidating the basic mechanisms of cardiac arrhythmia, but might also help to find better targets for pharmacological therapy and improved algorithms for electrical intervention designed to stop the arrhythmia. The goal of this chapter was to introduce the basis of the different types of modeling approaches that has been developed to handle these challenging tasks.

## 7.8  Computer Exercises: Reentry using Cellular Automata

The following computer exercise is based on a menu-driven Matlab program to simulate propagation in a two-dimensional network of deterministic cellular automata, as presented in Section 7.3. It represents a set of identical cellular automata lying on a rectangular lattice. Different menus allow the definition of the dimension of the lattice, of the state space of the cellular automaton (i.e., number of states during which one automaton can excite its neighbors, number of absolute and partially refractory states), of the threshold associated with each partially refractory state, and of the extent of neighborhood. The program also allows for different protocols involving the application of external stimuli on selected subsets of cellular automata. Subsets of cellular automata can also be disconnected from their neighbors to form inexcitable obstacles in the lattice. The program will be used to investigate different aspects of reentry in excitable tissue.

## Software

Start `Matlab`.

Type `mainaut1` at the `Matlab` prompt ($>>$) to start the program.

The main window will appear, with a main menu bar along the top. Some things to remember are:

- All parameters can be fixed through nested menus.

- Menu items are activated by clicking with the left mouse button.

- You must initialize the properties of the cellular automata before starting a simulation. This is done mostly using the various submenus under **define** in the main menu.

- When you have finished an initialization, you may save it in a file (by selecting **save**) such that it can be recalled.

- Details of the various menu options are included in a help file available on the book's web site, available via **www.springer-ny.com**. When you construct a protocol or initialize properties of the cellular automata, make sure you close down all (possibly nested) setup windows using the **OK** or **exit** button for each specific window before starting the cellular automata stimulation. This will ensure that all changes have been taken into account before the start of the simulation.

- The term **nb** appearing throughout the menu options and the description below means we are dealing with a "number."

We begin with some examples showing how the different menus can be used to solve specific problems. Then, we propose a set of exercises on the dynamics of excitable tissue.

## Examples

In this section, we present some exercises, with a description of the different steps to follow to reach the solution. We specify a given program submenu using the construct

> **/main menu option/submenu option/subsubmenu option**

For example, calling up the **lattice** submenu within the **define** main menu option is denoted by **/define/lattice**.

1. Program the cellular automaton as a simple three-state (one excited (E), one absolute refractory (AR), and one rest (R)) Wiener and Rosenblueth cellular automaton (see Section 7.3.1) using a neighborhood of radius $L = 1$ and set the threshold to 1 for excitation of the resting state, with no excitation possible from the other states. Set the protocol to initiate the propagation of a plane wave with an activation front parallel to one side of the lattice.

   The sequence of steps to initialize the model and the protocol are as follows:

   (a) Use **/define/lattice** to define a lattice size, say 10 by 10. Click **OK**.

   (b) Use **/define/state space** to define one excited state and one refractory state. This is done by erasing the numbers appearing in the boxes facing excited and refractory, and then typing 1 in each box. Click **OK**.

   (c) Use **/define/threshold** to set the threshold for excitation to one. Only the states **rest**, **ex1**, and **rp1** are defined for this model (recall step 2). The excited and refractory states (**ex1** and **rp1**) cannot be excited, so their threshold value should be zero (inexcitable). The threshold for excitation from the rest state should set to one (1). Click **OK**.

   (d) Use **/define/neighborhood** to define the cells within the neighborhood of the generic central cell **X**. Because we want the neighborhood to have radius $L = 1$, only the neighbors above, below, right, and left should be activated (they are the only ones within one unit of the center cell). Activate them and click **OK**.

   (e) Click **/state/reset** to set all the cells at their resting state.

   (f) Establish the simulation protocol. In this case, we want to stimulate all the cells lying along one border of the lattice.

   Select **/protocol/nb of stages** and set the number of stages to 1. Click **OK**.

   Select **/protocol/set** and click on the box holding the number 1. A new box appears with all the parameters for this stage of the protocol.

   - Set the number of step to say 10. This is the total number of time steps in this stage of the protocol

   - Set the period of the stimulation to 20. Only one stimulus will be applied, since the duration of this stage of the protocol is only 10 time steps.

   - Set the amplitude of stimulation to 1, which is the threshold to stimulate a cell from rest.

   - Click in the box holding the name **stimul** to select the cells to be stimulated. A new box showing the lattice appears. Click on the cells along one border of the lattice.

The cells that will be stimulated become black. Clicking again in a cell reverses its color to white, meaning that it will not be stimulated. When you have finished selecting the cells to be stimulated, click **OK**. Click **OK** also in the box holding the parameters of the protocol, and click **exit** in the box holding the stage.

(g) You are ready to run the simulation. Select **/run/start** and watch the simulation. A plane wave should travel across the lattice. By selecting again **/run/start**, you will rerun the simulation, but from the lattice state existing at the end of the previous simulation. If the plane wave travels too fast, you may click **/stepping/step** and then **/run/start**. One time step of simulation will be executed each time you click the left button of the mouse while holding it over the lattice. To see the correspondence between the color code and the cellular automaton states, click **/colormap**.

(h) You can also observe the temporal evolution of the state for different cells in the lattice. Select **/trace/choose cells**. A lattice appears on which you can select the cells with the same procedure that was used to choose the cells to stimulate. Click **OK**, and then rerun the simulation (**/run/start**). At the end, click **/trace/plot**. Graphics showing the temporal evolution of all the selected cells (state vs. time) appear. Selecting **/trace/plot** will always give a plot of the evolution of the cells for the last simulation in a new graphic window. In this way, you may compare the results of different protocols. These graphic windows can be closed by clicking on the box holding an **X** in their top right corner.

2. Find the minimum period of stimulation for which a plane wave is produced for each application of the stimulus.

- Return to the definition of the protocol. Select **/protocol/set** and click on the box holding the number 1. Change the period of the stimulation to 3. Close all the windows associated with the initialization of the protocol (clicking on **OK** and **exit** buttons) and rerun the simulation. You may reset the lattice to the resting state before starting the simulation by selecting **/state/reset**. At the end of the simulation, you can obtain new graphics of the evolution of the state for the selected cells by clicking on **/trace/plot**. What do you observe? Repeat after setting the period of stimulation to 2. What do you conclude?

3. Produce a sequence of 10 plane waves, with period 3, followed by a sequence of 10 plane waves with period 5.

   (a) Select **protocol/nb of stages** and set the number of stages to 2. Click **OK**.

   (b) Select **/protocol/set** and click on the box holding the number 1. Set the period of stimulation to 3, and the number of time steps to 30. Click **OK**.

   (c) Click on the box holding the number 2. For this stage of the protocol, set the period of stimulation to 5, and the number of time steps to 50. You must also define the amplitude of stimulation and select the cells to be stimulated in this stage of the protocol. In this case, these should be chosen identical to what was selected for stage 1. Close all the windows associated with the initialization of stage 2 with the **OK** buttons. Click **exit** in the box allowing the selection of the stages, set the lattice at the resting state, and run the simulation. The total simulation has now 80 time steps, corresponding to the simulation of stage 1, followed by stage 2.

4. Set a protocol that will induce the propagation of two fronts, turning respectively in the clockwise and counterclockwise directions, on a ring lying around the perimeter on the lattice

   (a) Select **protocol/nb of stages** and set the number of stages to 1. Click **OK**.

   (b) Select **/protocol/set** and click on the box holding the number 1.

- Set the number of time steps to 120.
- Set the period of stimulation to 3.
- Click on the button **stimul** and set only one cell in the middle of one of the borders to be stimulated.
- Click on the toggle **discon**. Disconnect the cells such that the disconnected cells define together the borders of a rectangle inside the lattice. The cells lying along the borders of the lattice must stay connected to define a ring of connected cells.
- Click **OK**.

   (c) Close all menus with **OK** and **exit**.

Run the simulation. Each stimulus should produce both a clockwise and an counterclockwise front turning around the circumference of the lattice and colliding at some location.

## Exercises

You will find below a series of questions to explore various behaviors of the cellular automaton. The questions are meant mostly to guide you through the more interesting aspects of the cellular automaton, but you are strongly encouraged to try any other combinations of interventions and/or parameters you can think of.

Ex. 7.8-1. **Plane wave.**

(a) Assuming that the cellular automaton can be excited only from its resting state, what is the link between the dimension of the state space (i.e., number of exciting states + number of refractory states + 1) and the limiting frequency of stimulation to get a plane wave at each stimulation? Does the speed of propagation of the activation front (corresponding to the displacement of the first excited state across the lattice) vary with the stimulation frequency?

[hint:] You can change the number of refractory states through /**define**/**state**, and check that they are inexcitable by looking at /**define**/**threshold**. You can change the period of stimulation in the protocol. You will find that the limiting period of stimulation to get a response at each stimulation is equal to the dimension of the state space, and that the speed of the activation front does not depend on the frequency of stimulation and equals 1 cell per timestep.

(b) How must the cellular automaton be modified (number of states, threshold, extent and amplitude of stimulation) to get a version where the speed of the activation front varies with the frequency of stimulation? In this case, is there a transient period before reaching a stable speed of propagation for each stimulation frequency? What is the relationship between the period of stimulation and the speed of propagation?

[hint:] Extend the neighborhood so that it contains $N > 1$ cells along each main direction. You may extend the lattice to a dimension 1 by a large number of columns to get a good representation of the propagation and see the evolution of the system both in time and space. Define the state space such that there are $N$ exciting states, and a number of refractory states $N_r \geq N - 1$. Set the threshold of each refractory state $R_j$, $j = 1, ..., N_r$, to $2 + N_r - j$. Set the threshold of the resting state to 1. Apply the stimulation on the first $N$ rows (or columns) of the lattice, with an amplitude of $N$. Observe the result while changing the frequency of stimulation and make a plot of the stable speed of propagation as a function of the period of stimulation. Observe the transient time needed to reach a stable pattern of propagation as a function of the period of stimulation, and the limiting value of the stimulation period to maintain a 1 : 1 entrainment (one plane wave produced at each stimulation).

(c) How would the results of the previous exercise be changed if the neighborhood was extended to include all the cells in the box of side $N$? Would the dimension of the lattice matter in this case?

Ex. 7.8-2. **Reentry on a Ring.**

(a) Design a protocol of stimulation to start a reentry on a ring constituted by a single line of cells around the periphery of the lattice.

> [hint:] Take the cellular automata you have used in the preceding exercise, but add a supplementary absolute refractory state after the last exciting state (using /**define**/**state**, and /**define**/**threshold**). Define a protocol with two stages. In the first stage, stimulate once on a number of cells corresponding to the dimension of the neighborhood, and set the number of time steps of the stage to the number needed for the clockwise and counterclockwise activation fronts to collide. Then, in the second stage, apply a second stimulus at an appropriate location and with an appropriate extent and amplitude to start the reentry. Are there many ways to choose the properties of the second stimulus, depending on the characteristics of the cellular automaton? Note that the width of the disconnected region must be greater than or equal to the extent of the neighborhood.

(b) Would it be possible to start the reentry by stimulation if the cellular automaton did not have a state-dependent threshold (all states inexcitable, except the resting state)?

Ex. 7.8-3. **Target Pattern.**

(a) Repeat the exercises on the plane wave for target patterns obtained by stimulating a center region of the lattice. Is the wave really circular? Are the speeds of propagation along the main axes and their dependence to the frequency of stimulation the same as for a plane wave?

(b) How can you change the neighborhood and/or the weights to obtain a propagating wavefront that looks more like an expanding circle?

(c) How can you change the neighborhood and/or the weights to obtain a propagating wavefront that looks like an ellipse, approximating the propagation in an anisotropic medium?

Ex. 7.8-4. **Reentry in a 2-D lattice.**

(a) Revert to the classical Wiener and Rosenblueth cellular automaton, with a given number $N_E$ of exciting states, $N_r$ of absolute refractory states, excitation from the resting state only, and $L = 1$ neighborhood. Two conditions established by Greenberg et al. to obtain self-sustained patterns of propagation are given in Section 7.3. Use these conditions to generate an initial condition (assigning by hand the state of the cells through the /state/modify menu) leading to persistent activity. What form do these solutions take? Are these solutions periodic, and if so, what is the period? What is the link between the period of the solutions and $N_E$ and $N_r$?

(b) Starting from a lattice at the resting state, is it possible to design a multistage protocol of stimulation leading to sustained reentry?

> [hint:] Consider a cellular automaton with $N_E = N_r = 1$ and a $2 \times 2$ lattice holding a pattern leading to sustained activity. Try to design a succession of stimuli applied to a lattice initially at rest that could create this pattern. You will realize that such a protocol is impossible.

(c) Suppose that you initialize randomly the state of the lattice. How disorganized could the solution become? Can you think of some measure of disorganization, and will it vary in time? Could it be related to what may happen in a continuous reaction–diffusion system?

> [hint:] In Ex. 7.8-4a, you have realized that any closed path of connected cells on which a cycle exists (as defined by the condition of Greenberg) is maintained and defined a periodic sequence. Two unconnected cycles will persist independently. Hence, the complexity can be defined by the number of cycles existing in the lattice, and this index should be constant. In a continuous medium, that would be equivalent to the number of vortices. However, generally, these vortices will move and interact, such that their number will not remain fixed.

(d) Set up the model to include a larger neighborhood (say $L = 3$ or more) and a threshold greater than 1 that is *not* state-dependent (only the rest state may be excited). If you increase $L$, you might want to increase the size of your lattice as well (say from $10 \times 10$ to $20 \times 20$). Do the initial conditions you developed to produce persistent patterns also work for this model? If not, find appropriate initial conditions to produce persistent patterns in this model. How are the patterns different or similar to those observed in Ex. 7.8-4a? What is the rotation period of spirals, if

any? Modify the weighting of contributions within the neighborhood to more accurately reflect the properties of a continuous
diffusion matrix. What changes are observed as a result of this
modification?

(e) Introduce a state-dependent threshold to the model. Define a
state space that has multiple refractory states. For example,
set $N_r = 4$. Define the first refractory state as inexcitable, and
assign a threshold of decreasing magnitude to refractory states
2 through 4, with the rest state having the lowest threshold. Use
$L = 3$ or more, so that you have enough connected neighbors to
include significant variations in the threshold.

   i. Reinvestigate the initiation of persistent patterns. In this
   case, try to generate spirals by a protocol of stimulation
   applied on the lattice initially at the resting state.

      [hint:] Generate a plane wave in one direction,
      and then a second plane wave, perpendicular to the
      first one. With appropriate timing between the two
      stimuli, as well as appropriate extent and amplitude
      for the second stimulus, this protocol can start a
      spiral pattern of propagation.

   ii. How does the spiral period behave in this model? Do the
   spirals rotate around a fixed pivot?

   iii. Try to generate the most complicated pattern of persistent
   activity you can. How does it compare to the patterns you
   were able to generate in the other models? Is the index of
   disorganization constant?

# 8

# Cell Replication and Control

## Michael C. Mackey, Caroline Haurie, and Jacques Bélair

## 8.1  Introduction

Though all blood cells are derived from hematopoietic stem cells, the regulation of this production system is only partially understood. Negative feedback control mediated by erythropoietin and thrombopoietin regulates erythrocyte and platelet production, respectively, and colony stimulating factor regulates leukocyte levels. The local regulatory mechanisms within the hematopoietic stem cells are also not well characterized at this point. Due to their dynamic character, cyclical neutropenia and other periodic hematological disorders offer a rare opportunity to more fully understand the nature of these regulatory processes. We review here the salient clinical and laboratory features of a number of periodic hematological disorders, and show through a detailed example (cyclical neutropenia) how mathematical modeling can be used to quantify and test hypotheses about the origin of these interesting and unusual dynamics. The emphasis is on the development and analysis of a physiologically realistic mathematical model including estimation of the relevant parameters from biological and clinical data, and numerical exploration of the model behavior and comparison with clinical data.

Hobart Reimann was an enthusiastic proponent of the concept of periodic diseases (Reimann 1963). None of these conditions has been as intensively studied as cyclical neutropenia, in which circulating neutrophil levels spontaneously oscillate from normal to virtually zero. This chapter reviews current knowledge about periodic hematological disorders, the hypotheses that have been put forward for their origin, and the analysis of these hypotheses through mathematical models. Some illustrative examples of these disorders are shown in Figure 8.1.

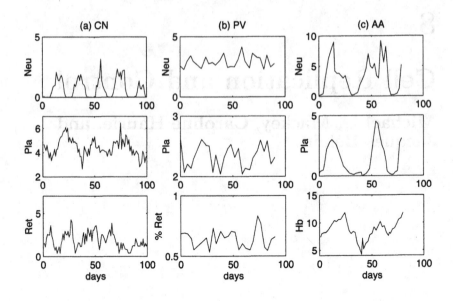

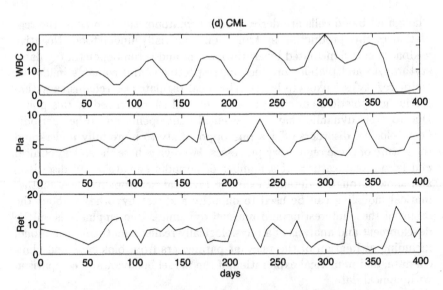

Figure 8.1. Representative patterns of circulating cell levels in four periodic hematological disorders considered in this chapter. Part (a) illustrates cyclical neutropenia (CN) (Guerry et al. 1973), (b) oscillations in polycythemia vera (PV) (Morley 1969), (c) oscillations in aplastic anemia (AA) (Morley 1979), and (d) periodic chronic myelogenous leukemia (CML) (Chikkappa et al. 1976). The density scales are: Neutrophils, $10^3$ cells/mm$^3$; white blood cells, $10^4$ cells/mm$^3$; platelets, $10^5$ cells/mm$^3$; reticulocytes, $10^4$ cells/mm$^3$; and Hb, g/dl. From Haurie, Mackey, and Dale (1998).

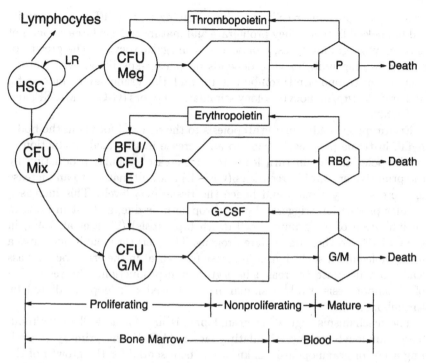

Figure 8.2. The architecture and control of hematopoiesis. This figure gives a schematic representation of the architecture and control of platelet (P), red blood cell (RBC), monocyte (M), and granulocyte (G) (including neutrophil, basophil, and eosinophil) production. Various presumptive control loops mediated by thrombopoietin, erythropoietin, and the granulocyte colony stimulating factor (G-CSF) are indicated, as well as a local regulatory (LR) loop within the totipotent hematopoietic stem cell population. CFU (BFU) refers to the various colony (burst) forming units (Meg = megakaryocyte, Mix = mixed, E = erythroid, and G/M = granulocyte/monocyte) which are the *in vitro* analogue of the *in vivo* committed stem cells. Adapted from Mackey (1996).

## 8.2 Regulation of Hematopoiesis

Although the regulation of blood cell production is complicated (Haurie et al. 1998), and its understanding constantly evolving, the broad outlines are clear.

Mature blood cells and recognizable precursors in the bone marrow ultimately derive from a small population of morphologically undifferentiated cells, the hemopoietic stem cells, which have a high proliferative potential and sustain hematopoiesis throughout life (Figure 8.2). The earliest hematopoietic stem cells are totipotent and have a high self-renewal capacity (Abramson et al. 1977; Becker et al. 1963; Lemischka et al. 1986), qualities that are progressively lost as the stem cells differentiate. Their progeny, the progenitor cells, or colony forming units (CFU), are committed to one cell lineage. They proliferate and mature to form large colonies of erythrocytes, granulocytes, monocytes, or megakaryocytes. The growth of colony forming units *in vitro* depends on lineage-specific growth factors, such as erythropoietin, thrombopoietin, and the granulocyte, monocyte, and granulocyte/monocyte colony stimulating factors (G-CSF, M-CSF, and GM-CSF).

Erythropoietin adjusts erythropoiesis to the demand for $O_2$ in the body. A fall in tissue $pO_2$ levels leads to an increase in the renal production of erythropoietin. This in turn leads to an increased cellular production by the primitive erythroid precursors (CFU-E) and, ultimately, to an increase in the erythrocyte mass and hence the tissue $pO_2$ levels. This increased cellular production triggered by erythropoietin is due, at least in part, to an inhibition of programmed cell death (apoptosis) (Silva et al. 1996) in the CFU-E and their immediate progeny. Thus, erythropoietin mediates a negative feedback such that an increase (decrease) in the erythrocyte mass leads to a decrease (increase) in erythrocyte production. The regulation of thrombopoiesis involves similar negative feedback loops mediated by thrombopoietin.

The mechanisms regulating granulopoiesis are not as well understood. The granulocyte colony stimulating factor, G-CSF, the primary controlling agent of granulopoiesis, is known to be essential for the growth of the granulocytic progenitor cells CFU-G *in vitro* (Williams et al. 1990). The colony growth of CFU-G is a sigmoidally increasing function of the granulocyte colony stimulating factor, G-CSF (Avalos et al. 1994; Hammond et al. 1992). One of the modes of action of the granulocyte colony stimulating factor, along with several other cytokines, is to decrease apoptosis (Koury 1992; Park 1996; Williams et al. 1990; Williams and Smith 1993). Neutrophil maturation time also clearly shortens under the action of the granulocyte colony stimulating factor (Price et al. 1996). Several studies have shown an inverse relation between circulating neutrophil density and serum levels of granulocyte colony stimulating factor (Kearns et al. 1993; Mempel et al. 1991; Takatani et al. 1996; Watari et al. 1989). Coupled with the *in vivo* dependency of granulopoiesis on the granulocyte colony stimulating factor, this inverse relationship suggests that the neutrophils regulate their own production through a negative feedback, as is the case of erythrocytes: An increase (decrease) in the number of circulating neutrophils would induce a decrease (increase) in the production of neutrophils through the adjustment of the granulocyte colony stimulating factor levels. Although mature neutrophils bear receptors for the granulocyte (G-CSF) and for the granulocyte/monocyte (GM-CSF) colony stimulating factors, the role of these receptors in governing neutrophil production is not yet known.

Little is known about how the self-maintenance of the hematopoietic stem cell population is achieved, this self-maintenance of hematopoietic stem cells depending on the balance between self-renewal and differentiation. Hematopoietic stem cells are usually in a dormant state but are triggered to proliferate after transplantation into irradiated hosts (Necas 1992), and the specific mechanisms regulating the differentiation commitment of hematopoietic stem cells are poorly understood (Ogawa 1993). However, mechanisms that could support autoregulatory feedback control loops controlling hematopoietic stem cell kinetics are starting to be investigated (Necas et al. 1988).

The selective responses of the erythrocytic, granulocytic, and megakaryocytic systems to increased demand of cell production indicate a relative autonomy of the peripheral control loops regulating these three cell lineages. The mechanisms regulating early hematopoiesis are, on the other hand, poorly understood, and strong connections may exist at this level between the regulation of the different blood lineages. In some of the periodic hematological disorders discussed here, such relations become visible though the occurrence of particular dynamical features common to all the blood lineages.

## 8.3   Periodic Hematological Disorders

We first introduce the main mathematical analysis technique we use to quantitatively assess the periodicity of clinical data.

### 8.3.1   Uncovering Oscillations

Fourier, or power spectrum, techniques are widely applicable when the data under study are evenly sampled but can give erroneous results when the data are unevenly sampled. While studying celestial phenomena, astrophysicists also encountered the problem of unevenly sampled data, and they developed an extension of the Fourier power spectrum, the Lomb periodogram, for evenly or unevenly sampled data (Lomb 1976). The statistical significance ($p$ value) of any peak can also be determined (Scargle 1982). Thus the Lomb technique is ideally suited to the detection of periodicity in hematological time series, since serial blood counts are usually sampled at irregular intervals. Appendix C contains more details on the Lomb periodogram, $P(T)$, including its definition in equation (C.5).

### 8.3.2   Cyclical Neutropenia

**General Features**

Cyclical neutropenia has been the most extensively studied periodic hematological disorder. Its hallmark is a periodic fall in the circulating neutrophil

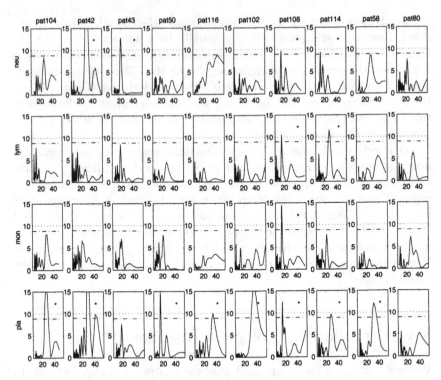

Figure 8.3. Lomb periodogram $P(T)$ [power $P$ versus period $T$ in days] of the blood cell counts of five cyclical neutropenia, three congenital neutropenic and two idiopathic neutropenic patients. The dotted lines in the Lomb periodogram give the $p = 0.10$ (lower dash–dot line) and $p = 0.05$ significance levels (upper dotted line); * indicates periodicity with significance $p \leq 0.10$. 'Neu': neutrophils, 'Lym': lymphocytes, 'Mon': monocytes, 'Pla': platelets. From Haurie, Mackey, and Dale (1999).

numbers from normal values to very low values. In humans it occurs sporadically or as an autosomal dominantly inherited disorder, and the period is typically reported to fall in the range of 19–21 days (Dale and Hammond 1988), though recent data indicate that the period may be as long as 46 days in some patients (Haurie et al. 1999) (see Figure 8.3).

Our understanding of cyclical neutropenia has been greatly aided by the discovery of an animal model, the grey collie (Figure 8.4). The canine disorder closely resembles human cyclical neutropenia with the exception of the period, which ranges from 11 to 15 days (Figure 8.5) (Haurie et al. 1999) and the maximum neutrophil counts, which are higher than for humans. For a review see Haurie, Mackey, and Dale (1998).

It is now clear that in both human cyclical neutropenia (Dale et al. 1972; Dale et al. 1972; Haurie et al. 1999; Hoffman et al. 1974) and the grey collie (Guerry et al. 1973; Haurie et al. 1999) there is not only a periodic

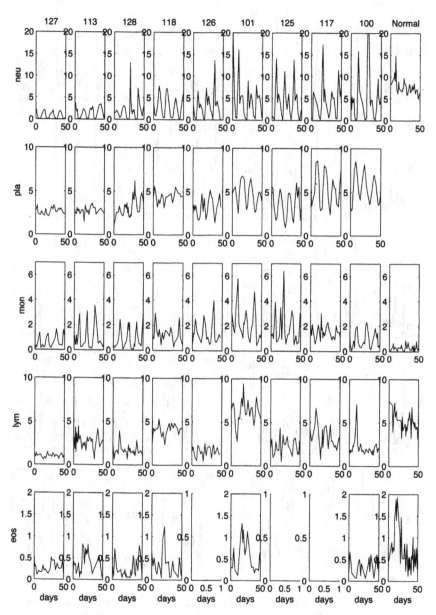

Figure 8.4. Differential blood counts in nine grey collies and one normal dog. Units: Cells $\times$ $10^{-5}$ per mm$^3$ for the platelets and Cells $\times$ $10^{-3}$ per mm$^3$ for the other cell types. 'Neu': neutrophils, 'Pla': platelets, 'Mon': monocytes, 'Lym': lymphocytes, 'Eos': eosinophils. From Haurie, Person, Mackey, and Dale (1999).

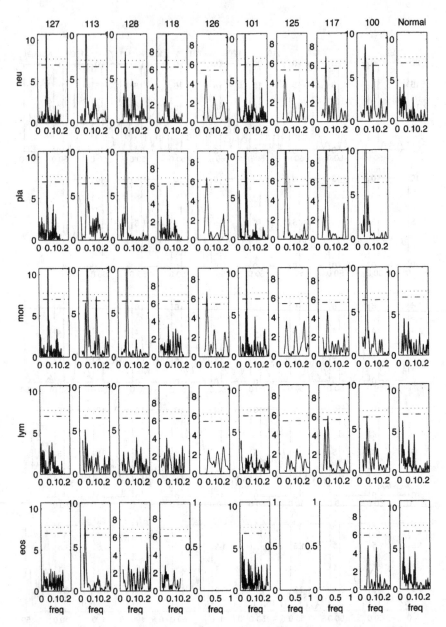

Figure 8.5. Lomb periodogram, equation (C.5), of the differential blood counts in nine grey collies. From Haurie, Person, Mackey, and Dale (1999).

fall in the circulating neutrophil levels, but also a corresponding oscillation of platelets, often the monocytes and eosinophils, and occasionally the reticulocytes and lymphocytes (see Figures 8.1A, 8.3, and 8.4). The monocyte, eosinophil, platelet, and reticulocyte oscillations are generally from normal to high levels, in contrast to the neutrophils, which oscillate from near normal to extremely low levels. Often (but not always), the period of the oscillation in these other cell lines is the same as the period in the neutrophils.

The clinical criteria for a diagnosis of cyclical neutropenia have varied widely. Using periodogram analysis, some patients classified as having cyclical neutropenia do not, in fact, display any significant periodicity, while other patients classified with either congenital or idiopathic neutropenia do display significant cycling (Haurie et al. 1999) as shown in Figure 8.3. Moreover the period of the oscillations detected through periodogram analysis in neutropenic patients may be as long as 46 days (Haurie et al. 1999).

## Origin

Transplantation studies show that the origin of the defect in cyclical neutropenia is resident in one of the stem cell populations of the bone marrow (Dale and Graw 1974; Jones et al. 1975; Jones et al. 1975; Weiden et al. 1974; Krance et al. 1982; Patt et al. 1973). Studies of bone marrow cellularity throughout a complete cycle in humans with cyclical neutropenia show that there is an orderly cell density wave that proceeds successively through the myeloblasts, promyelocytes, and myelocytes and then enters the maturation compartment before being manifested in the circulation (Brandt et al. 1975; Guerry et al. 1973). Further studies have shown that this wave extends back into the granulocytic progenitor cells (Jacobsen and Broxmeyer 1979) and erythrocytic progenitor cells (Dunn et al. 1977; Dunn et al. 1978; Hammond and Dale 1982; Jones and Jolly 1982), as well as in the erythrocytic burst forming units and granulocyte/monocyte colony forming units (Abkowitz et al. 1988; Hammond and Dale 1982), suggesting that it may originate in the totipotent hematopoietic stem cell populations.

## Fluctuations in Putative Regulators

In cyclical neutropenia, the levels of colony stimulating activity (related to the granulocyte colony stimulating factor) fluctuate inversely with the circulating neutrophil levels and in phase with the peak in monocyte numbers (Dale et al. 1971; Guerry et al. 1974; Moore et al. 1974). Erythropoietin levels oscillate approximately in phase with the reticulocyte oscillation (Guerry et al. 1974). It is unclear whether these correlations and inverse correlations between levels of circulating cells and putative humoral regulators are related to the cause of cyclical neutropenia, or are simply a secondary manifestation of some other defect.

## Effect of Phlebotomy and Hypertransfusion

The effect of bleeding and/or hypertransfusion on the hematological status of grey collies gives interesting results (Adamson et al. 1974). In the untreated grey collie erythropoietin levels cycle out of phase with the reticulocytes and virtually in phase with the neutrophil counts. After phlebotomy (bleeding of between 10% and 20% of the blood volume), the cycles in the neutrophils and reticulocytes continue as before the procedure, and there is no change in the relative phase between the cycles of the two cell types. Hypertransfusion (with homologous red cells) completely eliminates the reticulocyte cycling (as long as the hematocrit remains elevated), but has no discernible effect on the neutrophil cycle. Most significantly, when the hematocrit falls back to normal levels and the reticulocyte cycle returns, the phase relation between the neutrophils and the reticulocytes is the same as before the hypertransfusion. These observations suggest that the source of the oscillations in cyclical neutropenia is relatively insensitive to any feedback regulators involved in peripheral neutrophil and erythrocyte control, whose levels would be modified with the alteration of the density of circulating cells; and is consistent with a relatively autonomous oscillation in the hematopoietic stem cells (cf. Section 8.5).

### Effect of Cytokine Therapy

In both the grey collie (Hammond et al. 1990; Lothrop et al. 1988) and in humans with cyclical neutropenia (Hammond et al. 1989; Migliaccio et al. 1990; Wright et al. 1994) administration of the granulocyte colony stimulating factor leads to an increase in the mean value of the peripheral neutrophil counts by a factor of as much as 10 to 20, associated with a clear improvement of the clinical symptoms. However, the granulocyte colony stimulating factor does not obliterate the cycling in humans, but rather induces an increase in the amplitude of the oscillations and a decrease in the period of the oscillations in all the cell lineages, from 21 to 14 days (Hammond et al. 1989; Haurie et al. 1999).

## 8.3.3 Other Periodic Hematological Disorders Associated with Bone Marrow Defects

### Periodic Chronic Myelogenous Leukemia

Chronic myelogenous leukemia is a hematopoietic stem cell disease characterized by granulocytosis and splenomegaly (Grignani 1985). In 90% of the cases, the hematopoietic cells contain a translocation between chromosomes 9 and 22, which leads to the shortening of chromosome 22, referred to as the Philadelphia (Ph) chromosome. In most cases, the disease eventually develops into acute leukemia.

In 1967, Morley was the first to describe oscillations in the leukocyte count of patients with chronic myelogenous leukemia (Morley et al. 1967). Several other cases of cyclic leukocytosis in chronic myelogenous leukemia have now been reported, and these have been reviewed in Fortin and Mackey (1999). In the cases of periodic chronic myelogenous leukemia, the leukocyte count usually cycles with an amplitude of 30 to $200 \times 10^9$ cells/L and with periods ranging from approximately 30 to 100 days. The platelets and sometimes the reticulocytes also oscillate with the same period as the leukocytes, around normal or elevated numbers (Figure 8.1d). There have been no specific studies of hematopoiesis in patients with periodic chronic myelogenous leukemia.

## Polycythemia Vera and Aplastic Anemia

Polycythemia vera is characterized by an increased and uncontrolled proliferation of all the hematopoietic progenitors, and it involves, like chronic myelogenous leukemia, the transformation of a single hematopoietic stem cell. Two patients with polycythemia vera were reported with cycling of the reticulocyte, platelet, and neutrophil counts in one case (Figure 8.1b), and cycling only of the reticulocyte count in the other. The period of the oscillations was 27 days in the platelets, 15 days in the neutrophils, and 17 days in the reticulocytes (Morley 1969).

Finally, clear oscillations in the platelet, reticulocyte, and neutrophil counts (Figure 8.1c) were reported in a patient diagnosed with aplastic anemia (Morley 1979) and in a patient with pancytopenia (Birgens and Karl 1993), with periods of 40 and 100 days, respectively.

## Cytokine-Induced Cycling

The granulocyte colony stimulating factor, G-CSF, is routinely used in a variety of clinical settings, for example to treat chronic neutropenia or to accelerate recovery from bone marrow transplant and/or chemotherapy (Dale et al. 1993). The granulocyte colony stimulating factor may induce oscillations in the level of circulating neutrophils of neutropenic individuals (Haurie et al. 1999), and as will be seen later, in Section 8.5, this is of great significance in understanding cyclical neutropenia.

## Induction of Cycling by Chemotherapy or Radiation

Several reports describe induction of a cyclical neutropenia-like condition by the chemotherapeutic agent cyclophosphamide. In mongrel dogs on cyclophosphamide the observed period was on the order of 11 to 17 days, depending on the dose of cyclophosphamide (Morley et al. 1969; Morley and Stohlman 1970). In a human undergoing cyclophosphamide treatment, cycling with a period of 5.7 days was reported (Dale et al. 1973). Also, Gidáli, István, and Fehér (1985) observed oscillations in the granulocyte and iculocyte counts with three weeks periodicity in mice after mild irradiation. They observed an overshooting regeneration in the reticulocytes and the thrombocytes but not in the granulocytes. While the CFU-S returned to normal levels rapidly, the proliferation rate of CFU-S stayed abnormally elevated.

Five patients with chronic myelogenous leukemia receiving hydroxyurea showed oscillations in their neutrophils, monocytes, platelets, and reticulocytes with periods in the range of 30 to 50 days (Kennedy 1970). In one patient an increase of the hydroxurea dose led to a cessation of the oscillations. Chikkappa et al. (1980) report a cyclical neutropenia-like condition (period between 15 and 25 days) in a patient with multiple myeloma after three years of chemotherapy.

A $^{89}$Sr-induced cyclic erythropoiesis has been described in two congenitally anemic strains of mice, $W/W^v$ and $S1/S1^d$ (Gibson et al. 1984; Gibson et al. 1985; Gurney et al. 1981). $W/W^v$ mice suffer from a defect in the hematopoietic stem cells, and in $S1/S1^d$ mice the hematopoietic micro-environment is defective. The induction of cycling by $^{89}$Sr can be understood as a response to elevated cell death (Milton and Mackey 1989), as can the dynamic effects of chemotherapy.

### 8.3.4  Periodic Hematological Disorders of Peripheral Origin

Periodic autoimmune hemolytic anemia is a rare form of hemolytic anemia in humans (Ranlov and Videbaek 1963). Periodic autoimmune hemolytic anemia, with a period of 16 to 17 days in hemoglobin and reticulocyte counts, has been induced in rabbits by using red blood cell autoantibodies (Orr et al. 1968).

Cyclic thrombocytopenia, in which platelet counts oscillate from normal to very low values, has been observed with periods between 20 and 40 days and reviewed in Swinburne and Mackey (2000). The cases in which there was an implication of an autoimmune source for the disease had periods between 13 and 27 days, while patients with the amegakaryocytic version have longer periods. From the modeling work of Santillan et al. (2000) it seems clear that the periodicity of the autoimmune version is probably induced through a supercritical Hopf bifurcation.

# 8.4  Peripheral Control of Neutrophil Production and Cyclical Neutropenia

### 8.4.1  Hypotheses for the Origin of Cyclical Neutropenia

Given the interesting dynamical presentation of cyclical neutropenia in both its clinical and laboratory manifestations, it is not surprising that there have been a number of attempts to model this disorder mathematically. In this section we briefly review these attempts, since they focus the work of this section and simultaneously motivate the extensions that we have made.

The mathematical models that have been put forward for the origin of cyclical neutropenia fall into two major categories. Reference to Figure 8.2 will help place these in perspective. (See Dunn 1983; Fisher 1993 for other reviews.)

The first group of models builds upon the existence of oscillations in many of the peripheral cellular elements (neutrophils, platelets, and erythroid precursors, see Figure 8.2) and postulates that the origin of cyclical neutropenia is in the common hematopoietic stem cell population feeding progeny into all of these differentiated cell lines. A loss of stability in the stem cell population is hypothesized to be independent of feedback from peripheral circulating cell types (see below) and would thus represent a relatively autonomous oscillation driving the three major lines of differentiated hematopoietic cells.

Mackey (1978) analyzed a model for the dynamics of a stem cell population and concluded that one way the dynamic characteristics of cyclical neutropenia could emerge from such a formulation was via an abnormally large cell death rate within the proliferating compartment. This hypothesis allowed the *quantitative* calculation of the period of the oscillation that would ensue when stability was lost. This hypothesis has been expanded elsewhere (Mackey 1979; Milton and Mackey 1989) and allows a qualitative understanding of the observed laboratory and clinical effects of the granulocyte colony stimulating factor and chemotherapy discussed above (Mackey 1996). In spite of the resonance of this stem cell origin hypothesis in the clinical and experimental communities (Quesenberry 1983; Ogawa 1993) there has been little extension of this hypothesis in the modeling literature related to cyclical neutropenia.

The second broad group of these models identifies the origin of cyclical neutropenia with a loss of stability in the peripheral control loop, operating as a sensor between the number of mature neutrophils and the control of the production rate of neutrophil precursors within the bone marrow (Figure 8.2). This control has been uniformly assumed to be of a negative feedback type, whereby an increase in the number of mature neutrophils leads to a decrease in the production rate of immature precursors. The other facet of this hypothesis is a significant delay due to the maturation times required between the signal to alter immature precursor production and the actual alteration of the mature population numbers. Typical examples of models of this type that have specifically considered cyclical neutropenia are Kazarinoff and van den Driessche (1979); King-Smith and Morley (1970); MacDonald (1978); Morley, King-Smith, and Stohlman (1969); Morley and Stohlman (1970); Morley (1979); Reeve (1973); von Schulthess and Mazer (1982); Shvitra, Laugalys, and Kolesov (1983); Schmitz (1988); Wichmann, Loeffler, and Schmitz (1988); Schmitz, Loeffler, Jones, Lange, and Wichmann (1990); Schmitz, Franke, Brusis, and Wichmann (1993); Schmitz, Franke, Loeffler, Wichmann, and Diehl (1994); Schmitz, Franke, Wichmann, and Diehl (1995); all of which have postulated an alteration in the feedback on immature precursor production from the mature cell population numbers.

In the next section we show that it is highly unlikely that cyclical neutropenia is due to a loss of peripheral stability.

## 8.4.2    Cyclical Neutropenia Is Not Due to Peripheral Destabilization

**Development.** In the model development that follows, reference to the lower part of Figure 8.2, where the control of white blood cell production is outlined, will be helpful.

We let $x(t)$ be the density of white blood cells in the circulation (units of cells/$\mu$L blood), $\alpha$ be the random disappearance rate of circulating white blood cells (days$^{-1}$), and $\mathcal{M}_o$ be the production rate (cells/$\mu$L-day) of white blood cell precursors in the bone marrow.

The rate of change of the peripheral (circulating) white blood cell density is made up of a balance between the loss of white blood cells $(-\alpha x)$ and their production $(\mathcal{M}_o(\tilde{x}))$, or

$$\frac{dx}{dt} = -\alpha x + \mathcal{M}_o(\tilde{x}),  \tag{8.1}$$

wherein $\tilde{x}(t)$ is $x(t-\tau)$ weighted by a distribution of maturation delays, $\tilde{x}(t)$ is given explicitly by

$$\tilde{x}(t) = \int_{\tau_m}^{\infty} x(t-u)g(u)du \equiv \int_{-\infty}^{t-\tau_m} x(u)g(t-u)du,  \tag{8.2}$$

$\tau_m$ is the minimal maturation delay; and $g(\tau)$ is the density of the distribution of maturation delays as specified below in Section 8.4.2. Since $g(\tau)$ is a density, it is normalized by definition:

$$\int_0^{\infty} g(u)du = 1.  \tag{8.3}$$

To completely specify the semidynamical system described by equations (8.1) and (8.2), we must additionally give an initial function

$$x(t') \equiv \varphi(t')  \quad \text{for} \quad t' \in (-\infty, 0).  \tag{8.4}$$

**Distribution of Maturation Times.** A wide variety of analytic forms could be used for the density of the distribution of the maturation times in the bone marrow. We have chosen to use the density of the gamma distribution

$$g(\tau) = \begin{cases} 0, & \tau \leq \tau_m, \\ \frac{a^{m+1}}{\Gamma(m+1)}(\tau-\tau_m)^m e^{-a(\tau-\tau_m)}, & \tau_m < \tau, \end{cases}  \tag{8.5}$$

with $a, m \geq 0$, considered before (Blythe et al. 1984; Cooke and Grossman 1982; Gatica and Waltman 1988; Gatica and Waltman 1982) in a different context. This choice was predicated on two issues. First, we have found (see Section 8.4.2) that we can achieve a good fit of the existing data on cellular maturation times using equation (8.5). Secondly, the density of the gamma distribution has been used a number of times in the past (Kendall 1948; Powell 1955; Powell 1958) to fit distributions of cell cycle times. When the parameter $m$ in equation (8.5) is a nonnegative integer, then the corresponding equations (8.1) and (8.2) reduce to a system of linear ordinary differential equations coupled to a single nonlinear delayed equation with a discrete (not continuously distributed) delay (Fargue 1973; Fargue 1974; MacDonald 1989). This leads to analytic simplifications, though we do not use them here, since we have typically found noninteger values for the parameter $m$. We did, however, use this reduction to test the accuracy of our numerical simulations of the full model.

The parameters $m$, $a$, and $\tau_m$ in the density of the gamma distribution can be related to certain easily determined statistical quantities. The average of the *unshifted* density is given by

$$\tau_2 = \int_{\tau_m}^{\infty} \tau g(\tau) d\tau = \frac{m+1}{a}, \tag{8.6}$$

and thus the average maturation delay as calculated from equation (8.5) is given by

$$\langle \tau \rangle = \tau_m + \tau_2 = \tau_m + \frac{m+1}{a}. \tag{8.7}$$

The variance (denoted by $\sigma^2$) is given by

$$\sigma^2 = \frac{m+1}{a^2}. \tag{8.8}$$

Given the expressions (8.6), (8.7), and (8.8) in terms of the gamma distribution parameters $m$ and $a$, we may easily solve for these parameters in terms of $\tau_2$ and $\sigma^2$ to give

$$a = \frac{\tau_2}{\sigma^2} \tag{8.9}$$

and

$$m + 1 = \frac{\tau_2^2}{\sigma^2}. \tag{8.10}$$

**Parameter Estimation.** Several studies have shown that labeled neutrophils disappear from the circulation with a half-life $t_{1/2}$ of about 7.6 hours in humans (Dancey et al. 1976) and dogs (Deubelbeiss et al. 1975) with a range of 7 to 10 hours. Furthermore, this disappearance rate is unaffected in human (Guerry et al. 1973) and canine cyclical neutropenia (Dale et al. 1972) and is not altered by the administration of exogenous granulocyte colony stimulating factor (Price et al. 1996). Since the decay coefficient $\alpha$ of equation (8.1) is related to $t_{1/2}$ through the relation

$$\alpha = \frac{\ln 2}{t_{1/2}}, \tag{8.11}$$

we have taken values of $\alpha \in [1.7, 2.4]$ (days$^{-1}$) in all of the numerical work reported here.

Distributions of maturation times were determined from published data on the emergence of the number of labeled circulating neutrophils following pulse labeling by tritiated thymidine. The published graphed data were scanned and the postscript file viewed with Ghostview. Ghostview gives coordinates for the position of the points, which, using position of the axes, can be easily transformed to give the actual data points. The data were adjusted for the random death occurring at a rate $\alpha$ by using the method of Dancey, Deubelbeiss, Harker, and Finch (1976).

Assume that the neutrophils spend a period of time $u$ in the bone marrow, and $y$ in the blood. Then the fraction, $N(t)$, of labeled cells in the blood at time $t$ is the probability that the time in the marrow is less than $t$ and that the total time in the marrow and blood before death is greater than $t$. Let $g(u)$ be the density of the distribution of the maturation times in the marrow, and remember that $g(u)$ is the quantity that we wish to determine. Further note that because of the experimentally observed random destruction of neutrophils in the circulation, if the rate of random destruction is $\alpha$, then the density of the distribution of destruction rates is given by $\alpha e^{-\alpha y}$. With these observations, for $N(t)$ we finally have

$$N(t) = \int_0^t \int_{t-u}^\infty \alpha e^{-\alpha y} g(u) dy \, du = \int_0^t e^{-\alpha(t-u)} g(u) du. \tag{8.12}$$

Thus,

$$e^{\alpha t} N(t) = \int_0^t e^{\alpha u} g(u) du, \tag{8.13}$$

and differentiating both sides with respect to $t$ gives

$$\alpha e^{\alpha t} N(t) + e^{\alpha t} N'(t) = e^{\alpha t} g(t). \tag{8.14}$$

The final result for the density of marrow transit times is

$$g(t) = \alpha N(t) + N'(t). \tag{8.15}$$

Since we had discrete data points from the labeling data, we used the midpoint of two data points and the slope of the joining line in equation (8.15), and determined $g(t)$ at the midpoint. The mean and variance were calculated from the new density, and the corresponding $m$ and $a$ determined from equations (8.9) and (8.10) were used as the initial values in a nonlinear least squares fit to the data. The results of these determinations for a number of published data sets are summarized in Table 8.1. Figure 8.6

| Condition | $\langle\tau\rangle$ | $\sigma^2$ | $\tau_m$ | Ref | $a$ | $m$ |
|---|---|---|---|---|---|---|
| Normal Human | 9.70 | 16.20 | 3.8 | * | 0.36 | 1.15 |
| CN Human | 7.57 | 12.01 | 1.2 | ♣ | 0.53 | 2.38 |
| 30 μg G-CSF Human | 6.27 | 4.60 | 2.4 | ◇ | 0.84 | 2.26 |
| 300 μg GCSF Human | 4.86 | 2.30 | 2.0 | ◇ | 1.24 | 2.56 |
| Normal Dog | 3.68 | .198 | 3.0 | ♡ | 3.43 | 1.34 |
| Gray Collie Apogee | 3.21 | 0.042 | 2.6 | ♠ | 14.52 | 7.86 |
| Gray Collie Nadir | 3.42 | 0.157 | 2.6 | ♠ | 5.22 | 3.28 |

Table 8.1. Distribution of maturation time parameters deduced from published data. The units of $\langle\tau\rangle$ and $\tau_m$ are in days, $\sigma^2$ is in days$^2$, and $a$ is in days$^{-1}$. For the references, * = Perry et al. (1966), ♣ =Guerry et al. (1973), ◇ = Price et al. (1996), ♡ = Deubelbeiss et al. (1975), and ♠ = Patt et al. (1973). See the text for details.

shows the raw data as well as the fits to the data using the density of the gamma distribution.

### The Steady State and Stability

**The Steady State.** The equilibrium solution for the functional differential equation (8.1)–(8.2) occurs when

$$\frac{dx}{dt} = 0 = -\alpha x + \mathcal{M}_o(\tilde{x}), \qquad (8.16)$$

so the steady state $x^*$ is defined implicitly by the solution of the equation

$$\alpha x^* = \mathcal{M}_o(x^*). \qquad (8.17)$$

Given the presumptive monotone decreasing nature of the negative feedback production rate inferred from the biology, there can be but a unique value for the steady-state white blood cell density $x^*$. It is important to note that $x^*$ is independent of the distribution of the maturation times. However, the stability of $x^*$ is dependent on the density $g(\tau)$, as we now show.

**Stability.** One of the primary considerations of this section has to do with the stability of the unique steady state, defined implicitly by equation (8.17), and how that stability may be lost. We examine the stability of $x^*$ in the face of very small deviations away from the steady state. Though the mathematical development may seem much different in this model, it is fundamentally the same as the procedure for examining the stability of steady states in ordinary differential equations.

Throughout this analysis, an important parameter that will appear is the *slope* of the production function $\mathcal{M}_o$ evaluated at the steady state, denoted by $\mathcal{M}'_{o*}$. Because of our arguments concerning the negative feedback nature of the peripheral control mechanisms acting on neutrophil production, we know that this slope must be nonpositive (i.e., negative or zero).

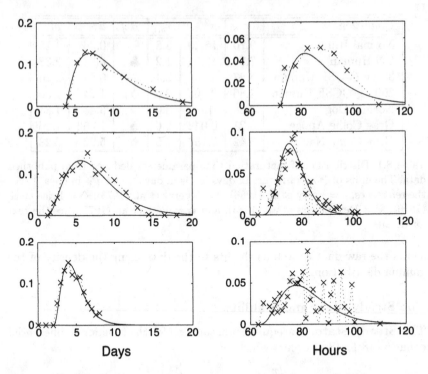

Figure 8.6. Densities of distributions of maturation times and the least square fits to the data achieved using the density of the gamma distribution. The three left-hand panels are for humans and show, from top to bottom, a normal human, data from a cyclical neutropenia patient, and a normal human receiving 300 μg granulocyte colony stimulating factor. The three right-hand panels are for dogs and correspond to (top to bottom) a normal dog, a grey collie at the apogee of the cycle, and a grey collie at the nadir of the cycle. See Table 8.1 for the parameters used to fit the data and the references for the source of the data. From Hearn, Haurie, and Mackey (1998).

To examine the local stability, we write out equation (8.1) for small deviations of $x$ from $x^*$. In the first (linear) approximation this gives

$$\frac{dx}{dt} \approx -\alpha x + \mathcal{M}_{o*} + (\tilde{x} - x^*)\mathcal{M}'_{o*}, \qquad (8.18)$$

wherein

$$\mathcal{M}_{o*} \equiv \mathcal{M}_o(\tilde{x} = x^*) \qquad (8.19)$$

and

$$\mathcal{M}'_{o*} \equiv \frac{d\mathcal{M}_o(\tilde{x})}{d\tilde{x}}|_{\tilde{x}=x^*}. \qquad (8.20)$$

Utilizing equation (8.17) and defining the deviation from equilibrium as $z(t) = x(t) - x^*$, we can rewrite equation (8.18) in the form

$$\frac{dz}{dt} = -\alpha z + \mathcal{M}'_{o*} \int_{-\infty}^{t-\tau_m} z(u)g(t-u)du. \qquad (8.21)$$

As before, we assume that the deviation $z$ from the steady state has the form $z(t) \bowtie \exp(\lambda t)$, substitute this into equation (8.21), carry out the indicated integrations, and finally obtain

$$\lambda + \alpha = \mathcal{M}'_{o*} \left(\frac{a}{\lambda + a}\right)^{m+1} e^{-\lambda \tau_m}. \qquad (8.22)$$

Equation (8.22) for the eigenvalues $\lambda$ may have a variety of solutions. If an eigenvalue $\lambda$ is real, then a simple graphical argument shows that the eigenvalue will be negative and contained in the open interval $(-\alpha, -a)$.

Alternatively, the eigenvalue solutions of equation (8.22) may be complex conjugate numbers, in which case, as seen before, the most interesting thing to know is when the real part of the eigenvalue is identically zero. This will define the boundary between a locally stable steady state when Re $\lambda < 0$ and a locally unstable steady state with Re $\lambda > 0$.

To investigate this possibility, we take $\lambda = \mu + i\omega$ and substitute this into equation (8.22) to give, with $\mu = 0$,

$$i\omega + \alpha = \mathcal{M}'_{o*} \left(\frac{a}{i\omega + a}\right)^{m+1} e^{-i\omega \tau_m}, \qquad (8.23)$$

or rewriting,

$$\left[(i\omega + \alpha)\left(1 + i\frac{\omega}{a}\right)^{m+1}\right] = \mathcal{M}'_{o*} e^{-i\omega \tau_m}. \qquad (8.24)$$

This equation can be manipulated to give a set of parametric equations in $\alpha$ and $\mathcal{M}'_{o*}$. We start by setting

$$\tan \theta = \frac{\omega}{a}. \qquad (8.25)$$

Using de Moivre's formula in equation (8.24) gives

$$(\alpha + i\omega)(\cos[(m+1)\theta]) + i\sin[(m+1)\theta]) =$$
$$\mathcal{M}'_{o*} \cos^{m+1} \theta (\cos \omega \tau_m - i \sin \omega \tau_m). \qquad (8.26)$$

Equating the real and imaginary parts of equation (8.26) gives the coupled equations

$$\alpha - \mathcal{M}'_{o*} R \cos \omega \tau_m = \omega \tan[(m+1)\theta] \qquad (8.27)$$

and

$$\alpha \tan[(m+1)\theta] + \mathcal{M}'_{o*} R \sin \omega \tau_m = -\omega, \qquad (8.28)$$

where

$$R = \frac{\cos^{m+1}\theta}{\cos[(m+1)\theta]}. \tag{8.29}$$

Equations (8.27) and (8.28) are easily solved for $\alpha$ and $\mathcal{M}'_{o*}$ as parametric functions of $\omega$ to give

$$\alpha(\omega) = -\frac{\omega}{\tan[\omega\tau_m + (m+1)\tan^{-1}(\omega/a)]} \tag{8.30}$$

and

$$\mathcal{M}'_{o*}(\omega) = -\frac{\omega}{\cos^{m+1}[\tan^{-1}(\omega/a)]\sin[\omega\tau_m + (m+1)\tan^{-1}(\omega/a)]}. \tag{8.31}$$

To show that the stability boundary defined implicitly by equations (8.30) and (8.31) delimits a transition from a locally stable steady state to a locally unstable steady state as $\mathcal{M}'_{o*}$ decreases, we must show that the real part of the eigenvalue is negative on one side of the boundary and positive on the other. Thus, the real part of $d\lambda/d\mathcal{M}'_{o*}$, or equivalently of $(d\lambda/d\mathcal{M}'_{o*})^{-1}$, must be negative when $\lambda = i\omega$.

Implicit differentiation of equation (8.22) yields

$$\left(\frac{d\lambda}{d\mathcal{M}'_{o*}}\right)^{-1} = \left(\frac{\lambda+a}{a}\right)^{m+1} e^{\lambda\tau_m} + \mathcal{M}'_{o*}\frac{m+1}{\lambda+a} + \mathcal{M}'_{o*}\tau_m, \tag{8.32}$$

and the use of equation (8.22) in (8.32) gives

$$\left(\frac{d\lambda}{d\mathcal{M}'_{o*}}\right)^{-1} = \mathcal{M}'_{o*}\left(\frac{1}{\lambda+\alpha} + \frac{m+1}{\lambda+a} + \tau_m\right). \tag{8.33}$$

Evaluating (8.33) at $\lambda = i\omega$ and eliminating complex numbers in the denominators, we have

$$\left(\frac{d\lambda}{d\mathcal{M}'_{o*}}\right)^{-1} = \mathcal{M}'_{o*}\left(\frac{\alpha-i\omega}{\alpha^2+\omega^2} + \frac{(m+1)(a-i\omega)}{a^2+\omega^2} + \tau_m\right), \tag{8.34}$$

with

$$Re\left(\left(\frac{d\lambda}{d\mathcal{M}'_{o*}}\right)^{-1}\right) = \mathcal{M}'_{o*}\left(\frac{\alpha}{\alpha^2+\omega^2} + \frac{(m+1)a}{a^2+\omega^2} + \tau_m\right). \tag{8.35}$$

If $\mathcal{M}'_{o*}$ is negative (as in our case), then the right-hand side of equation (8.35) is negative, indicating that for increases in $\mathcal{M}'_{o*}$ to more positive values at the boundary where $\mu \equiv 0$, the real part of the eigenvalue $\lambda$ is crossing from positive to negative.

Thus, we conclude that the locus of points defined by equations (8.30) and (8.31) defines the location in $(\alpha, \mathcal{M}'_{o*})$ parameter space where a supercritical Hopf bifurcation takes place and a periodic solution of period

$$T_{\text{Hopf}} = \frac{2\pi}{\omega} \tag{8.36}$$

occurs.

## Implications of the Local Stability Analysis

In Figure 8.7 we have parametrically plotted $\mathcal{M}'_{o*}(\omega)$ versus $\alpha(\omega)$ ($\omega$ is the parameter) [equations (8.30) and (8.31)] to give the stability boundaries for a normal human and a human with cyclical neutropenia using the data of Table 8.1. (Ignore the lines corresponding to G-CSF for the time being). The two vertical dashed lines correspond to the normal range of $\alpha$ values as discussed in Section 8.4.2; the lower dashed line is the stability boundary for the cyclical neutropenia case, and the solid line is for the normal human. Regions above a given stability boundary in $(\alpha, \mathcal{M}'_{o*})$ parameter space correspond to a locally stable steady-state neutrophil level, while regions below are unstable. For values of $(\alpha, \mathcal{M}'_{o*})$ exactly on a given line there is a bifurcation to a periodic solution with Hopf period $T_{\text{Hopf}}$ as discussed above.

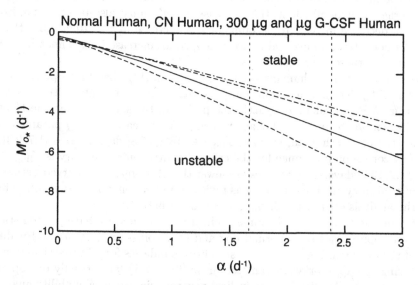

Figure 8.7. A parametric plot of the regions of linear stability and instability based on data for normal humans (solid line) (Perry et al. 1966), humans with cyclical neutropenia (CN) (lower dashed line) (Guerry et al. 1973), and normal humans administered granulocyte colony stimulating factor (G-CSF) (upper dashed line is for 30 $\mu$g, and the dash–dot line is for 300 $\mu$g) (Price et al. 1996). In this and all subsequent stability diagrams, points $(\alpha, \mathcal{M}'_{o*})$ above a given stability line correspond to linear stability of the steady state, and those below correspond to an unstable steady state. See the text for details. From Hearn, Haurie, and Mackey (1998).

**Implications for the Origin of Cyclical Neutropenia.** The first point to be noted is the following: If the model for granulopoiesis is stable for a normal human, then a simple alteration of the characteristics of the maturation time distribution to correspond to the value for cyclical neutropenia

(Table 8.1) is incapable of singlehandedly inducing an instability. Furthermore, note that the unique steady state of the model as given implicitly by equation (8.17) is *independent* of any alterations in the distribution of maturation times. However, the dynamically varying neutrophil levels in cyclical neutropenia are often depressed relative to the normal state (Section 8.3.2), thus implying that a simple alteration of the distribution of maturation times could not be the sole source of cyclical neutropenia dynamics alone.

Examination of Figure 8.7 shows that if the dynamic behavior of cyclical neutropenia is to be a result of an instability in this model, then in addition to the known alterations in the distribution of maturation times, there must be a concomitant decrease in $\mathcal{M}'_{o*}$ to more negative values such that $(\alpha, \mathcal{M}'_{o*})$ falls in the zone of parameter space where $x^*$ is unstable. Since one of the hallmarks of cyclical neutropenia is an oscillation about a reduced average neutrophil count, this decrease in $\mathcal{M}'_{o*}$ must also be accompanied by a decrease in $\mathcal{M}_{o*}$ to account for the decrease in $x^*$. (Remember that $\alpha$ is not altered in cyclical neutropenia, so an increase in $\alpha$ cannot be the source of these depressed levels.)

Suppose that in humans such a decrease in $\mathcal{M}'_{o*}$ has taken place, i.e., that $\mathcal{M}'_{o*}$ has become sufficiently negative for an unstable situation to occur. We can calculate exactly the period of the solution when the Hopf bifurcation to unstable behavior occurs. In the case of the $g$ parameters for the normal human, we have $T_{\text{Hopf}} \in [18.2, 17.8]$ days for $\alpha \in [1.7, 2.4]$. The corresponding range for the cyclical neutropenia boundary is $T_{\text{Hopf}} \in [14.2, 13.8]$ days. These values are lower than the smallest observed periods in clinical cyclical neutropenia as reviewed in Section 8.3.2 and as found in the analysis of Haurie, Mackey, and Dale (1999).

Turning to the case of canine cyclical neutropenia, we have plotted stability boundaries for a normal dog and grey collies at the peak and nadir of their cycle in Figure 8.8. The stability boundaries for all three situations (using the appropriate parameters from Table 8.1) fall virtually on top of one another. As with human cyclical neutropenia, the local stability analysis suggests that in contrast with the hypothesis of Schmitz, Loeffler, Jones, Lange, and Wichmann (1990), the origin of canine cyclical neutropenia is not a consequence of alterations in the distribution of marrow maturation times for neutrophil precursors alone. Rather, as in the human case, a shift in $\mathcal{M}'_{o*}$ to more negative values would be required to effect the requisite instability.

Assume for the grey collie that such a shift in $\mathcal{M}'_{o*}$ to values sufficiently negative to destabilize the system has taken place. What, then, are the predicted Hopf periods at the onset of the ensuing oscillation? Based on the data for normal dogs presented in Table 8.1, for $\alpha \in [1.7, 2.4]$ the local stability analysis of Section 8.4.2 predicts that $T_{\text{Hopf}} \in [8.5, 8.2]$ days. For the grey collie maturation distribution data taken at the nadir of the cycle this range is reduced to $T_{\text{Hopf}} \in [8.0, 7.6]$ days, while the collie data from

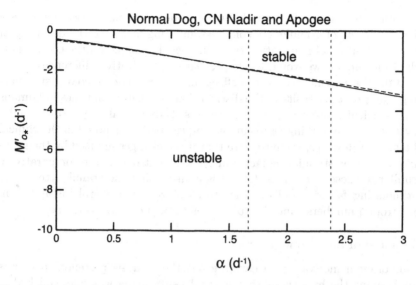

Figure 8.8. A parametric plot of the regions of linear stability and instability based on data for normal dogs taken from Deubelbeiss, Dancey, Harker, and Finch (1975) and from grey collies at the apogee and nadir of their oscillation as taken from Patt, Lund, and Maloney (1973). Note that the three stability boundaries are virtually indistinguishable from one another. CN = cyclical neutropenia. From Hearn, Haurie, and Mackey (1998).

the apogee predicts $T_{Hopf} \in [7.4, 7.1]$ days. All of these estimates are below the reported ranges for the period of canine cyclical neutropenia discussed in Section 8.3.2 and in Haurie, Person, Mackey, and Dale (1999).

Thus, for both human and grey collie cyclical neutropenia we conclude that there is no evidence from the linear stability analysis that the dynamics of cyclical neutropenia are due to an instability in the peripheral control of granulopoiesis caused by a change in the distribution of cell maturation times.

**Assessing the Effects of Granulocyte Colony Stimulating Factor.** The second point that we can address with the aid of the local stability analysis of Section 8.4.2 is the effect of granulocyte colony stimulating factor on the stability of the system in normal humans. In Figure 8.7 we have plotted the stability boundaries for the data of Table 8.1 corresponding to the alterations in normal humans induced by 30 μg and 300 μg granulocyte colony stimulating factor reported by Price, Chatta, and Dale (1996). (Note that if the individuals in this study weighed 70 kg, then the dosage was either 0.43 μg/kg–day or 4.3 μg/kg–day, respectively.) It is clear from Figure 8.7 that the region of parameter space in which the normal human control system is stable is actually *decreased* by the administration of granulocyte colony stimulating factor, since the stability boundaries for both dosages of granulocyte colony stimulating factor lie above the stability boundary for a normal human. Unfortunately, we have been unable to locate any data for the effects of granulocyte colony stimulating factor on the density $g$ of the distribution of maturation times in dogs, but based on the comparable data for humans we would not expect large quantitative differences.

If data were available for the effects of the granulocyte colony stimulating factor on the density of the distribution of maturation times in humans with cyclical neutropenia, we could assess the potential role of the granulocyte colony stimulating factor in altering the period as noted in the clinical literature. However, we must note that if the changes induced by the granulocyte colony stimulating factor in cyclical neutropenia are comparatively similar to those in normals, then it is unlikely that the granulocyte colony stimulating factor could ever act to stabilize a peripheral instability in neutrophil numbers, since its role seems to be a destabilizing one.

### Discussion and Conclusions

Our original motivation in carrying out the analysis presented here was to examine the hypothesis that cyclical neutropenia was due to a loss of stability in the peripheral control of neutrophil production. Based on the considerations of Section 8.4.2 that are independent of the precise nature of the control function assumed, we conclude that any alterations of parameters in this peripheral control system consistent with the extant laboratory and clinical data on cyclical neutropenia are unable to reproduce either the characteristics of clinical cyclical neutropenia or its laboratory counterpart in the grey collie. Further, we conclude that the dynamic effects of granulocyte colony stimulating factor treatment of cyclical neutropenia are probably not primarily due to the alterations of the peripheral control dynamics.

Rather, we conclude that the dynamics of cyclical neutropenia are due to a destabilization of the hematopoietic stem cell population as originally proposed by Mackey (1978) and Mackey (1979).

## 8.5    Stem Cell Dynamics and Cyclical Neutropenia

In trying to understand and model the properties of cyclical neutropenia as discussed in Section 8.3.2, one of the most crucial clues is the observation of the effect of continuous cyclophosphamide and busulfan administration in normal dogs (Morley and Stohlman 1970; Morley et al. 1970). Though in most animals these drugs led to a pancytopenia whose severity was proportional to the drug dose, in some dogs low doses led to a mild pancytopenia, intermediate doses gave a cyclical neutropenia-like behavior with a period between 11 and 17 days, and high drug levels led either to death or gross pancytopenia. When the cyclical neutropenia-like behavior occurred it was at circulating white blood cell levels of one-half to one-third normal. To this we must add the observation that patients undergoing hydroxurea therapy sometimes develop cyclical neutropenia-like symptoms (Kennedy 1970), as do patients receiving cyclophosphamide (Dale et al. 1973).

Both cyclophosphamide and busulfan selectively kill cells within the DNA synthetic phase of the cell cycle, and the fact that both drugs are capable of inducing cyclical neutropenia-like behavior strongly suggest that the origin of cyclical neutropenia as a disease is due to an abnormally large death rate (apoptosis) in the proliferative phase of the cell cycle of a population of cells – the hematopoietic stem cells – more primitive than the granulocyte/monocyte colony forming units, CFU-GM, and the erythrocytic burst forming units, BFU-E. Here we interpret the effects of an increase in the rate of irreversible apoptotic loss from the proliferating phase of the hematopoietic stem cells ($\gamma$ in Figure 8.9) on blood cell production (Mackey 1978).

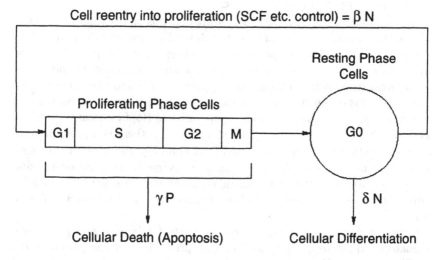

Figure 8.9. A schematic representation of the control of hematopoietic stem cell regeneration. Proliferating phase cells ($P$) include those cells in $G_1$, S (DNA synthesis), $G_2$, and M (mitosis), while the resting phase ($N$) cells are in the $G_0$ phase. Local regulatory influences are exerted via a cell-number-dependent variation in the fraction of circulating cells, $\delta$ is the **normal** rate of differentiation into all of the committed stem cell populations, while $\gamma$ represents a loss of proliferating phase cells due to apoptosis. See Mackey (1978), Mackey (1979) for further details.

The dynamics of this hematopoietic stem cell population are governed (Mackey 1978; Mackey 1979) by the pair of coupled differential delay equations

$$\frac{dP}{dt} = -\gamma P + \beta(N)N - e^{-\gamma\tau}\beta(N_\tau)N_\tau \qquad (8.37)$$

and

$$\frac{dN}{dt} = -[\beta(N) + \delta]N + 2e^{-\gamma\tau}\beta(N_\tau)N_\tau, \tag{8.38}$$

where $\tau$ is the time required for a cell to traverse the proliferative phase, and the resting to proliferative phase feedback rate $\beta$ is taken to be

$$\beta(N) = \frac{\beta_0\theta^n}{\theta^n + N^n}. \tag{8.39}$$

An examination of equation (8.38) shows that this equation could be interpreted as describing the control of a population with a delayed mixed feedback-type production term $[2e^{-\gamma\tau}\beta(N_\tau)N_\tau]$ and a destruction rate $[\beta(N) + \delta]$ that is a decreasing function of $N$.

This model has two possible steady states. There is a steady state corresponding to no cells, $(P_1^*, N_1^*) = (0, 0)$, which is stable if it is the only steady state and which becomes unstable whenever the second positive steady state $(P_2^*, N_2^*)$ exists.

The stability of the nonzero steady state depends on the value of $\gamma$. When $\gamma = 0$, this steady state cannot be destabilized to produce dynamics characteristic of cyclical neutropenia. On the other hand, for $\gamma > 0$, increases in $\gamma$ lead to a decrease in the hematopoietic stem cell numbers and a consequent decrease in the cellular efflux (given by $\delta N$) into the differentiated cell lines. This diminished efflux becomes unstable when a critical value of $\gamma$ is reached, $\gamma = \gamma_{\text{crit},1}$, at which a supercritical Hopf bifurcation occurs. For all values of $\gamma$ satisfying $\gamma_{\text{crit},1} < \gamma < \gamma_{\text{crit},2}$, there is a periodic solution of equation (8.38) whose period is in good agreement with that seen in cyclical neutropenia. At $\gamma = \gamma_{\text{crit},2}$, a reverse bifurcation occurs and the greatly diminished hematopoietic stem cell numbers as well as cellular efflux again becomes stable. All of these properties are illustrated in Figure 8.10.

Separate estimations of the parameter sets for human and grey collie hematopoietic stem cell populations give predictions of the period of the oscillation at the Hopf bifurcation that are consistent with those observed clinically and in the laboratory.

Numerical simulations, shown in Figures 8.11 and 8.12, of equations (8.37) and (8.38) bear out the results of the above local stability analyses. As expected, an increase in $\gamma$ is accompanied by a decrease in the average number of circulating cells. For certain values of $\gamma$ an oscillation appears. Over the range of $\gamma$ in which an oscillation occurs, the period increases as $\gamma$ increases. However, the amplitude of the oscillation first increases and then decreases. (Similar observations hold for the model of autoimmune hemolytic anemia as the control parameter $\gamma$ is increased.) When all the parameters in the model are set to the values estimated from laboratory and clinical data, no other types of bifurcations are found. Although these simulations also indicate the existence of multiple bifurcations and chaotic behaviors, these more complex dynamics are observed only for

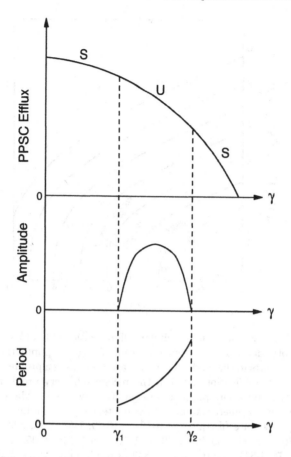

Figure 8.10. Schematic representation of the combined analytic and numerically determined stability properties of the hematopoietic stem cell model. See the text for details. From Mackey (1996).

nonphysiological choices of the parameters. Thus the observed irregularities in the fluctuations in blood cell numbers in cyclical neutropenia cannot be related to chaotic solutions of equation (8.38). These results suggest that cyclical neutropenia is likely related to defects, possibly genetic, within the hematopoietic stem cell population that lead to an abnormal ($\gamma > 0$) apoptotic loss of cells from the proliferative phase of the cell cycle.

## 8.5.1 Understanding Effects of Granulocyte Colony Stimulating Factor in Cyclical Neutropenia

Recent clinical and experimental work has focused on the modification of the symptoms of hematological disorders, including periodic hematopoiesis, by the use of various synthetically produced cytokines (Sachs 1993; Sachs

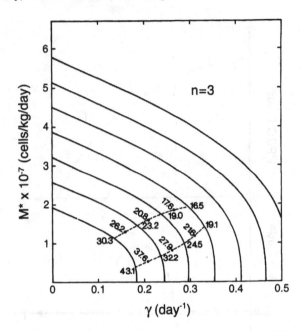

Figure 8.11. Variation of the total cellular differentiation efflux $(\delta N)$ as a function of the apoptotic death rate $\gamma$ from the proliferating cell population in humans $(n = 3)$. Parameters in the model were estimated assuming a proliferating fraction of 0.1, and an amplification of 16 in the recognizable erythroid, myeloid, and megakaryocytic precursors populations. See Mackey (1978), Mackey (1979) for details. The hematopoietic stem cell parameters corresponding to each curve from the top down are $(\delta, \beta_0, \tau, \theta \times 10^{-8}) = (0.09, 1.58, 1.23, 2.52)$, $(0.08, 1.62, 1.39, 2.40)$, $(0.07, 1.66, 1.59, 2.27)$, $(0.06, 1.71, 1.85, 2.13)$, $(0.05, 1.77, 2.22, 1.98)$, $(0.04, 1.84, 2.78, 1.81)$, and $(0.03, 1.91, 3.70, 1.62)$ in units $(\text{days}^{-1}, \text{days}^{-1}, \text{days}, \text{cells/kg})$. The dashed solid lines indicate the boundaries along which stability is lost in the linearized analysis, and the numbers indicate the predicted (Hopf) period (in days) of the oscillation at the Hopf bifurcation. From Mackey (1978).

and Lotem 1994; Cebon and Layton 1984), e.g., the recombinant colony stimulating factors rG-CSF and rGM-CSF, whose receptor biology is reviewed in (Rapoport et al. 1992), and Interlukin-3. These cytokines are now known to interfere with the process of apoptosis or to lead to a decrease in $\gamma$ within the context of the hematopoietic stem cell model of Section 8.5.

Human colony stimulating factors increase both the numbers and proliferation rate of white blood cell precursors in a variety of situations (Bronchud et al. 1987; Lord et al. 1989; Lord et al. 1991). Furthermore, colony stimulating factor in mice is able to stimulate replication in both stem cells and early erythroid cells (Metcalf et al. 1980).

It is known that in aplastic anemia and cyclical neutropenia there is an inverse relationship between plasma levels of granulocyte colony stimulating factor and white blood cell numbers (Watari et al. 1989). Further, it has

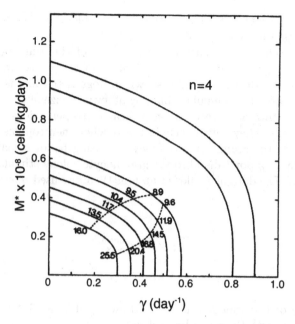

Figure 8.12. As in Figure 8.11, but all calculations done with parameters appropriate for dogs. Reproduced from Mackey (1978).

been shown (Layton et al. 1989) that the $t_{1/2}$ of the granulocyte colony stimulating factor in the circulation is short, on the order of 1.3 to 4.2 hours, so the dynamics of the destruction of the granulocyte colony stimulating factor are unlikely to have a major role in the genesis of the dynamics of cyclical neutropenia.

In the grey collie it has been shown that at relatively low doses of the granulocyte colony stimulating factor the mean white blood cell count is elevated (from 10 to 20 times), as is the amplitude of the oscillations (Hammond et al. 1990), while higher dosages (Lothrop et al. 1988; Hammond et al. 1990) lead to even higher mean white blood cell numbers but eliminate the cycling. Another interesting observation is that in the collie, granulocyte colony stimulating factor administration results in a decrease in the period of the peripheral oscillation. The elevation of the mean white blood cell levels and the amplitude of the oscillations, as well as an enhancement of the oscillations of platelets and reticulocytes, at low levels of the granulocyte colony stimulating factor has also been reported in humans (Hammond et al. 1990; Migliaccio et al. 1990; Wright et al. 1994), and it has been also noted that the fall in period observed in the collie after granulocyte colony stimulating factor administration occurs in humans with a fall in period from 21 to about 14 days. Finally, it should be mentioned that treatment with granulocyte colony stimulating factor in patients with agranulocytosis has also lead to a significant increase in the mean white blood cell counts and, in some patients, to the induction of white blood cell oscillations with periods ranging from 7 to 16 days.

Our major clue to the nature of the effects of the granulocyte colony stimulating factor comes from its prevention of apoptosis, and the work of Avalos et al. (1994), who have shown in dogs that there is no demonstrable alteration in the number, binding affinity, or size of the granulocyte colony stimulating factor receptor on cyclical neutropenia dogs as compared to normal dogs. They thus conclude that cyclical neutropenia "is caused by a defect in the granulocyte colony stimulating factor signal transduction pathway at a point distal to the granulocyte colony stimulating factor binding ... ." The data of Avalos et al. (1994) can be used to estimate that

$$\gamma_{max}^{CN} \approx 7 \times \gamma_{max}^{norm}. \tag{8.40}$$

The results of Hammond, Chatta, Andrews, and Dale (1992) in humans are consistent with these results in dogs.

Less is known about the effect of the granulocyte/monocyte colony stimuling factor, GM-CSF, but it is known that administration of GM-CSF in humans gives an elevation of the mean white blood cell level but only by relatively modest amounts, 1.5 to 3.9 times (Wright et al. 1994), but either dampens the oscillations of cyclical neutropenia or eliminates them entirely. The same effect has been shown (Hammond et al. 1990) in the grey collie. It is unclear whether the period of the peripheral cell oscillations has a concomitant decrease, as is found with the granulocyte colony stimulating factor. The abnormal responsiveness of precursors to granulocyte colony stimulating factor in grey collies and humans with cyclical neutropenia (Hammond et al. 1992; Avalos et al. 1994) is mirrored in the human response to the granulocyte/monocyte colony stimulating factor (Hammond et al. 1992).

Thus, the available laboratory and clinical data on the effects of colony stimulating factors in periodic hematopoiesis indicate that (1) there is extensive intercommunication between all levels of stem cells; and (2) within the language of nonlinear dynamics, colony stimulating factors may be used to titrate the dynamics of periodic hematopoiesis to the point of inducing a reverse Hopf bifurcation (disappearance of the oscillations). In the course of this titration, there may also be a shift in the period.

The behavior in periodic hematopoiesis when colony stimulating factor is administered is qualitatively consistent with the hematopoietic stem cell model discussed in Section 8.5, since it is known that colony stimulating factor interferes with apoptosis, and thus administration of colony stimulating factor is equivalent to a decrease in the apoptotic death rate $\gamma$.

# 8.6    Conclusions

Delayed feedback mechanisms constitute a core element in the regulation of blood cell populations. These delayed feedback mechanisms can produce oscillations whose period typically ranges from 2 to 4 times the delay, but which may be even longer. Thus it is not necessary to search for illusive and mystical entities (Beresford 1988), such as ultradian rhythms, to explain the periodicity of these disorders.

The observations in this chapter emphasize that an intact control mechanism for the regulation of blood cell numbers is capable of producing behaviors ranging from no oscillation to periodic oscillations to more complex irregular fluctuations, i.e., chaos. The type of behavior produced depends on the nature of the feedback, i.e., negative or mixed, and on the value of certain underlying control parameters, e.g., peripheral destruction rates or maturation times. Pathological alterations in these parameters can lead to periodic hematological disorders.

# 8.7    Computer Exercises: Delay Differential Equations, Erythrocyte Production and Control

## Objectives

The purpose of these exercises is to gain some familiarity with the behavior of the solutions of differential delay equations by using both analytical and numerical approaches. We are going to do this within the context of a simple model for erythrocyte production and control. For the numerical work, you will use XPP* for these exercises.

## A Simple Model for the Regulation of Red Blood Cell Production

Consider the control of erythrocyte, or red blood cell, production as represented schematically in Figure 8.13.

A fall in circulating erythrocyte numbers leads to a decrease in hemoglobin levels and thus in arterial oxygen tension. This decrease in turn triggers the production of renal erythropoietin, which increases the cellular production within the early committed erythrocyte series cells, and thus the cellular efflux from the erythrocytic colony forming unit, CFU-E, into the identifiable proliferating and nonproliferating erythroid precursors,

---

*See Introduction to XPP in Appendix A.

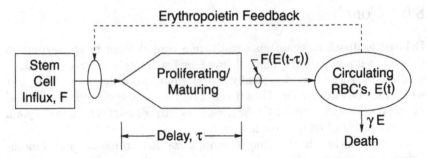

Figure 8.13. A schematic representation of the regulation of red blood cell production. Here the major features and parameters are defined for the simple model that leads to equation (8.41). From Mackey (1996).

and ultimately augments circulating erythrocyte numbers (i.e., negative feedback).

To formulate this sequence of physiological processes in a mathematical model, we let $E(t)$ (cells/kg) be the circulating density of red blood cells as a function of time, $\beta$ (cells/kg-day) be the stem cell influx under erythropoietin control, $\tau$ (days) be the time required to pass through recognizable precursors, and $\gamma$ (days$^{-1}$) be the loss rate of red blood cells in the circulation. We can then write the rate of change of erythrocyte numbers as a balance between their production and their destruction:

$$\frac{dE(t)}{dt} = \beta(E(t - \tau)) - \gamma E(t). \tag{8.41}$$

Once a cell from the hematopoietic stem cell compartment is committed to the erythroid series, it undergoes a series of nuclear divisions and enters a maturational phase for a period of time ($\tau \simeq 5.7$ days) before release into the circulation, and the argument in the production function is therefore $E(t - \tau)$, and not $E(t)$. Thus, changes that occur at time $t$ were actually initiated at a time $t - \tau$ in the past. We adopt the usual convention of $E_\tau(t) = E(t - \tau)$, and also do not explicitly denote the time unless necessary. Then the simple model (8.41) for red blood cell dynamics takes the alternative form

$$\frac{dE}{dt} = \beta(E_\tau) - \gamma E. \tag{8.42}$$

To define an appropriate form for the production function $\beta$, we use in vivo measurements of erythrocyte production rates in rats and other mammals including humans. The feedback function saturates at low erythrocyte numbers, and is a decreasing function of increasing red blood cell levels. A convenient function that captures this behavior, has sufficient flexibility to be able to fit the data, and is easily handled analytically is given by

$$\beta(E_\tau) = \beta_0 \frac{\theta^n}{E_\tau^n + \theta^n}, \tag{8.43}$$

where $\beta_0$ (units of cells/kg-day) is the maximal red blood cell production rate that the body can approach at very low circulating red blood cell numbers, $n$ is a positive exponent, and $\theta$ (units of cells/kg) is a parameter. These three parameters have to be determined from experimental data related to red blood cell production rates.

Combining equations (8.42) and (8.43), we have the final form for our model of red blood cell control given as

$$\frac{dE}{dt} = \beta_0 \frac{\theta^n}{E_\tau^n + \theta^n} - \gamma E. \tag{8.44}$$

As discussed in Chapter 9, we have to specify an initial condition in the form of a *function* defined for a period of time equal to the duration of the time delay. Thus we will select

$$E(t') = \phi(t'), \qquad -\tau \leq t' \leq 0. \tag{8.45}$$

Usually we consider only initial functions that are constant, but it must be noted that some differential delay equations can display multistable behavior in which there are two or more coexisting locally stable solutions, depending on the initial function.

Ex. 8.7-1. A *steady state* (or stationary) solution for the model (8.44) is defined by the requirement that the red blood cell number not be changing with time. This means that

$$E(t) = E(t - \tau) = E_\tau(t) = \text{a constant, the steady state} = E^*, \tag{8.46}$$

and

$$\frac{dE}{dt} = 0 \quad \text{so} \quad \beta(E^*) = \beta_0 \frac{\theta^n}{E^{*n} + \theta^n} = \gamma E^*. \tag{8.47}$$

We cannot solve (8.47) to get an analytic form for $E^*$, but a simple graphical argument shows that there is only one value of $E^*$ satisfying (8.47). This value of the steady state occurs at the intersection of the graph of $\gamma E^*$ with the graph of $\beta(E^*)$.

In this first problem, you must determine the stability of $E^*$ using a linear expansion in (8.44).

a. Expand the function $\beta$ around $E^*$ to obtain, with $z(t) = E(t) - E^*$, the linear differential delay equation

$$\frac{dz}{dt} = \beta'(E^*)z_\tau - \gamma z. \tag{8.48}$$

b. Assuming that $z(t) \simeq e^{\lambda t}$ in equation (8.48), derive the equation

$$\lambda = \beta'(E^*)e^{-\lambda \tau} - \gamma \tag{8.49}$$

that $\lambda$ must satisfy.

Letting $\lambda = i\omega$, show that the relation connecting $\tau$, $\beta'(E^*)$, and $\gamma$ that must be satisfied in order for the eigenvalues to have real part identically zero is given by

$$\tau = \frac{\cos^{-1}\left(\frac{\gamma}{\beta'(E^*)}\right)}{\sqrt{\beta'(E^*)^2 - \gamma^2}}. \tag{8.50}$$

Ex. 8.7-2. In this exercise, you will numerically integrate equation (8.44). In the code written to simulate equation (8.44), the erythrocyte numbers $E$ have been scaled by the numerical value of the parameter $\theta$ in the erythropoietin feedback function (8.43), since XPP gets very cranky whenever dependent variables exceed 100 in absolute value.

(a) Rewrite equation (8.44) in the dependent variable, and define a new variable $x = E/\theta$ to transform equation (8.44) into

$$\frac{dx}{dt} = \frac{\beta_0}{\theta} \frac{1}{x_\tau^n + 1} - \gamma x. \tag{8.51}$$

The code for equation (8.44) is written in aiha1.ode.

(b) Open up XPP with the code for equation (8.44) by typing xppaut aiha1.ode, and turn off the bell.

The choice of Method, which determines the numerical algorithm used to integrate the differential delay equation, is essential. Differential delay equations can be pretty tricky to deal with, and many people have come to grief by using an adaptive step size method to integrate them. (Can you think of why this might happen?) So, we want to make sure that the one that is being used is *not* adaptive. Check that (R)unge-Kutta x method is selected.

Having gotten through the above, you are ready to try a numerical simulation.

You should see a periodic variation in the scaled erythrocyte numbers $(x)$ versus time, and this is because $\gamma$ was picked to be inside the zone of instability of $x^* \equiv E^*/\theta$.

What is the period and amplitude of this periodic variation?

(c) Now you can compare the predictions, from the linear analysis, of the stability boundaries, and at the same time determine qualitatively how the period and amplitude of the oscillatory erythrocyte numbers within the unstable range of $\gamma$ change as $\gamma$ is increased.

Use a (R)ange of gamma, with Steps: 5 to begin with and the starting and ending values slightly below and above the values of $\gamma_1$ and $\gamma_2$, respectively, employing different values in an exploratory mode.

Next you can try to change some of the other parameters in the equation to see how they affect the solution behavior, as well as its stability, and see how your intuition matches with what you see numerically. Can you find any values of parameters such that numerically there seems to be a secondary bifurcation?

Ex. 8.7-3. The analysis of equation (8.44) can be continued by using an approximation yielding an analytic solution.

As in Chapter 9, we let $n \to \infty$ in the nonlinear Hill function (8.43), so the nonlinearity becomes progressively closer to a step function nonlinearity. Equation (8.44) then becomes

$$\frac{dE(t)}{dt} = -\gamma E(t) + \begin{cases} F_0, & 0 \le E_\tau < \theta, \\ 0, & \theta \le E_\tau. \end{cases} \tag{8.52}$$

The nonlinear differential delay equation (8.52) can be alternatively viewed as a pair of ordinary differential delay equations, and which one we have to solve at any given time depends on the value of the retarded variable $E_\tau$ with respect to the parameter $\theta$. (This method of solution is usually called the *method of steps*.)

As initial function for equation (8.52) of the type in (8.45), pick one that satisfies $\phi(t') > \theta$ for $-\tau \le t' \le 0$ and specify that $\phi(0) \equiv E_0$, a constant.

(a) Solve the equation

$$\frac{dE}{dt} = -\gamma E \qquad \theta < E_\tau, \quad E(t=0) \equiv E_0, \tag{8.53}$$

to obtain

$$E(t) = E_0 e^{-\gamma t}, \tag{8.54}$$

valid until a time $t_1$ determined by the condition $\theta = E(t_1 - \tau)$. Show that the value of $t_1$ is given by

$$t_1 = \frac{1}{\gamma} \ln\left\{\frac{E_0 e^{\gamma \tau}}{\theta}\right\}. \tag{8.55}$$

From this value of $t_1$, show that the value of $E$ at $t = t_1$ can be calculated as

$$E(t = t_1) \equiv E_1 = \theta e^{-\gamma \tau}. \tag{8.56}$$

(b) To proceed for times greater that $t_1$, solve the other differential equation given in (8.52), namely,

$$\frac{dE}{dt} = -\gamma E + F_0 \qquad E_\tau \le \theta, \quad E(t_1) = E_1, \tag{8.57}$$

to get

$$E(t) = E_1 e^{-\gamma(t-t_1)} + \frac{F_0}{\gamma}\left[1 - e^{-\gamma(t-t_1)}\right], \tag{8.58}$$

which is a solution valid until a time $t_2$ defined by $\theta = E(t_2 - \tau)$. Compute the value of $t_2$ as

$$t_2 = \frac{1}{\gamma} \ln \left\{ \left( \frac{E_0}{\theta} \right) \left[ \frac{E_1 - (F_0/\gamma)}{\theta - (F_0/\gamma)} \right] e^{2\gamma\tau} \right\}, \qquad (8.59)$$

and the value of the solution at time $t_2$, $(E(t = t_2) \equiv E_2)$ to obtain

$$E_2 = \frac{F_0}{\gamma} + \left( \theta - \frac{F_0}{\gamma} \right) e^{-\gamma\tau}. \qquad (8.60)$$

(c) In the computation of the third portion of the solution, you must once again solve equation (8.53) subject to the endpoint conditions determined in the last calculation. Show that

$$E(t) = E_2 e^{-\gamma(t - t_2)} \qquad (8.61)$$

is the expression to use, and determine that

$$t_3 = \frac{1}{\gamma} \ln \left\{ \left( \frac{E_0 E_2}{\theta^2} \right) \left[ \frac{E_1 - (F_0/\gamma)}{\theta - (F_0/\gamma)} \right] e^{3\gamma\tau} \right\}, \qquad (8.62)$$

so that $E(t_3) \equiv E_3$ is given by

$$E_3 = \theta e^{-\gamma\tau}. \qquad (8.63)$$

What can you conclude by comparing equations (8.56) and (8.63)? Calculate the period of the periodic solution just derived and show that it is given by

$$T = 2\tau + \frac{1}{\gamma} \ln \left\{ \left[ \frac{\frac{F_0}{\gamma\theta}}{\frac{F_0}{\gamma\theta} - 1} - e^{-\gamma\tau} \right] \left[ \frac{F_0}{\gamma\theta} - e^{-\gamma\tau} \right] \right\}. \qquad (8.64)$$

Ex. 8.7-4. From the previous exercises it seems that first-order differential delay equations with negative feedback have only one bifurcation between a stable steady state and a stable limit cycle when looked at numerically.

However, the situation is quite different if one looks at a system with a *mixed* feedback replacing the negative feedback. Mixed feedback is a term that has been coined to indicate that the feedback function has the characteristics of positive feedback over some range of the state variable, and negative feedback for other ranges.

These systems have a host of bifurcations, and you can explore these numerically by using XPP to study the prototype equation

$$\frac{dx}{dt} = -\gamma x + \beta \frac{x_\tau}{1 + x_\tau^n}, \qquad (8.65)$$

a variant of which was originally proposed as a model for the regulation of white blood cell production (Mackey and Glass 1977).

(a) How many steady states does equation (8.65) have?
(b) Compute analytically when the steady state becomes unstable in equation (8.65). What is the Hopf period at that point?
(c) Explore the range of behavior that the numerical solutions can take as one changes various parameters. (Note that you can eliminate one of the parameters by, for example, scaling the time by $\gamma$ so there is only the three–parameter set $(\beta, \tau, n)$ to deal with.)

# 9

# Pupil Light Reflex: Delays and Oscillations

## John Milton

## 9.1 Introduction

For the experimentalist, the analysis of "real-life" dynamical systems is problematic. The obvious strategy of comparing prediction to observation by varying a control parameter is faced by the practical issues of first identifying the relevant control parameter(s) and then devising ways to alter it. Furthermore, all real systems are continually subjected to the influence of random, uncontrolled fluctuations, herein referred to as "noise." The observed dynamics reflect the interplay between noise and the intrinsic dynamics of the system. Finally, the possibility that a dynamical system can generate very complex time series necessitates a careful consideration of the way in which we measure the data. All measurement devices are limited by their resolution, and hence, in the frequency domain, some of the frequency characteristics of the signal generated by the dynamical system will be filtered out.

In this chapter we consider the analysis of dynamical systems in which the value of a state variable, $x(t)$, depends on its value at some time, $\tau$, in the past, $x(t - \tau)$. We refer to $\tau$ as the latency or time delay. Our findings are discussed with reference to studies on the human pupil light reflex.

## 9.2 Where Do Time Delays Come From?

Since conduction velocities are finite, all real spatially extended dynamical systems have time delays. One way to think about a time delay is to regard it as a transport delay, or dead time. Figure 9.1 shows the result of delivering a transient stimulus to a dynamical system and measuring its response. In the absence of a time delay, the response of the dynamical system lags behind that of the stimulus (Figure 9.1A). A characteristic of a dynamical system that possesses a lag is that the time course of the stimulus overlaps

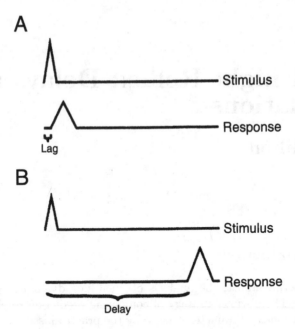

Figure 9.1. Comparison between A) a lag and B) a time delay. See text for discussion.

that of the response. However, in the case of a time delay there is a clear separation between the stimulus and the response (Figure 9.1B). The larger the gap between stimulus and response, the greater the necessity of explicitly including a time delay in the mathematical model (i.e., small delays can be approximated as lags, but not large ones). The presence of a time delay means that what happened in the past has important consequences for what happens in the present (see Problems 1 and 2 at the end of the chapter).

Time delays arise, for example, in the nervous system because of axonal conduction and integration times, in cell biology because of cell maturation times, and in molecular biology because of the time required for transcription and translation. Thus, mathematical models of these physiological systems take the form of a delay differential equation, for example

$$\frac{dx(t)}{dt} = -\alpha x(t) + f(x(t - \tau)),\qquad(9.1)$$

where $\alpha$ is a constant. It should be noted that in order to solve equation (9.1) it is not sufficient to specify an initial value of $x(0) = X_o$ (which is what we would do to solve, for example, the first-order ordinary differential equation $dx(t)/dt = -\alpha x(t)$). Instead, we must specify an initial function, i.e., all of the values of $x(s)$ that lie on the interval $[-\tau, 0]$, i.e., $x(s), s \in [-\tau, 0]$. This is the way in which the dynamical system remembers its past. Since the number of points on an interval is infinite, equation (9.1) is also referred to as an infinite-dimensional differential equation or a functional differential equation.

In this chapter we restrict our attention to the case of a single, discrete time delay. This situation most commonly arises in the analysis of feedback control mechanisms such as those that arise in the description of the growth of animal (Gourley and Ruan 2000; Nicholson 1954) and blood cell populations (Mackey, Haurie, and Bélair, Chapter 8), physiological and metabolic control mechanisms (an der Heiden and Mackey 1982; Glass and Mackey 1988; Mackey and Glass 1977), and neural feedback control mechanisms (Eurich and Milton 1996; Foss and Milton 2000; Milton 1996; Milton, Longtin, Beuter, Mackey, and Glass 1989).

## 9.3   Pupil Size

Pupillary size reflects a balance between two opposing muscle groups located in the iris (Loewenfeld 1993). Pupil constriction is due to the increased tension in the pupillary constrictor. The pupillary constrictor muscle is circularly arranged. It is innervated by the parasympathetic nervous system, and the motor nucleus, called the Edinger–Westphal nucleus, is located in the midbrain. There are two mechanisms for pupillary dilation: (1) "active" reflex dilation is due to contraction of the pupillary dilator; and (2) "passive" reflex dilation is to due to inhibition of the activity of the Edinger–Westphal nucleus (Loewenfeld 1958). The pupillary dilator muscle is radially arranged and is innervated by the sympathetic nervous system. Its motor nuclei are located in the hypothalamus.

It is naive to think of pupil size as simply a balance between constricting and dilating forces. Obviously, pupillary size is not uniquely determined by this balance. For every dilating force, there is a constricting force that balances it to give the same pupil size. In other words, a measurement of pupil size is not sufficient to uniquely determine the activity in the dilating and constricting muscles. It is only through observations of the dynamics of pupil change that these forces can be estimated (Terdiman, Smith, and Stark 1969; Usui and Hirata 1995).

The term **pupil light reflex** refers to the changes in pupil size that occur in response to a light pulse, or pulses (Figure 9.2). Recently, it has been shown that the pupils of mice whose retina lack rods and cones respond to bright light (Lucas, Douglas, and Foster 2001). The photoreceptor for this nonvisual pathway is a melanopsin-containing retinal ganglion cell (Hattar, Liao, Takao, Berson, and Yau 2002). These specialized ganglion cells integrate light energy over very long time scales (typical latencies of 10 to 20 seconds) and respond poorly to brief stimuli (Berson, Dunn, and Takao 2002). Thus this nonvisual pathway for monitoring ambient light levels is

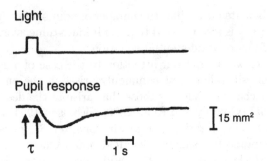

Figure 9.2. A light pulse can be used to measure the time delay $\tau$ for the pupil light reflex.

thought to be most relevant for the circadian light entrainment mechanism and is unlikely related to the responses of pupil size to light stimuli typically studied by pupil aficionados. However, light is not the only input that influences pupil size. For example, pupil size is also affected by changes in lens accommodation (Hunter, Milton, Lüdtke, Wilhelm, and Wilhelm 2000), changes in the level of consciousness (O'Neill, Oroujeh, Keegan, and Merritt 1996; O'Neill, Oroujeh, and Merritt 1998; Wilhelm, Lüdtke, and Wilhelm 1998; Wilhelm, Wilhelm, Lüdtke, Streicher, and Adler 1998; Yoss, Moyer, and Hollenhurst 1993), and the effects of cardiac pulses (Calcagnini, Lino, Censi, Calcagnini, and Cerutti 1997; Daum and Fry 1982) and respiration (Borgdorff 1975; Daum and Fry 1981; Yoshida, Yana, Okuyama, and Tokoro 1994). Here we focus on the oscillations in pupil size that occur when the gain in the reflex arc is high (i.e., pupil cycling).

Figure 9.2 shows the changes in pupil size that occur following a single light pulse. There are no structures present in the iris musculature that are analogous to the muscle spindles and Golgi tendon organs in skeletal muscle. This means that there is no feedback in the pupil light reflex that is directly related to pupil size. Rather, the input is the amount of light falling on the retina (Stark 1968; Stark and Sherman 1957). The retinal light flux, $\phi$, is equal to

$$\phi = IA, \tag{9.2}$$

where $I$ is the retinal illuminance (e.g., lumens $mm^{-2}$) and $A$ is the pupil area ($mm^2$). The iris acts very much like the aperture in a camera. If the retinal light flux, $\phi$, is too high, the pupil light reflex decreases $A$ and hence $\phi$. On the other hand, if $\phi$ is too low, $\phi$ is increased by increasing $A$. In other words the pupil light reflex acts like a negative feedback control mechanism. Furthermore it is a time-delayed negative feedback control mechanism. Pupil size does not change immediately in response to a change in illumination, but begins to change after a delay, or latency, of $\approx 300$ msec (Figure 9.2).

Table 9.1. Spontaneous dynamical behaviors exhibited by the pupil.

| Type of Dynamical Behavior | Description |
| --- | --- |
| Regular oscillations Simple waveforms | Pupil cycling<br>    Edge-light; Stern (1944)<br>    Electronic feedback<br>        Continuous negative feedback; Longtin et al. (1990); Stark (1962)<br>        Piecewise constant negative feedback; Milton and Longtin (1990); Milton et al. (1988); Reulen et al. (1988) |
| Complex waveforms | Hippus (narcoleptics) Yoss et al. (1993)<br>Cycling with piecewise constant "mixed" feedback; Longtin and Milton (1988); Milton et al. (1989) |
| Irregular oscillations | Intermittent irregular pupil cycling in demyelinative optic neuropathy; Milton et al. (1988) |
| Noise-like fluctuations | "Open-loop" hippus; Stark et al. (1958) |

There are a number of reasons why biomathematicians have been fascinated by the pupil. First, the pupil exhibits a wide range of dynamical behaviors (Table 9.1). For example, irregular variations in pupil area ("hippus") occur spontaneously, whereas regular oscillations ("pupil cycling") can be induced by focusing a small light beam at the pupillary margin. Second, the pupil light reflex is one of the few neural reflexes in the human that can be monitored noninvasively, for example, by the use of infrared video cameras; and that can be readily manipulated, for example, pharmacologically or by changing ambient light levels. Finally, it it possible to insert a known feedback into the reflex by using a technique known as clamping (Milton et al. 1988; Stark 1962; Reulen et al. 1988). This means that there exists the possibility of directly comparing theoretical predictions to experimental observations in a precise way that is simply not possible for other human neural feedback control mechanisms.

## 9.4  Pupil Light Reflex

The first attempt to model the pupil light reflex in terms of a delay differential equation was made by Longtin and Milton. A schematic representation of this model is shown in Figure 9.3. Here we present a simpler derivation

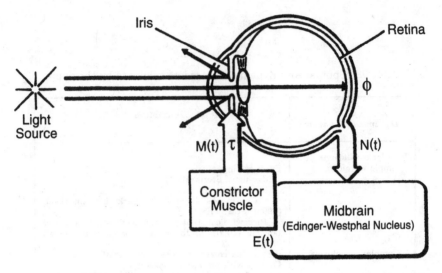

Figure 9.3. Schematic representation of the pupil light reflex. See text for discussion. Adapted from Longtin and Milton (1989b).

of their model that enables certain of the physiological aspects of the pupil light reflex to be better discussed. A detailed and more rigorous development of this model is presented elsewhere (Longtin and Milton 1989b; Longtin and Milton 1989a; Longtin, Milton, Bos, and Mackey 1990; Milton 1996).

## 9.5    Mathematical Model

There is a logarithmic compression of light intensities at the retina (Weber–Fechner law). This means that the output of the retina to a given light flux, measured in terms of the frequency of neural action potentials in the optic nerve, $N(t)$, is of the form

$$N(t) = \eta \ln\left[\frac{\phi(t - \tau_{\mathrm{r}})}{\overline{\phi}}\right], \qquad (9.3)$$

where $\eta$ is a positive constant, $\overline{\phi}$ is the threshold retinal light level (i.e., the light level below which there is no response), and $\tau_{\mathrm{r}}$ is the time required for retinal processing. The notation $\phi(t - \tau_{\mathrm{r}})$ indicates that $N(t)$ is a function of the retinal light flux, $\phi$, measured at a time $\tau_{\mathrm{r}}$ in the past.

Figure 9.4 shows a plot of $N(t)$ as a function of $\phi$. The retina is able to detect light levels ranging from a low of just a few photons to a high corresponding to direct sunlight. This is an $\approx 10^{11}$-fold range in $\phi$! However, pupil area ranges from a minimum of $\approx 10$ mm$^2$ to a maximum of $\approx 50$ mm$^2$, i.e., a 5-fold range. In a dimly lighted room using a light-emitting diode light source, pupil area can be varied over the range of 12–40 mm². Thus under typical laboratory conditions we have access to most of the range of possible pupil sizes, but only to a relatively small range of the possible illuminance. Thus in the discussion that follows we assume that the retinal illuminance is constant and focus on the effects of changes in pupil area on $\phi$.

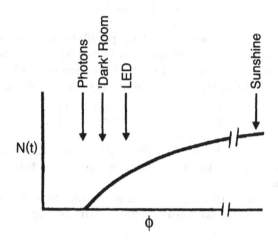

Figure 9.4. Spike frequency in the optic nerve described by equation (9.3) as a function of the retinal light flux, $\phi$.

The essential nonlinearities in the pupil light reflex arise in either the retina or the iris musculature (Howarth, Bailey, Berman, Heron, and Greenhouse 1991; Inoue 1980; Semmlow and Chen 1977). Surprisingly, the midbrain nuclei merely act as a linear lowpass filter and introduce a time delay (Inoue 1980). Thus the output of the Edinger–Westphal nucleus, $E(t)$, is

$$E(t) = \eta' \ln\left[\frac{\phi(t - (\tau_r + \tau_m))}{\bar{\phi}}\right], \tag{9.4}$$

where $\tau_m$ is the time delay introduced by the midbrain nuclei and $\eta'$ is a constant.

When the action potentials reach the neuromuscular junction of the pupillary constrictor muscle they initiate a complex sequence of events (e.g., transmitter release and diffusion across the synaptic cleft, release of $Ca^{2+}$, actin–myosin interactions: The net result is that the tension produced by the constrictor muscle changes. There is a direct relationship between neural spike frequency and muscle activity, but an inverse relation between muscle tension and pupil area. Longtin and Milton (1989b) proposed that the relationship between neural activity and pupil area could be determined in two steps: Step 1: determine the relationship between $E(t)$ and muscle activity, $x$; Step 2: determine the relationship between $x$ and pupil area, $A$. As we will see, the advantage of this approach is that the relationship between $E(t)$ and $x$ need not be precisely formulated. Thus we assume that

$$E(t) = M\left(x, \frac{dx}{dt}, \frac{d^2x}{dt^2}, \cdots\right) \tag{9.5}$$

$$\approx k\left(\frac{dx}{dt} + \alpha x\right), \tag{9.6}$$

where $\alpha$ represents the rate constant for pupillary movements and $k$ is a constant. A justification for the use of a first-order approximation is the success that this model has in describing the dynamics of pupil cycling (discussed later in this chapter). Combining equations (9.4) and (9.6) gives

$$\frac{dx(t)}{dt} + \alpha x(t) = \beta \ln\left[\frac{\phi(t-\tau)}{\bar{\phi}}\right], \tag{9.7}$$

where $\beta = \eta'/k$ and $\tau$ is the total time delay in the pupil reflex arc, i.e., the sum of $\tau_r$, $\tau_m$, and the time taken by the events that occur at the neuromuscular junction.

In order to write equation (9.7) in terms of pupil area, it is necessary to have a function, $f(x)$, that relates neural activity to pupil area, i.e.,

$$A = f(x).$$

Then we can define

$$g(A) = f^{-1}(x),$$

and rewrite equation (9.7) as

$$\frac{dg}{dA}\frac{dA}{dt} + \alpha g(A) = \beta \ln\left[\frac{\phi(t-\tau)}{\bar{\phi}}\right]. \tag{9.8}$$

Intuitively, we know that $f(x)$ must satisfy two requirements: (1) Pupil area must be positive for all $x$ and be bounded by finite limits; and 2) it should reflect the role played by the elastomechanical properties of the iris. One possible choice is shown in Figure 9.5A (Longtin and Milton 1989b; Longtin and Milton 1989a):

$$A = f(x) = \frac{A_{m_1}\theta^n}{\theta^n + x^n} + A_{m_2}, \tag{9.9}$$

where $A_{m_2}$ is the minimum pupil area, $A_{m_1} + A_{m_2}$ is the maximum pupil area, and $n, \theta$ are positive constants. A plot of $g(A) = f^{-1}(x)$ is shown in Figure 9.5B.

Over typical ranges of light intensities, the pupil light reflex functions in a linear range (Loewenfeld and Newsome 1971). If we restrict our attention to this range, then we can replace $g(A)$ by the dotted straight line shown in Figure 9.5B,

$$g(A) \approx \Psi' - \hat{k}A, \tag{9.10}$$

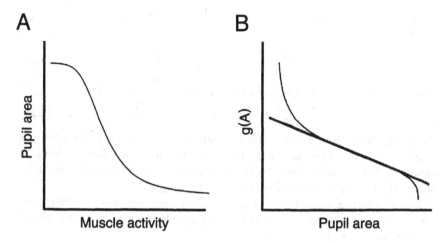

Figure 9.5. (A) Plot of pupil area versus muscle activity described by equation (9.9). (B) Plot of the inverse of equation (9.9), $g(A)$, versus pupil area. See text for discussion.

and hence equation (9.8) becomes

$$-\hat{k}\frac{dA(t)}{dt} - \alpha\hat{k}A(t) + \alpha\Psi' = \beta\ln\left[\frac{\phi(t-\tau)}{\bar{\phi}}\right].$$

The arguments presented above suggest that when $I$ is constant, the retinal light flux, $\phi$, is confined to a fairly small interval (i.e., it can vary only by about three-fold). Thus $\ln\phi$ does not vary much either, and hence we can write

$$\beta\ln\phi(t-\tau) = \beta\ln IA(t-\tau) \approx \beta\ln\phi_o + \Delta A(t-\tau),$$

where $\Delta$ is a constant that gives the slope of the straight linear approximation.* Using this approximation we can write

$$\frac{dA(t)}{dt} + \alpha A(t) = \Psi_o + dA(t-\tau), \tag{9.11}$$

where $\Psi_o > 0$ and $d < 0$ are constants.

## 9.6 Stability Analysis

At the critical point $A^*$ of equation (9.11) we have that $dA(t)/dt = 0$, $A(t) = A(t-\tau)$, and

$$A^* = \frac{\Psi_o}{\alpha - d}. \tag{9.12}$$

---

*More precisely, $\Delta$ is the slope of the straight line approximation evaluated at the fixed point $A^*$.

If we define $u(t) = A(t) - A^*$, then we can linearize equation (9.11) about $A^*$ to get

$$\frac{du(t)}{dt} + \alpha u(t) = du(t - \tau).$$ (9.13)

The characteristic equation, obtained by taking $u = \exp(\lambda t)$, is

$$\lambda + \alpha = de^{-\lambda \tau}.$$ (9.14)

Equation (9.14) is a transcendental equation and hence has an infinite number of roots. The fact that there are an infinite number of roots to equation (9.14) corresponds to the fact that an initial function must be specified in order to determine the solution uniquely.

Stability means that the real parts of all of the eigenvalues are negative. Instability occurs when the real part of any one eigenvalue becomes positive. Thus, if we write the eigenvalue as

$$\lambda = \gamma + if,$$

we are interested in the case where $\gamma = 0$, since this gives us the boundary at which a change in stability occurs. In this case equation (9.14) can be written as

$$\frac{\alpha}{d} + i\frac{f}{d} = e^{-if\tau}.$$ (9.15)

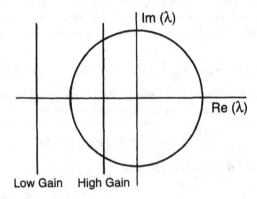

Figure 9.6. Graphical method to evaluate the stability of equation (9.15). See text for discussion.

We can use a graphical method (MacDonald 1989) to gain an insight into the stability of equation (9.15) (Figure 9.6). The left-hand side of equation (9.15) is a straight line in the complex plane, and the right-hand side is a circle in the complex plane. The stability depends on two parameters: (1) the ratio $\alpha/d$, which is related to the gain of the feedback loop (Longtin and Milton 1989b; Longtin and Milton 1989a), and (2) the time delay $\tau$. There are two cases:

Case 1. *Low gain ($\alpha > d$):* In this case there is no intersection between the left- and right-hand sides of equation (9.15). This means that there is no root and hence that there can be no change in stability. Since we have chosen $\alpha > 0$, the pupil light reflex is stable for all possible time delays.

Case 2. *High gain ($\alpha < d$):* In this case there will be a root, since there is an intersection of the graphs described by the left- and right- hand sides of equation (9.15). Thus, provided that the gain and $\tau$ are sufficiently large, there can be a change in stability.

A remarkable fact about a first-order delay differential equation with negative feedback is that when the critical point loses its stability, a stable limit cycle arises, and moreover, this is the only type of stable solution that arises. What is even more interesting is that we can determine the frequency of this limit cycle that arises at the point of instability. To see this we use the Euler relationship

$$e^{-if\tau} = \cos f\tau - i \sin f\tau.$$

If we substitute this into equation (9.15) and equate real and imaginary parts, we obtain

$$\cos f\tau = \frac{\alpha}{\beta} \tag{9.16}$$

and

$$\sin f\tau = -\frac{f}{\beta}. \tag{9.17}$$

Dividing equation (9.16) by equation (9.17) yields

$$\frac{\sin f\tau}{\cos f\tau} = -\frac{f}{\beta} = \tan f\tau. \tag{9.18}$$

From the known properties of the tan function we see that

$$\frac{\pi}{2} < f\tau < \pi,$$

or

$$\frac{\pi}{2\tau} < f < \frac{\pi}{\tau}.$$

In terms of the period, $T$, of the oscillation,

$$T = \frac{2\pi}{f};$$

thus

$$4\tau > T > 2\tau. \tag{9.19}$$

In general, the period of the oscillation generated by a delayed negative feedback control mechanism is $\geq 2\tau$. It is useful to remember these relationships between $\tau$ and $T$, since they provide a clue to determine whether a time delay plays an important role in the generation of an experimentally observed oscillation.

## 9.7  Pupil Cycling

The gain in the pupil light reflex has been estimated to be $\approx 0.12 - -0.16$ (a value of one corresponds to the gain at which the pupil light reflex becomes unstable) (Stark 1959; Stark 1968). Thus under normal physiological conditions we do not expect to see spontaneous, regular oscillations in pupil area. However, the gain in the pupil light reflex can be easily increased by using the simple experimental trick of focusing a narrow light beam at the pupillary margin (Stern 1944). It is instructive to follow the changes in pupil area as a function of the relative positions of the light beam and the iris margin (Figure 9.7). When the pupil has attained its minimal value, the iris blocks light from reaching the retina, and hence the pupil begins to dilate. When pupil area increases sufficiently (horizontal dotted line), the iris no longer shades light from the retina. However, the pupil does not immediately begin to constrict, but continues dilating for a time equal to the time delay for pupillary constriction. Then the pupil begins to constrict, and the iris margin again shields the retina from the light beam. The pupil continues to constrict for a time equal to the time delay for pupillary dilation and then dilates again. If there were no time delays, the position of the iris margin would be fixed by the position of the light beam.

In edge-light pupil cycling, the gain is zero for all pupil areas except when the iris margin corresponds to the position of the light beam. At this point the gain is infinite, since a small change in pupil area dramatically affects the retinal light flux $\phi$. Thus the regular oscillations in pupil size seen under these conditions are also referred to as high-gain pupil oscillations.

The period of pupil cycling under normal lighting conditions is about 900 msec (Loewenfeld 1993). Under these conditions the time delay is $\approx 300$ msec. From equation (9.19) the predicted period of pupil cycling, $T_{\text{pc}}$, is

$$600 \text{ msec} < T_{\text{pc}} < 1200 \text{ msec.} \tag{9.20}$$

Thus the observed period of pupil cycling is in excellent agreement with that predicted from our stability analysis.

Measurements of the period of pupil cycling are used clinically as a simple test to detect pathology within the reflex arc. Pupil cycling occurs even when the sympathetic supply to the iris is cut surgically (Milton, Longtin, Kirkham, and Francis 1988) or blocked pharmacologically. Thus pupil cycling is primarily sensitive to changes in the parasympathetic arc of the pupil light reflex, the latency of the pupil light reflex, and the mechanical properties of the iris. For example, in diseases that affect the optic nerve such as multiple sclerosis, the latency of the reflex is increased (Loewenfeld 1993). This is reflected by an increase in the period of pupil cycling.

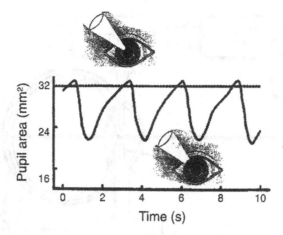

Figure 9.7. Pupil cycling obtained by focusing a narrow light beam at the pupillary margin.

However, the clinical usefulness of measuring pupil cycling to detect optic nerve demyelination is limited by the fact that partial demyelination of axons more typically produces conduction block, and thus it is difficult to sustain regular pupil cycling (Milton, Longtin, Kirkham, and Francis 1988).

Technical problems have limited the usefulness of measurements of pupil cycling (Loewenfeld 1993). Despite its simplicity, it is quite difficult to measure, even with the use of a slit lamp. For example, many patients are distracted by the positioning of a light beam at the edge of their pupil and often will look directly into the beam. Moreover, if the amplitude of the oscillations is small or variable, it can be quite difficult to reliably count the number of cycles over, for example, 30–60 seconds in order to estimate the period.

One technique that overcomes these difficulties is clamping (Figure 9.8). Clamping refers to a technique whereby known feedback is inserted into the reflex arc noninvasively (Longtin and Milton 1988; Milton et al. 1988; Reulen et al. 1988; Stark 1962). First, the reflex arc is "opened" by focusing a light beam onto the center of the pupil that has a diameter smaller than the smallest possible diameter of the pupil (typically $\approx 1$ mm) (Stark and Sherman 1957). Thus, the retinal light flux is

$$\phi = A_{\mathrm{b}}I,$$

where $A_{\mathrm{b}}$ is the cross-sectional areas of the incident light beam, and again, $I$ is the retinal illuminence. Under these conditions changes in pupil size do not affect $\phi$. The feedback loop is then closed again by measuring pupil area with a pupillometer and then using the measured area to change the retinal illuminence $I$.

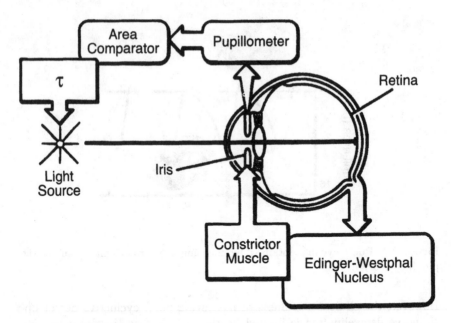

Figure 9.8. The clamped pupil light reflex. See text for discussion. Adapted from Milton and Longtin (1990).

Typically, an analog signal proportional to pupil area is sent to a computer and the computer uses a rule to change the retinal illuminence $I$ based on $A$ ("area comparator" in Figure 9.8). A rule that represents an idealization to edge-light pupil cycling is (see Figure 9.9)

$$H(A_\tau) = \begin{cases} \text{light ON} & \text{if } A_\tau > A_{\text{ref}}, \\ \text{light OFF} & \text{if } A_\tau < A_{\text{ref}}. \end{cases} \tag{9.21}$$

This rule corresponds to piecewise constant negative feedback. The advantage of using clamping to induce pupil cycling is that the light beam is focused onto the center of the pupil rather than at the pupillary margin. Thus pupil cycling is easier to sustain.

For pupil cycling we have, using equations (9.8) and (9.21),

$$\frac{dg}{dA}\frac{dA}{dt} + \alpha g(A) = H(A_\tau). \tag{9.22}$$

If we take

$$g(A) \approx A,$$

then our model for pupil cycling becomes

$$\frac{dA}{dt} + \alpha A = \begin{cases} A_{\text{on}} & \text{if } A_\tau > A_{\text{ref}}, \\ A_{\text{off}} & \text{otherwise}, \end{cases} \tag{9.23}$$

where $A_{\text{on}}, A_{\text{off}}$ are the asymptotic values attained by the pupil, respectively, when the light is on or off indefinitely.

The observations in Figure 9.2 indicate that the rate constant for pupil constriction, $\alpha_c$, is not the same as that for pupil dilation, $\alpha_d$. Thus in order to completely specify the solution of Equation (9.23) it is necessary to specify seven parameters: $A_{ref}, \tau_c, \tau_d, \alpha_c, \alpha_d, A_{on}, A_{off}$. In the discussion that follows we will assume that $\tau_c = \tau_d$ (Loewenfeld 1993; Milton and Longtin 1990); thus there are six parameters to be determined. The time delay, $\tau$, can be measured from the response of the pupil to a light pulse (Figure 9.2) and the parameter, $A_{ref}$ is set by the experimenter. Next we describe how the remaining four parameters can be determined from the solution of equation (9.23).

By inspecting equation (9.23) we see that the solution can readily be constructed by piecing together exponentials. If we assume that the solution has settled onto the limit cycle oscillation, then we can write the solution as

$$A(t) = \begin{cases} A_{on} + [A(t_o) - A_{on}] \exp(-\alpha_c(t - t_o)), & \text{for } A(s - \tau) > A_{ref}, \\ A_{off} + [A(t_o) - A_{off}] \exp(-\alpha_d(t - t_o)), & \text{for } A(s - \tau) \leq A_{ref}, \end{cases}$$
(9.24)

where $s \in (t_o, t)$.

The period $T$ of the oscillation is equal to (see Problem 3 at the end of the chapter)

$$T = 2\tau + t_1 + t_2,$$
(9.25)

$$= 2\tau + \alpha_c^{-1} \ln \left[ \frac{A_{max} - A_{on}}{A_{ref} - A_{on}} \right] + \alpha_d^{-1} \ln \left[ \frac{A_{min} - A_{off}}{A_{ref} - A_{off}} \right].$$
(9.26)

The maximum amplitude of the oscillation, $A_{max}$, can be determined from equation (9.24) by choosing $A(t_o) = A_{ref}$,

$$A_{max} = A_{ref} S_{max} + I_{max},$$
(9.27)

where

$$S_{max} = \exp(-\alpha_d \tau),$$
$$I_{max} = A_{off}(1 - \exp(-\alpha_d \tau)).$$

In a similar manner we can write for the minimum amplitude of the oscillation, $A_{min}$,

$$A_{min} = A_{ref} S_{min} + I_{min},$$
(9.28)

where

$$S_{min} = \exp(-\alpha_c \tau),$$
$$I_{min} = A_{on}(1 - \exp(-\alpha_c \tau)).$$

The importance of equations (9.27) and (9.28) is that they show that the four remaining parameters can be chosen by measuring $A_{max}$ and $A_{min}$ during pupil cycling as a function of $A_{ref}$ (see Figure 9.9B).

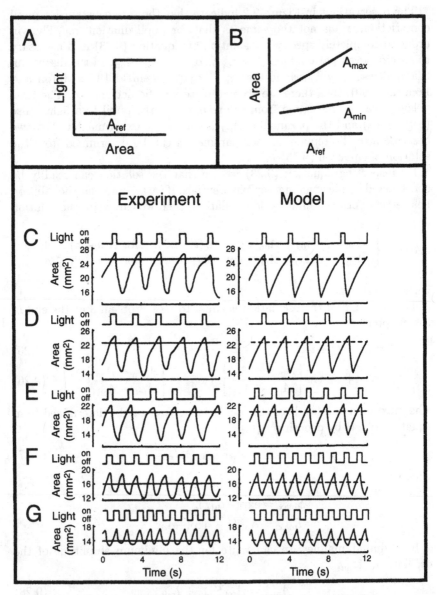

Figure 9.9. Comparison of pupil cycling in the clamped pupil light reflex to that predicted by equation (9.23) using the area comparator shown in (A). The values of the parameters were determined by plotting the minimum and maximum amplitudes of the oscillation versus $A_{ref}$ in (B) as described in the text. The value of $A_{ref}$ is presented by the horizontal dotted line and was set at (C) 25.0 mm$^2$, (D) 22.5 mm$^2$, (E) 20.1 mm$^2$, (F) 16.2 mm$^2$, and (G) 14.0 mm$^2$. For this subject the pupil latency was 280 msec. There was a machine delay of 100 msec and thus the total time delay was 380 msec.

Figure 9.9 compares pupil cycling observed for different values of $A_{\text{ref}}$ to that predicted by equation (9.23). The predicted amplitude and period of the oscillations agree to within 5–10%. The essential difference between the predictions of the model and experimental observations is the presence of the slope discontinuity in the model; i.e., the waveform predicted by the model is not the same as observed experimentally. These observations are consistent with the fact that linear approximations, e.g., equation (9.6), are typically quite good for estimating the period and amplitude of oscillations, but poor for estimating the waveform of the oscillations (Stark 1962). In other words, the effects of the nonlinearities in the reflex arc are expected to appear in the waveform of the oscillations.

The slope discontinuity does not appear when the model for the pupil light reflex is of second order or higher (Bressloff, Wood, and Howarth 1996). Surprisingly, these extensions suggest that dynamics more complex than a limit oscillation may occur, for example, multistability (Campbell et al. 1995b; Campbell et al. 1995a; Foss et al. 1996; Foss and Milton 2000; Foss et al. 1997). Multistability means that multiple attractors coexist. Analytical studies of a harmonic oscillator with delayed piecewise constant negative feedback indicate that qualitatively different limit attractors having the same period exist (an der Heiden et al. 1990; an der Heiden and Reichard 1990; Bayer and an der Heiden 1998; Milton et al. 1998). These limit cycle attractors differ with respect to the shape of the waveform. Preliminary experimental measurement indicate that different waveforms can indeed be detected during pupillary cycling (Milton, Bayer, and an der Heiden 1998).

Despite the temptation to extend a model for pupil cycling beyond that of equation (9.23), it must be kept in mind that the approximation of pupillary movements during constriction and dilation as simple exponential processes (equation (9.6)) is actually quite robust. For example, suppose we replace $H(A_\tau)$ in equation (9.21) by the more complex feedback, referred to as piecewise constant mixed feedback,

$$H(A_\tau) = \begin{cases} \text{light OFF} & \text{if } A_\tau < \theta_1, \\ \text{light ON} & \text{if } \theta_1 \le A_\tau < \theta_2, \\ \text{light OFF} & \text{otherwise,}. \end{cases}$$

where $\theta_1, \theta_2$ are two different values of the area threshold. Equation (9.22) with this choice of $H(A_\tau)$ is of particular interest since it is a piecewise constant version of the Mackey–Glass equation (an der Heiden and Mackey 1982; Mackey and Glass 1977). It can be shown analytically that this piecewise constant delay equation produces a variety of complex oscillations, including chaos (an der Heiden and Mackey 1982). Experimentally excellent agreement is seen between the simpler bifurcations and those predicted to occur (Milton et al. 1989). However, the more complex waveforms are difficult to observe experimentally due to the effects of parametric noise in the pupil light reflex. Thus it is not possible to confirm the prediction that chaotic dynamics arise in the paradigm.

## 9.8   Localization of the Nonlinearities

The order, i.e., the highest derivative, of the differential equation that describes the pupil light reflex can be determined experimentally by evaluating the transfer function. Mathematically, the transfer function of a linear system is defined as the ratio of the Laplace transform of the output to the Laplace transform of the input when all initial conditions are zero. Experimentally, the transfer function of the pupil light reflex can be measured by sinusoidally modulating the light input at frequency $f_s$ and then measuring the amplitude and phase lag of the resultant oscillations in pupil area under open loop conditions as a function of $f_s$. When this experiment was done it was found that the open-loop transfer function was of third order (Stark 1959). This means that the linear properties of the pupil light reflex are described by a third-order delay differential equation of the form

$$\frac{d^3\phi(t)}{dt^3} + \delta\frac{d^2\phi(t)}{dt^2} + \beta\frac{d\phi(t)}{dt} + \alpha\phi(t) = F(\phi(t-\tau)),$$

where $\alpha, \beta, \delta$ are constants. However, the models we have discussed for the pupil light reflex predict that the transfer function is of first order. It is quite likely that other properties of the pupil light reflex need to be considered in order to account for the experimentally measured transfer function.

The essential nonlinearities in the pupil light reflex arise in either the retina or the iris musculature (Howarth, Bailey, Berman, Heron, and Greenhouse 1991; Inoue 1980; Semmlow and Chen 1977). We have already considered two nonlinearities: (1) the logarithmic compression of light intensities by the retina (Weber–Fechner law, equation (9.3)); and (2) the asymmetry in the rates of pupillary constriction and dilation. A method that can be used to better pinpoint the location of other important nonlinearities in the pupil light reflex is pupil flicker fusion. This test refers to the response of the pupil to trains of square wave light pulses. Clinical observations suggest that pupil flicker fusion is most sensitive to alterations in the mechanical properties of the iris and its musculature (Loewenfeld 1993). Two additional observations support this impression: (1) midbrain neurons are capable of responding to higher frequencies of photic stimulation than the pupil (Gamlin, Zhang, and Clarke 1995; Smith, Masek, Ichinose, Watanabe, and Stark 1970); and (2) in the pigeon, the pupil responds to higher frequencies of square wave light pulses than is possible for human pupils (Loewenfeld 1993). This observation is thought to be due to the fact that the pupillary dilator muscle in pigeon is striated muscle, whereas in the human it is a myoepithelial cell.

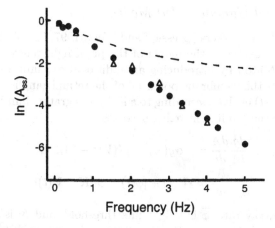

Figure 9.10. Plot of the natural logarithm of the steady-state amplitude, $A_{ss}$, of pupillary oscillations for pupil flicker fusion as a function of the frequency periodic square-wave light pulses (width 100 msec) (•) and electrical stimulation ($\Delta$). To facilitate comparison the amplitudes have been normalized to the values observed for the lowest pulse train frequency (0.05 Hz). Dotted line gives the prediction of equation (9.24).

Figure 9.10 shows the change in the steady-state amplitude of the pupil oscillation as a function of the frequency of a train of brief (100 msec) square wave light pulses. As frequency increases, there is an exponential decrease in the steady-state amplitude. The pupil flicker fusion obtained by direct electrical stimulation of the retina ($\Delta$) is identical to that obtained with light stimulation (•). When an active electrode is placed on a subject's temple near the canthus of the eye and the indifferent electrode on the ipsilateral elbow, neural elements of the retina are stimulated that are located central to the photoreceptors and peripheral to the optic nerve fibers, i.e., possibly the ganglion cells (Welpe 1967). The similarity in pupil flicker fusion elicited by photic and electrical stimuli suggests that the retinal photoreceptors are not a significant nonlinearity in the pupil light reflex (see also comments surrounding equation (9.5)).

The linearity of the semilog plot shown in Figure 9.10 should not be interpreted as evidence indicating that pupil flicker fusion is a linear phenomenon. The dotted line in Figure 9.10 shows the predicted pupil flicker fusion using the piecewise constant model of the pupil light reflex given by equation (9.24). Indeed, it can be shown, using the principle of linear superposition, that there is no linear model of the pupil light reflex that produces the exponential decrease in $A_{ss}$ shown in Figure 9.10. Thus the observations in Figure 9.10 suggest that the remaining nonlinearities reside either at the level of the retinal ganglion cells or the iris musculature.

## 9.8.1 Retinal Ganglion Cell Models

Bressloff and his coworkers (Bressloff and Wood 1997; Bressloff, Wood, and Howarth 1996) extended the model of the pupil light reflex developed by Longtin and Milton by introducing an additional dynamic variable, $V(t)$, that represents the membrane potential of the retinal ganglion cells. By assuming that $V(t)$ evolves according to a leaky-integrator shunting equation, the model for the pupil light reflex becomes

$$\frac{dg}{dA}\frac{dA}{dt} = -\alpha g(A) + \gamma f(V(t - \tau)),$$

$$\frac{dV}{dt} = \epsilon V(t) + [\phi(t) - \overline{\phi}][K - V(t)], \tag{9.29}$$

where $\epsilon$ is a decay rate, $\overline{\phi}$ is a light flux threshold, and $K$ is a membrane reversal potential. Consequently, the effective decay rate depends on both the membrane potential $V(t)$ and the retinal flux $\phi(t)$. This model accounts for the frequency-dependent shift in the average pupil size in response to sinusoidally modulated light better than equation (9.11) (Loewenfeld 1993; Howarth 1989; Howarth et al. 1991; Varjú 1969).

## 9.8.2 Iris Musculature Effects

It is quite likely that the properties of the iris musculature also need to be considered in order to account for a third-order open-loop transfer function. This conclusion is supported by the observation that the response of the pupil to a step input of light depends on both the light intensity and the initial size of the pupil ("pupil size effect")(Krenz and Stark 1984; Krenz and Stark 1985; Loewenfeld and Newsome 1971; Sun and Stark 1983; Sun, Krenz, and Stark 1982; Sun, Tauchi, and Stark 1982).

The term **pupil-size effect** refers to the following phenomenon. When an initially large pupil in the dark is exposed to a step input of dim light, the pupil transiently constricts, then dilates ("pupillary escape").On the other hand, if the step input of light is sufficiently intense, the pupil constricts, but does not subsequently dilate ("pupillary capture"). For step inputs of intermediate brightness, oscillations are superimposed onto escape and capture. In contrast, an initially small pupil is captured by either dim or bright light. The essential role played by pupil size is emphasized by the observations that identical results are obtained whether initial pupil size is adjusted by changing ambient light levels or accommodation (Sun and Stark 1983). The clinical importance of the pupil-size effect (a normal phenomenon) arises in the differential diagnosis of an afferent pupillary defect (a sign of optic nerve disease) (Loewenfeld 1993).

The effects of the iris musculature on the pupil light reflex enter through the choice of the function $M$, i.e., equation (9.6). For example, we would obtain a second-order delay differential equation for the pupil light reflex, by taking

$$E(t) \approx k \left( \frac{d^2x}{dt^2} + \beta \frac{dx}{dt} + \alpha x \right),$$

so that equation (9.11) becomes

$$\frac{dg}{dA} \frac{d^2A}{dt^2} + B(g, dg/dt) \frac{dA}{dt} + \alpha g(A)A = f(\phi_\tau),$$

where

$$B(g, dg/dt) = \frac{d^2g}{dA^2} + \beta \frac{dg}{dA},$$

and $\alpha, \beta$ are constants. It is not difficult in this way to construct models that produce dynamics that qualitatively resemble those of the pupil-size effect (see Ex. 9.13-1). Following this, a third-order model for the pupil light reflex could be obtained by including the effects of the retinal ganglion cells (e.g., equation (9.29)).

## 9.9   Spontaneous Pupil Oscillations?

Although the small gain in the pupil light reflex suggests that spontaneous oscillations in pupil size should not occur, regular oscillations in pupil size have been observed. Figure 9.11A shows pupil size as a subject stares at a fixed target, located at infinity, without blinking under constant lighting conditions. Parasympathetic activity to the pupil is directly proportional to light intensity and thus remains constant. However, sympathetic activity slowly decreases, and thus the pupil slowly decreases in size. As can be seen, at a critical point very regular oscillations in pupil size appear.

The etiology of these oscillations is currently unknown. The current interpretation is that these oscillations arise because the subject has become drowsy, or at the very least inattentive. As such, the study of these oscillations is currently attracting a great deal of attention for the possible development of an inexpensive alternative to polysomnography for diagnosing and following the response to treatment of patients with certain sleep disorders (O'Neill, Oroujeh, Keegan, and Merritt 1996; O'Neill, Oroujeh, and Merritt 1998; Wilhelm, Lüdtke, and Wilhelm 1998; Wilhelm, Wilhelm, Lüdtke, Streicher, and Adler 1998; Yoss, Moyer, and Hollenhurst 1993). Thus it is possible that these pupil oscillations represent the input of a "sleep oscillator" into pupil reflex arc.

However, the subject in Figure 9.11 denied feeling drowsy. The observations in Figure 9.11B raise the possibility that these oscillations are generated as a consequence of intrinsic mechanisms within the reflex arc itself. The application of a very small step increase in retinal illuminance to further decrease pupil size completely abolishes the oscillation. Moreover, the oscillations quickly return when the step of light is turned off.

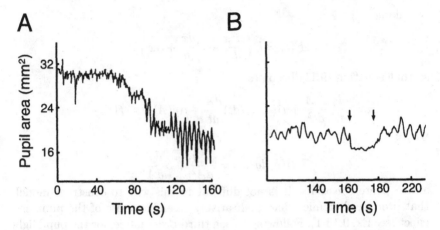

Figure 9.11. (A) Time-dependent changes in pupil area as an alert subject stares at an object without blinking. Prior to these measurement each eye received two drops of local anesthetic to minimize blinking (0.5% proxymetacrine plus artificial tears. (B) Effect of a small step increase in retinal illuminance ($\approx 3$ trolands) on spontaneous pupil oscillations. See text for discussion.

These observations suggest that pupil size is the critical determinant for the occurrence of the oscillations, not simply a "sleepy" state.

It is possible that the mechanism for these oscillations could be intrinsic to the iris musculature. For example, the sum of the length–tension diagrams for the two opposing muscles in the iris is nonlinear (Usui and Hirata 1995), and in particular, a "cubic nonlinearity" can arise if the length–tensions of the two opposing muscles are properly arranged. The presence of such a nonlinearity in the iris musculature would provide a simple explanation for a limit cycle-type oscillation under certain conditions.The computer exercise Ex. 9.13-1 shows that phenomena very similar to those observed for the pupil-size effect can be generated by the van der Pol equation: a nonlinear ordinary differential equation possessing a cubic nonlinearity.

## 9.10   Pupillary Noise

Figure 9.12 shows that small-amplitude, irregular fluctuations in pupil diameter can be detected for an alert subject under conditions of open-loop constant illumination. The source of these fluctuations is thought to be multiplicative (i.e., parametric) noise that is injected into the pupil light reflex at the level of the midbrain nuclei (Stanten and Stark 1966; Stark, Campbell, and Atwood 1958). There is ample evidence to suggest that the observed dynamics reflect an interplay between noise and the intrinsic properties of the reflex arc. For example, the amplitude, and possibly the

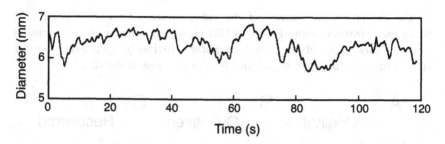

Figure 9.12. Spontaneous fluctuations in pupil diameter monitored for an alert subject under conditions of constant illumination.

frequency content, of these irregular fluctuations in pupil size are modified by the properties of the iris musculature, in particular by the "expansive range nonlinearity" (Longtin and Milton 1989b; Stanten and Stark 1966; Usui and Stark 1978). The smoothness of the waveforms of the oscillations observed during pupil cycling reflects the powerful band-pass filtering properties of the limit cycle attractor (Kosobud and O'Neill 1972; Stark 1959). Spontaneous fluctuations in the waveforms of pupil cycling have been suggested to reflect noise-induced switches between coexistent attractors (Milton, Bayer, and an der Heiden 1998) as has been observed for human postural sway (Eurich and Milton 1996). The pupillary dynamics measured when the pupil light reflex is clamped with piecewise constant mixed feedback are dominated by the effects of parametric noise that causes switches between attractors (Milton et al. 1989; Milton and Foss 1997).

Chapter 8 deals with the effects of noise on the dynamics of time-delayed feedback control mechanisms. Here we focus on two practical aspects for the study of pupillary noise: (1) How can small amplitude fluctuations be measured accurately?; and (2) What are the effects of noise on parameter estimation?

## 9.10.1 Noisy Pupillometers

In present-day pupillometers, the pupil is imaged by the charge-coupled device (CCD) of a video camera, and an analog voltage proportional to pupil size, is digitized by an analog-to-digital (A/D) board in order to enter the data into a computer. It is not known whether the physiologically relevant measurement of pupil size is diameter or area (Loewenfeld 1993). Both the form, and consequently the dynamics produced by equations (9.11) and (9.29), would be greatly altered if they were written in terms of diameter rather than area.

There is a technical problem inherent in the measurement of small fluctuations in pupil diameter. The problem arises because the magnitude of the small fluctuations in pupil size approach the limits of resolution of the pupillometer. Resolution is limited because of the finite size of the pixel on the charge-coupled device of the video camera. This error is referred to as quantization error. Figure 9.13B shows the effects of quantization error on measurements on pupil diameter. Obviously, quantization error significantly distorts the signal and results in a loss of detail.

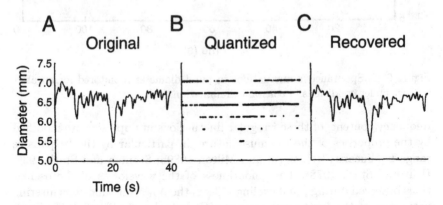

Figure 9.13. (A) Pupil diameter measured with a pupillometer with high resolution and low noise level. (B) The original signal was then quantized with added Gaussian distributed noise before quantization into six bins to simulate a noisy pupillometer with relatively poor resolution. (C) The signal was recovered using a lowpass filter. Details of the filter design are given in Hunter et al. (2000).

One solution is to increase the resolution of video-pupillometers by using charge-coupled devices having higher numbers of pixels. However, a less expensive and easier solution is to use a "noisy pupillometer" (Hunter et al. 2000). Much of the signal detail by adding noise to the signal at the point before it is digitized is recovered using a technique referred to as dithering (see computer exercise Ex. 9.13-2). The essential requirement for dithering is that the time series be oversampled in the presence of noise. The major frequency components of the open-loop irregular fluctuations in pupil size are $\leq 0.3$ Hz (Longtin, Milton, Bos, and Mackey 1990; Stark, Campbell, and Atwood 1958), and the sampling frequency of video-pupillometers is $\geq 25$ Hz. Adding a small-amplitude noisy component to the "true" pupil will result in a switching between two different quantization levels on the A/D board. If the frequency content of the noise is much higher than that for the primary signal, then over short intervals the average population of the two quantization levels is a measure of how close the true signal is to the threshold that separates bins on the A/D board. Thus it follows that the true signal can be recovered by lowpass filtering (equivalent to averaging) the quantized data.

Figure 9.13 illustrates this procedure (see also Exercise Ex. 9.13-2). An analog input signal proportional to pupil diameter (Figure 9.13A) is discretized in the presence of noise at 25 Hz (Figure 9.13B). The quantized signal is then lowpass filtered (for details see Hunter et al. 2000) to yield a time series that closely resembles the original time series (compare Figures 9.13A and 9.13C). The lowpass filtering smoothes the original time series, i.e., the high-frequency components have been removed. Thus measurements of pupil diameter cannot be used to detect the presence of low-amplitude, high-frequency fluctuations in pupil size. Obviously, pupillometers that directly measure pupil area would be better suited for this task.

## 9.10.2  Parameter Estimation

There is a fundamental uncertainty in the accuracy by which a parameter can be estimated. To illustrate how serious a problem this can be, consider the question of estimating the delay for light offset, $\tau_d$, for pupil cycling (Milton et al. 1993).

If $\tau_c \neq \tau_d$, then equation (9.27) becomes

$$A_{max} = A_{off} [1 - \exp(-\alpha_d \tau_d)] + A_{ref} \exp(-\alpha_d \tau_d). \tag{9.30}$$

Thus $\tau_d$ can be calculated, provided that $\alpha_d$ is known. A variety of strategies for measuring $\alpha_d$ can be derived by analyzing equation (9.24) (Milton, Ohira, Steck, Crate, and Longtin 1993). For example, when the pupil area equals $A_{ref}$ we have

$$A_{ref} = A_{off} + K_d \exp(-\alpha_d \tau_d), \tag{9.31}$$

where $K_d$ is a constant. But we also have

$$\frac{dA_d}{dt} = -\alpha_d K_d \exp(-\alpha_d \tau_d), \tag{9.32}$$

and hence

$$\frac{dA_d}{dt} = \alpha_d A_{off} - \alpha_d A_{ref}, \tag{9.33}$$

where $d(A_d)/dt$ is $dA/dt$ measured from pupil dilation. Thus a plot of $d(A_d)/dt$ versus $A_{ref}$ will be linear with slope $-\alpha_d$. This is observed experimentally (data not shown).

It can be shown that $\Delta \tau_d$, the uncertainty in $\tau_d$, is (Milton et al. 1993)

$$\Delta \tau_d = -\left[ \frac{1}{\alpha_d S_{max}} \right] \Delta S_{max} - \left[ \frac{1}{\alpha_d^2} \ln \frac{1}{S_{max}} \right] \Delta \alpha_d. \tag{9.34}$$

In two trials it was found that $\alpha_d$ was, respectively, 0.45 sec$^{-1}$ and 0.40 sec$^{-1}$ and that $S_{max}$ was 0.85 and 0.87. Although the variation in $\alpha_d$ and $S_{max}$ is $\leq 10\%$, the uncertainly in calculating $\tau_d$ is 100 msec! This uncertainty in the value of a parameter underscores the difficulty faced by those who wish to construct more detailed models for the pupil light reflex.

# 9.11    Conclusions

In 1984, Lawrence Stark (1984) proposed that the pupil light reflex was *the* paradigm of a human neural feedback control mechanism. He suggested that under appropriate conditions, phenomena generated by the pupil light reflex would likely be observed in other neural reflexes as well. How far have we progressed in our understanding of human neural feedback control mechanisms?

Two thousand years before the birth of Christ, the Chinese had figured out the use of belladonna alkaloids to produce a fixed, dilated pupil. The motivations for this work were sex (big pupils were attractive) and politics (monitoring changes in pupil size to determine whether someone was lying does not work if pupil size is fixed).At about the time of the birth of Christ, the Romans had figured out the use of anticholinergic drugs to constrict the pupil in order control the pressure in the eye following surgical removal of a cataract. The anatomy of the pupil light reflex was worked out in about 980 A.D. The recent treatise by Loewenfeld on the human pupil (Loewenfeld 1993) is in two volumes that together measure nearly five inches in thickness. The second volume contains only the references. In fact, there was a third volume of references that the publisher refused to publish!

However, despite nearly four thousand years of experimental and theoretical investigations, I feel that the pupil light reflex still has lessons to teach us. In this chapter I have tried to point to many unanswered questions that beg to be solved. My hope is that some readers will pick up the gauntlets I have thrown down and complete the study of this important neural control mechanism.

# 9.12    Problems

1. Consider the time-delayed first-order differential equation

$$\frac{dx(t)}{dt} + x(t - \tau) = 0,$$

   where $\tau$ is the time delay. There is a fixed point at $x(t) = 0$. For what value of the time delay does the fixed point become unstable?

2. The Pulfrich phenomenon illustrates the effects of a delay on neural dynamics (Lit 1960; Sokol 1976). Make a simple pendulum by attaching a paper clip to a string and allow the pendulum to oscillate to and fro in a frontal plane. Cover one eye with a filter that reduces the light intensity. The target will now appear to move in an elliptic path. The explanation for this stereo phenomenon is that the covered eye

is more weakly stimulated than the uncovered eye, and this results in a delay of the transmission of visual signals to the visual cortex from the covered eye. This disparity in transmission time between the two eyes is interpreted by the brain as a disparity in space, thus producing the stereo illusion. Based on this background,

- When the filter is placed over the right eye, do you expect that the rotation goes clockwise or counterclockwise? What about when the filter is placed over the left eye? Verify these predictions experimentally.

The Pulfrich phenomenon has been extensively studied (see, for example, Lit (1960) and references therein). An ambitious reader who wishes to construct a Pulfrich apparatus should consult Sokol (1976).

3. equation (9.26) shows that the period of the oscillations in pupil area that occur during pupil cycling can be broken into four contributions. Sketch the solution of equation (9.24) and identify the portion of the period accounted for by each of the terms in equation (9.26).

## 9.13 Computer Exercises: Pupil-Size Effect and Signal Recovery

### Data File

**pupil.dat** Data file of pupil area samples at 60 Hz with a high-resolution pupillometer

### Software

There are 5 Matlab[†] scripts you will use for these exercises:

**plot_vdpol.** An example of how to do fourth-order Runge Kutta integration of the van der Pol equation for specified parameters (calls the function **vdpol_pars** to set the parameters and the function **vdpol** to compute the derivatives).

**vdpol.** A function that returns the derivatives of the van der Pol equation (called from the script **plot_vdpol**).

**vdpol_pars.** A function to set the default parameters for the van der Pol equation (called from the script **plot_vdpol**).

**quantize_data.** A script that loads the data in **pupil.dat** and creates the quantized data in **quantized.dat** and **quantized_nse.dat**.

---

[†]See Introduction to Matlab in Appendix B.

**recover.** A script that loads the quantized data and applies a low-pass filter to it to try and recover the true signal. The plots compare the results with pupil.dat.

Many thanks to John D. Hunter for his work on the Matlab programs used in these computer exercises.

## Exercises

**Ex. 9.13-1. Dynamics in "pupil-size effect."** A mathematical model that captures qualitatively the dynamics seen in the "pupil-size effect" is the van der Pol equation

$$\frac{d^2x(t)}{dt^2} - k\left[d^2 - x^2(t)\right]\frac{dx(t)}{dt} + a^2x(t) = bu_0(t),$$

where $a, b, k, d$ are positive constants and $u_0(t)$ is a step function, i.e.,

$$u_0(t) = \begin{cases} 0 & \text{if } t < 0, \\ 1 & \text{if } t > 0. \end{cases}$$

Assume that the initial conditions are $x(0) = 5$ and $dx/dt|_{t=0} = 0$ and take $d^2 = k = 0.05$ and $a = 1$. Use a computer program to calculate $x(t)$ when $b$ has the values 10, 4.5, 3, 0, $-3$, $-5$, $-10$, and compare these results to the discussion in Section 7.7.2. (The student can use the m-file **plot_vdpol.m** to integrate this equation for a single value of $b$. The student does not need to alter the other two files, **vdpol.m** and **vdpo.l_pars.m**.) Now investigate the stability of the fixed points for this equation and see whether you can understand the results on the computer simulations.

**Ex. 9.13-2. Signal recovery.** The effects of dithering on the recovery of a signal distorted by quantization effects is quite amazing (Figure 9.13). A data file of pupil size versus time sampled at 60 Hz using a high-resolution pupillometer is provided in the file **pupil.dat**. By using the m-file **quantize_data.m** you can mimic the effects of using a pupillometer with a much lower resolution. Run this program to create two new data files from pupil.dat: **quantized.dat** is the original pupil data quantized into discrete values to show what a continuous signal would look like sampled by a machine with poor resolution. The data file **quantized_nse.dat** is the same as **quantized.dat** except that Gaussian-distributed white noise has been added at the time of the quantization. The m-file **recover.m** loads the quantized data files and performs a low-pass filtering to recover the true signal. The plots compare the results with **pupil.dat**. Using these m-files as well as m-files that you write yourself, answer the following questions:

(a) Compute the RMS error between the recovered quantized signals that are generated by **recover.m**. The root mean square is defined as sqrt(mean(x.^2)).

(b) Write a script that computes the RMS error of the recovered noisy signal as a function of the low-pass corner frequency (the variable cf in **recover.m**). The corner frequency is the frequency at which the filter starts to seriously attenuate the filtered signal, typically defined as 3 decibels of attenuation. The file **recover.m** contains an example of how to define a low-pass filter in Matlab from the corner frequency, stop frequency, and attenuation parameters. Plot the RMS error (error is original signal minus recovered signal) as a function of the corner frequency. You may want to compare your results for the corner frequency with the example provided in the figure pupil_dither.ps

(c) Write a script that varies the noise intensity before quantization (see, for example, **quantize_data.m**). Plot the RMS error of the recovered signal as a function of the noise intensity.

(d) What are good choices for the low-pass corner frequency and the noise intensity? Is there a relationship between the ideal noise intensity and the quantization interval?

For more details concerning the optimal choice of the noise intensity and the cutoff intensity for the low-pass filter, the interested reader is referred to Gammaitoni (1995) and Hunter et al. (2000).

## 9.14 Computer Exercises: Noise and the Pupil Light Reflex

This section proposes simulations of the delay-differential equation

$$\frac{dA}{dt} = -\alpha A + \frac{c}{1 + \left[\frac{A(t-\tau)^n}{\theta}\right]} + k \tag{9.35}$$

in the presence of noise, either on the parameter $c$ or the parameter $k$. The noise is not Gaussian white noise, but rather colored (yet still Gaussian) Ornstein–Uhlenbeck noise. It is defined by two parameters, the noise intensity $D$ and the correlation time $t_{cor}$.

The integration time step is determined as a fraction of the time delay $\tau/N$. A good choice for the problem here is to choose $N = 100$. You can alter all parameters of the simulation; in particular, you can increase the value of the parameter $n$ to produce a Hopf bifurcation in the deterministic equation. This increases the slope of the feedback function around the fixed point, which becomes unstable if this slope is too high.

## Software

There are two Matlab[‡] programs, which run similarly to the programs **langevin.m** and **fhnnoise.m** in Chapter 6 on the effect of noise on non-linear dynamics (see page 184), except for the fact that here we have a delay-differential equation. The integration method is Euler–Maruyama of order one (i.e., integral Euler–Maruyama).

> For the program **pupilc.m**, in which the noise is on the parameter $c$, other parameters are set to (but can be modified): $\alpha = 3.21$, $\theta = 50.0$, $k = 0.0$, $\tau = 0.3$, and the average value of $c$ is $\bar{c} = 200$.

> For the program **pupilk.m**, in which noise is on the parameter $k$, parameters are as in **pupilc.m**, except that $c = 200$, and the average value of $k$ is $\bar{k} = 0.0$.

## Exercises

**Ex. 9.14-1. Hopf bifurcation in the absence of noise as $n$ increases.** In the absence of noise (noise intensity equals zero), pinpoint the Hopf bifurcation in the equation as $n$ is increased. Other parameters are $\tau = 0.3, \langle c \rangle = 200, \theta = 50, \alpha = 3.21, k = 0$.

**Ex. 9.14-2. Critical slowing down.** Satisfy yourself that even in the absence of noise, the transients are more difficult to get rid of as the bifurcation point is approached from either side (again varying $n$ as above). This is an example of "critical slowing down."

**Ex. 9.14-3. Effect of weak noise on the dynamics.** Investigate the effect of a bit of noise on $c$ or $k$ on the dynamics in the vicinity of the bifurcation point. You should find that oscillations are now visible even when the bifurcation parameter is set below its deterministic bifurcation value.

**Ex. 9.14-4. Effect of the noise correlation time.** Investigate the effect of the correlation time of the noise on the solutions. You can also study what happens when $t_{cor}$ is varied while keeping constant the variance $D/t_{cor}$ of the Ornstein–Uhlenbeck noise.

**Ex. 9.14-5. Stochastic bifurcation point.** Compute densities (via histograms) for the solution for different values of $n$. Try to estimate the value of $n$ for which the distribution becomes bimodal (this is the stochastic bifurcation point), and compare it to the deterministic bifurcation point $n_o$.

**Ex. 9.14-6. Limit of large $n$ in the absence of noise.** What happens in the limit where $n$ is very large (e.g., $> 30$) in the absence of noise?

---

[‡]See Introduction to Matlab in Appendix B.

Why is the piecewise constant negative feedback case different from the smooth negative feedback case? You should find that the solution looks, at any given time, like an increasing or decreasing exponential. With noise, you should now find that the period fluctuates more strongly than the amplitude, i.e., the opposite of the case for low $n$.

# 10

# Data Analysis and Mathematical Modeling of Human Tremor

**Anne Beuter**
**Roderick Edwards**
**Michèle S. Titcombe**

## 10.1 Introduction

Oscillatory behavior is a common form of normal biological function. The motor system contains many forms of oscillatory behaviors, and surprisingly, both the causes and the mechanisms of these rhythmicities are often poorly understood. For example, when reading the literature on human tremor, one is struck by the limitations of our current knowledge regarding its origin and mechanisms, even though this common neurological sign has been studied for centuries (Galen, second century A.D., cited in Garrison 1969; de la Boë (Sylvius) 1675; Sauvages, 1763, cited in Capildeo 1984, p. 286; Parkinson 1817; Charcot 1877; Eshner 1897; Binet 1920; Walshe 1924).

Section 10.2 attempts to summarize what is currently known about tremor and is organized in five parts, including (1) the definition, classification, and measurement of tremor; (2) the physiology of tremor; (3) the characteristics of tremor observed in patients with Parkinson's disease; (4) the conventional methods used to analyze tremor; and (5) a summary of the initial attempts that have been proposed to model human tremor. Section 10.3 outlines some linear time series analysis techniques that have been used to characterize tremor data in terms of their amplitude, frequency, amplitude fluctuations, and how closely they resemble a sinusoidal oscillation. This section also discusses whether to record the displacement, velocity, or acceleration of tremor. Details of concepts used in time series analysis (e.g., the power spectrum) can be found in Appendix C. Section 10.4 describes how one can use deviations of tremor time series from linear stochastic processes to discriminate between types of tremor. Section 10.5

discusses mathematical models pertaining to Parkinsonian tremor. Finally, Section 10.6 reviews how to apply techniques of time series analysis to pathological tremor data.

## 10.2  Background on Tremor

### 10.2.1  Definition, Classification, and Measurement of Tremor

Tremor is defined as an approximately rhythmical movement of a body part (face, jaw, palate, eyes, trunk, extremities). Typical tremor of the extremities is involuntary, irregular, and continuous, and its frequency and amplitude may fluctuate. Normal physiological or enhanced physiological tremor of the extremities often looks noisy and irregular with broad power spectra, while pathological tremors, in general, tend to be more regular, slower, less noisy, and with spectra containing sharper peaks. The power spectrum provides information on the frequency content of the recorded tremor signal: A broad spectrum indicates that many frequencies contribute to the signal, and a concentrated peak in the spectrum indicates that there is one main frequency to the tremor (see Appendix C for more detail on power spectra).

Since 1993 there have been two main classifications of tremor in use. The first one is based on the state of activity of the body part when tremor is observed, and the second is based on the etiology of the underlying disease or condition. As reported by Bain (1993) and Deuschl et al. (1998), the classification of tremor by state of activity includes:

1. Rest tremor occurring when relevant muscles are not activated and the body part is fully supported against gravity; and

2. Action tremor occurring when relevant muscles are activated, which includes postural, kinetic, isometric, intention, and task-specific tremors.

Postural tremor is apparent during the maintenance of a posture that is opposed to gravity. Kinetic tremor is observed during any kind of movement. Intention tremor occurs when there is an exacerbation of kinetic tremor toward the end of a goal-directed movement. Isometric tremor occurs when a voluntary muscle activity is opposed by a rigid stationary object. Finally, there also exist several task-specific tremors that occur during the performance of highly skilled activities such as writing, playing a musical instrument, using a specific tool, and speaking.

The second classification of tremor is based on etiology and includes:

1. enhanced physiological tremor, i.e., an increase in tremor amplitude caused by various factors (listed below);

2. tremor observed in:

(a) hereditary, degenerative and idiopathic diseases such as Parkinson's disease (described in Section 10.2.3), or

(b) benign essential tremor, a hereditary condition occurring in 0.3 to 4.0% of the population, one half of the cases with an autosomal dominant trait and causing rest or action tremor of the head, neck, voice, and upper extremities, or

(c) dystonic tremor, which is hereditary or idiopathic and is associated with jerky spasms during the performance of highly skilled acts;

3. neuropathic tremor, which occurs in some patients with peripheral neuropathy, such as Charcot–Marie–Tooth or Guillain–Barré syndromes, in the form of action tremor and sometimes rest tremor;

4. midbrain tremor, also called rubral or Holmes's tremor, is often seen in multiple sclerosis or brainstem vascular disease in the form of rest and action tremors;

5. cerebellar tremor, which typically takes the form of the kinetic and especially intention tremor and is often associated with other motor signs such as dysmetria;

6. drug induced tremors, including the well-known effect of prolonged heavy ingestion of alcohol (some of the other drugs are listed below);

7. orthostatic tremor, which corresponds to *rapid* tremor of the legs and trunk appearing only on standing and stopping, or when the subject walks (severe cases) or leans against a support;

8. psychogenic tremor in the form of rest or action tremor that can change in amplitude or frequency and fluctuate quickly;

9. tremor in metabolic diseases (e.g., hyperthyroidism).

Although tremor is one of the most common pathologic signs in neurology, its characteristics are generally too nonspecific, and there is too much overlap within these characteristics to associate a specific neuropathology with each form of tremor (Elble and Koller 1990; Gresty and Buckwell 1990; Findley and Cleeves 1989).

Some 37 factors have been reported to have an influence on tremor. These include the part of the body tested, the position of the extremity, the amount of work done by the limb both prior to and during testing, the weight of the extremity used, voluntary tensing of the measured extremity, the presence of hypertonus or of neural damage, the subject's age, the state of consciousness, emotional state, toxic metabolic chemical substances, peripheral sensory impressions, respiration, stuttering, time of day, and room temperature (Wachs and Boshes 1966, p. 67). Circumstances that exacerbate or induce tremor (Hallett 1991) include:

1. drugs such as neuroleptics, amphetamines, cocaine, lithium carbonate, beta-adrenergic agonists, methylxanthine (caffeine), anticonvulsants, and tricyclic antidepressants;

2. withdrawal from alcohol;

3. toxic substances (such as mercury, manganese, lead, arsenic, copper, toluene, dioxin);

4. emotion (anxiety, stress, fear, anger);

5. exercise;

6. fatigue;

7. hypoglycemia;

8. hypothermia;

9. thyrotoxicosis (manifestations associated with an excess of thyroid hormones);

10. pain;

11. trauma.

What is not clear is whether all of these circumstances induce changes in tremor that are similar in terms of time- and frequency-domain characteristics. Further information on factors influencing tremor can be found in the *Consensus Statement of the Movement Disorder Society on Tremor* (Deuschl et al. 1998).

There are several qualitative and quantitative measures of tremor. Qualitative measures include tremor-rating scales such as the Webster Scale (1968) and the Unified Parkinson's Disease Rating Scale (1987), which tend to be limited in reliability and validity (Elble and Koller 1990). Hand coordination tests such as drawing spirals can also be used, but they do not provide direct measures of tremor. Quantitative measures of tremor examine variables such as displacement, velocity, acceleration, electromyography (surface recording or single motor unit studies), and force.

Since the 1880s numerous techniques have been used to record tremor (Elble and Koller 1990). Initially, displacements were measured with a tambour applied to the limb and transmitted to a smoked rotating drum. Angular displacements were measured using potentiometers mounted in goniometers or attached to a joy stick. Force was measured using ultrasensitive semiconductor strain gauges. To eliminate the limited range of motion of most systems, some researchers have used photochronographic techniques and cinephotography. Since its introduction in 1956, the accelerometer has become the most popular technique used in tremor research and neurology clinics. The smallest accelerometers are piezoresistive devices including a small seismic mass at the end of a cantilever beam on which strain gauges are attached in a Wheatstone bridge arrangement. A voltage proportional to acceleration is generated, amplified, and filtered. Sensitivity of accelerometers can reach $0.01$ mV/cm/s$^2$ (Elble and Koller 1990). Miniature accelerometers have also been used to record tremor by telemetry (actigraph) and to gain better insight into the fluctuations of tremor over long periods of time (van Someren et al. 1996; van Someren et al. 1998).

Electromyography (EMG) is also a popular technique for recording tremor, especially when information regarding tremor frequency and motor unit synchronization is desired. Surface electromyography is adequate to record activity of large muscles located right under the skin. In principle, the amplitude of electromyographic recordings is inversely proportional to the distance between the motor unit and the electrode, but it is attenuated by the skin and other soft tissues. Recording electromyography from small neighboring muscles such as hand muscles or from deep muscles causes crosstalk between muscles and is better done using wire and needle electrodes, although artifacts caused by voluntary movement may still be a problem. Electromyographic signals must be rectified and filtered, and the resulting signal is supposed to be proportional to force, and therefore acceleration, and fluctuations in electromyographic signals are associated with fluctuations in force causing tremor. However, these relationships contain nonlinearities, and the relationship between tremor amplitude and surface electromyography is still unclear, especially when the signals are weak (Norman et al. 1999; Timmer et al. 1998a). When motion is limited or absent, tremor frequency can also be estimated by recording single motor unit spike trains or sensory afferent discharges (Elble and Koller 1990).

Beuter et al. (1994) proposed a new technique using semiconductor lasers to quantify displacement with a high resolution ($2\sigma = 15$ microns) that requires no contact between the recording system and the subject. The laser system transducing displacement is placed at the midpoint of the recording range from the target, which is a lightweight card attached to the fingernail of the index finger. Later, Norman et al. (1999) compared recordings of tremor from a displacement laser, a velocity laser, an accelerometer, and from electromyography. The laser transducing velocity is a low-power helium–neon laser in which the beam is split, with one part directed at the target and the other (reference beam) directed at a rotating disk inside the laser. Backscattered light from the rotating disk is used to determine the sign of the velocity signal. Vibration of the target is detected and converted into a calibrated output voltage that is proportional to the target velocity.

As we have just seen, the transducers used to record tremor are varied. In addition, the anatomical location of recording, the types of tremor recorded, and the duration of the recording vary widely across studies (Beuter and de Geoffroy 1996). The consequence of this lack of standardization is that the validity of the results and the general conclusions drawn about discriminating different types of tremor have been severely limited.

## 10.2.2  Physiology of Tremor

According to Elble (1986) normal tremor contains two distinct oscillatory components. The first component is produced by the underdamped limb combined with the stretch reflex (see Figure 10.1). It has a frequency determined by the mass and stiffness of the limb. The passive mechanical properties of body parts are a source of oscillation when they are perturbed by external or internal forces (Elble and Koller 1990). External mass loads tend to decrease the frequency of this tremor, while elastic loads tend to increase it. As Elble (1986) indicates, this tremor component is not associated with electromyographic bursts of activity in the involved muscles, but there is reflex modulation of motoneuron firing when tremor becomes large (e.g., fatigue, emotion, or drugs).

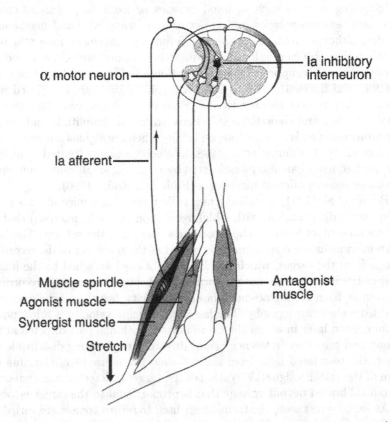

Figure 10.1. When the muscle is stretched, the Ia afferent fibers increase their firing rate. They make monosynaptic excitatory connections to α motor neurons innervating the agonist (or homonymous) muscle from which they arise and to motor neurons innervating synergist muscles. However, they inhibit motor neurons to antagonist muscles through an inhibitory interneuron. Adapted from Kandel, Schwartz, and Jessell (1991).

The exact role of the stretch reflex in this tremor component is still unclear. While some researchers believe that the stretch reflex enhances tremor, others think that it suppresses it. Timmer et al. (1998b) find that reflexes only modify peaks in the power spectra that are present from other causes. Stiles and Pozos (1976) have called this component of tremor the mechanical-reflex component of tremor. As Elble (1986) points out, this component has been considered as a segmental phenomenon, but it may involve transcortical and transcerebellar feedback pathways as well. Cardioballistics, breathing, external perturbations and irregularities in unfused motor unit contractions continually force this system and produce tremor in everybody.

The second component is called the 8 to 12 Hz component (Elble 1986). Elble associates this component with intense synchronous modulation of motor unit spike trains at 8–12 Hz, regardless of mean motor unit firing frequency. As opposed to the first component, this 8–12 Hz component is resistant to frequency changes. Elble (1986) indicates that inertial, elastic and torque loads, as well as limb cooling, produce less than 1–2 Hz frequency change. However, its amplitude can be modified by manipulating either the stretch reflex or limb mechanics. The synchronous modulation of motor unit spike trains and the apparent stability of frequency led Elble to suggest that a neuronal oscillator influenced by sensory information with an unknown location in the central nervous system is responsible for this tremor component.

As opposed to the mechanical reflex component, the 8–12 Hz component varies between subjects, and in some cases this component is hardly visible if at all. It also seems to be influenced by age (Bain 1993) and is consistently observed in deafferented limbs (Marsden et al. 1967), which may suggest that some component of physiological tremor arises independently from sensory feedback.

The resonant frequencies of body parts vary in the range 3–5 Hz (elbow) and 17–30 Hz (finger). The resonant frequency of the wrist is similar (8–12 Hz) to that of the second component of physiological tremor. Resonant frequencies of other parts of the upper limb are even lower and correspond to frequencies commonly found in pathological tremor (e.g., Parkinsonian tremor, which is discussed in the next section). These observations suggest that recording tremor in the finger would avoid superimposing the mechanical and neural components of tremor.

It is not clear whether physiological tremor exists for a purpose. To this question Bernstein (1967) suggests that the 8–12 Hz component is generated by a central mechanism that regulates and coordinates motor control. This pacemaker would act as an internal reference necessary for coordinating neuronal activities, but Bernstein does not indicate where this pacemaker might be located nor what its characteristics may be. Elble and Koller (1990) also cite studies that suggest that finger movements tend to occur in phase with the 8–12 Hz tremor (Travis 1929) or at times of peak momentum produced by tremor (Goodman and Kelso 1983), suggesting that central oscillations can modulate the timing of motor events. Elble and Koller (1990) note that the origin of tremor has been attributed to various anatomical structures including the cerebral cortex (Alberts 1972), the thalamus and basal ganglia (Narabayashi 1982), the inferior olive and midbrain tegmentum (Lamarre et al. 1975), the cerebellum (Vilis and Hore 1980), the Renshaw cells in the spinal cord (Elble and Randall 1976), the stretch reflex (Lippold 1970), and the muscles (Rondot et al. 1968). These structures do not function in isolation and probably interact in the generation of normal and pathological tremors.

Although tremor appears to be a normal and simple manifestation of the nervous system, it is in fact a deeply complex physiological phenomenon that is influenced by central and peripheral structures (McAuley and Marsden 2000; Elble 1996) and by pharmacological, mechanical, environmental, and psychological conditions or events. An attempt to represent this complex system is presented in Figure 10.2.

## 10.2.3  Characteristics of Tremor in Patients with Parkinson's Disease

Parkinson's disease is a neurodegenerative disorder characterized by bradykinesia (i.e., slowing down of movements), rigidity, tremor, and postural instability. It is associated with a degeneration of dopaminergic neurons connecting the substantia nigra and the striatum in a subcortical region of the brain. However, other neurotransmitters and other subcortical regions of the brain are also affected in this disease. Patients with Parkinson's disease tend to exhibit masked facies, stooped posture, and shuffling gait (Stiles and Pozos 1976). The majority of patients with Parkinson's disease exhibit rest tremor with a frequency between 4 and 6 Hz. The onset of tremor is usually unilateral. It involves rotational and translational movements ("pill-rolling tremor"). Rest tremor is inhibited by voluntary activation of the muscles and augmented when the subject is walking or distracted. Early in the disease, anticholinergics can be employed for patients with a predominance of tremor. Levodopa and dopaminergic agonists are commonly used for all symptoms of Parkinson's disease, but their effect on rest tremor varies widely across subjects.

Postural tremor (5–12 Hz) is also seen in about 60% of patients with Parkinson's disease even when no rest tremor is observed. In fact, in 10 to 20% of these patients postural tremor will be the only form of tremor exhibited throughout the course of the illness (Elble and Koller 1990). Electromyographic bursts in antagonistic muscles of patients with Parkinson's disease are either synchronous or alternating, and double or triple motor unit discharges are observed in postural tremor (Elek et al. 1991). Cogwheeling corresponds to palpable tremor on passive manipulation of the

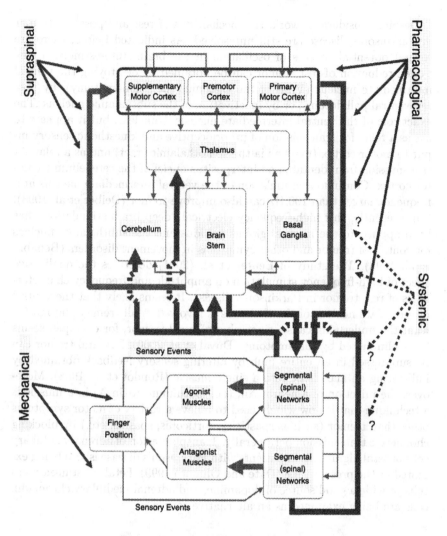

**Deterministic or Stochastic?**

Figure 10.2. Structures thought to be implicated in the generation of tremor with speculation of the interactions between levels. See also Figures 6.1 and 7.1, page 111 and page 123, of Elble and Koller (Elble and Koller 1990) for principal neural pathways presumed to be implicated in cerebellar tremor and Parkinsonian tremor, respectively. The term "systemic" refers to other physiological systems such as respiration, circulation, and circadian rhythms whose effect on the motor system is not completely understood.

limb. However, it is not specific to Parkinson's disease and can be felt, for example, in normal subjects with enhanced physiological tremor.

Despite considerable work, the mechanisms of rest and postural tremor in Parkinson's disease are still unresolved. As indicated before, there are several potential sources of oscillation in the brain. Interestingly, a sustained reduction of rest tremor can be obtained by destroying the ventral intermediate nucleus (Vim) of the thalamus (i.e., thalamotomy), by posteroventral pallidotomy, or by inactivation of the subthalamic nucleus. The exact role of the ventral intermediate nucleus is unclear, but it appears to act as a relay for transmission of proprioceptive and kinesthetic sensory input to motor cortex (maybe via the spinothalamic tract) and as a relay for transmission from dentate and interpositus nuclei of the cerebellum to motor cortex. Continuous stimulation of the ventral intermediate nucleus at a frequency larger than 100 Hz can also improve tremor (Deiber et al. 1993). More recently, this high-frequency electrical "deep brain stimulation" has been applied to the internal globus pallidus and the subthalamic nucleus for control of tremor and other symptoms in movement disorders (Benabid et al. 1998). The study of Beuter et al. (2001) describes the qualitative effect of high-frequency stimulation on amplitude and frequency characteristics of rest tremor in Parkinson's disease. It seems likely that the ventral intermediate nucleus of the thalamus is involved in all tremors, no matter what the underlying cause, since physiological tremor, for example, seems to be eliminated by thalamotomy (Duval et al. 2000). Postural tremor can be suppressed in an entire limb by altering sensory feedback obtained by infiltrating an anesthetic into a single muscle (Rondot et al. 1968). Moreover, the injection of a neurotoxin (i.e., botulinum toxin A) in a muscle is a technique that is now widely used to relieve a variety of motor symptoms other than tremor (such as spasticity, torticollis, spasms, etc.) by blocking cholinergic transmission peripherally. Transplantation of adrenal medullary cells or fetal nigral cells in patients with Parkinson's disease is also being explored at the present time (Date and Ohmoto 1999). Fetal substantia nigra cells provide a good source of dopamine, and adrenal medullary chromaffin cells are being examined as an alternative source.

## 10.2.4  Conventional Methods Used to Analyze Tremor

Conventional methods used to analyze tremor include amplitude and frequency analyses. Amplitudes are often given in root mean square, amplitude distributions, peak-to-peak amplitude, average absolute amplitude, or total displacement. Frequencies are often presented in half-power frequency, total power in consecutive bands, highest-band power, highest-peak frequency, semi-interquartile range, number of movement reversals, median frequency, etc. See Appendix C for more information on concepts of time series analysis, such as root mean square and peak frequency in power spectra.

In general, amplitude and frequency tend to be inversely related. Eshner (1897) showed that tremor frequency decreased from 10 to 6 Hz as amplitude increased from control to its largest level. Studying the transition from physiological tremor to shivering tremor in normal subjects, Stuart et al. (1966) noted a decrease of frequency from 10.3 to 8.3 Hz over a 50-fold increase of acceleration amplitude, and more recently, Stiles (1975) found similar results. In general, frequency is reported to be more stable than amplitude, although Friedlander (1956) makes a historical review of reported frequencies of normal postural tremor since 1876, which indicates wide discrepancies between studies.

Marsden et al. (1969b) were among the first to examine simultaneously the oscillations of the two hands and their coherency. This coherency indicates the proportion of the power in the oscillations of one hand that might be explained by oscillations of the other hand (Marsden et al. 1969b). A coherency of 1 indicates a completely deterministic linear relation between the hands, while a coherency of 0 indicates complete statistical independence. These authors combined pharmacological and mechanical experiments on the two arms to show that ballistocardiac oscillations, which should affect both arms similarly, probably contributed only a small amount to tremor in the fingers (i.e., about 10%). Other studies based on the cross-spectrum and coherency indicate that both physiological and pathological tremors are driven by two separate central generators, one for each side, since coherency between the hands is low, with the dramatic exception of orthostatic tremor, where very high coherencies were found (Köster et al. 1998; Timmer et al. 2000). The cross-spectrum and coherency are described in more detail in Section 10.3.6 and in Appendix C.

Two other important issues should be considered in analyzing tremor. First, can diagnosis be improved by considering additional features of tremor beyond amplitude and frequency? Recently, many additional features of tremor time series have been considered, for the purpose of discriminating different types of tremor as well as discriminating normal from pathological tremor in early stages of pathology when tremor amplitude is small. Beyond amplitude and frequency, the analysis has been extended to power spectral distribution, morphology, and dynamics (Timmer et al. 1993, Timmer et al. 2000, Beuter and Edwards 1999, Edwards and Beuter 2000). Certain measures for characterizing tremor are discussed in Sections 10.3 and 10.4. Second, what type of tremor recording is most effective for discrimination (displacement/velocity/acceleration; rest/postural/kinetic)? For example, despite the fact that well-developed Parkinsonian tremor is primarily a rest tremor, for patients in early stages of Parkinson's disease, postural tremor in the index finger without visual feedback allowed much better discrimination from control subjects than either rest tremor in the hand or postural tremor with visual feedback in the index finger (Beuter and Edwards 1999; Edwards and Beuter 2000). The question of recording displacement, velocity, or acceleration is discussed in Section 10.3.1.

## 10.2.5  Initial Attempts to Model Human Tremor

As indicated by Beuter and de Geoffroy (1995), a number of hypotheses have been examined by researchers to explain tremor in Parkinson's disease going from an abnormal rise in the threshold of the Renshaw cells to an increased delay in spinal feedback. Some investigators have suggested that Parkinsonian tremor is caused by a central generator, which could be located in the inferior olive, while others have proposed that this pathological tremor appears in the spinal cord under the influence of abnormal supraspinal influences. The current state of knowledge suggests that both the central nervous system and sensory feedback pathways participate in the generation of tremor and that Parkinson's disease may be considered as a dynamical disease (Beuter and Vasilakos 1995). What is not clear, however, is the exact nature of the respective contribution of central and peripheral mechanisms. Overall, few attempts to model human tremor have been published. Interestingly, most studies focus on tremor in Parkinson's disease and use a coarse scale to model the generation of tremor.

In 1962, Austin and Tsai (1962) proposed a simple differential equation model of tremor in Parkinson's disease. In this model they hypothesized that the downstream paths of excitation are relatively excessive as a result of multiple pathological lesions involving predominantly inhibitory regions of the brain. They used two ordinary differential equations to describe the rate of increase in the ratio of facilitation to inhibition caused by the disease and to the amount of tremor already present. They consider that the pathologically exaggerated downstream excitation level acting on spinal interneurons is mainly responsible for Parkinsonian tremor. Thus in their model the oscillatory tendency of spinal interneurons is in turn transferred to motoneurons. They found some agreement between their model and data recorded on six patients who received intravenous injections of adrenalin or sodium amytal to simulate the effect of emotion or fatigue, respectively. Unfortunately, the authors do not explain how they recorded tremor nor what type of tremor they recorded. Furthermore, their model does not consider the time delays inherent to any physiological system. The model of Austin and Tsai is discussed in more detail in Section 10.5.1.

Later, Gurfinkel and Osovets (1973) were the first to mention the presence of large-amplitude fluctuations associated with a reduction by about half of the frequency in Parkinsonian tremor. They hypothesized that the stability in the maintenance of a posture is obtained by periodic changes in the strength of the muscle at the frequency of physiological tremor. The oscillations in strength have an amplitude that is a function of the deviation angle from the set position and the coefficient of amplification in a feedback loop. The appearance of Parkinsonian tremor is regarded as a transition corresponding to an abrupt increase in the coefficient of amplification in the controlling chain. They explain the transition by a spontaneous rise in amplification in a feedback loop, but because they have insufficient available data, they do not elaborate on this point. The authors also neglect all delays present in the central and peripheral feedback loops and do not consider the possibility that both forms of tremor (physiological and pathological) may be present in the same patient.

In 1976, Stein and Oguztoreli (1976) examined how mechanical and reflex factors could contribute to tremor. They assumed that physiological tremor is a linear phenomenon that has a small amplitude and used a second-order system based on a mass–spring system. The differential equations they used generated oscillations corresponding to the frequency of physiological tremor. Parkinsonian tremor was obtained by using longer supraspinal reflexes. In addition, the authors examined briefly the possibility that nonlinearities present in muscle receptors could limit the magnitude of the oscillations. In their model, at a specific gain the reflex oscillation became dominant, while at a higher gain, it was the mechanical oscillation that dominated.

The model proposed by Fukumoto (1986) is the first model to include both an oscillator located in the central nervous system (4–6 Hz) and a peripheral feedback loop located in the stretch reflex (8–12 Hz). While the central loop is often believed to play a role in pathological tremor, the peripheral loop is often associated with physiological tremor. The author hypothesized that Parkinsonian tremor is caused by a reduced contractive force of intrafusal muscle fibers due to fatigue (assumed to be caused by chronic stimulation of the gamma system) and modeled it with an autoregressive (AR) model. An autoregressive model is a model with linear dynamics in which optimal model parameters are chosen, using the autocorrelation function, to minimize the prediction error for a given time series (see Appendix C). In Fukumoto's model, eight independent variables were used, and the manipulation of only one of them (i.e., the intensity parameter of intrafusal fiber force) allowed the author to reproduce both physiological and Parkinsonian tremors. Although this model represented a new contribution to the field when it was proposed, it is limited by the fact that it is based on *linear* differential equations, and deals only with frequency while ignoring the morphology of the oscillation. There is now evidence that Parkinsonian tremor fluctuates in intensity (Edwards and Beuter 1996).

Gantert et al. (1992) were the first to consider nonlinear dynamics approaches to tremor time series analysis. These techniques do not propose a specific model, but test for consistency with a class of models. They measured acceleration of the stretched hand and applied three methods to the tremor time series: the correlation dimension, which provides a test of stochasticity; a test for linearity in the form of an autoregressive moving average (ARMA) model; and fitting the data to a linear state space model. For physiological tremor the correlation dimension suggested that the dynamics are stochastic and that the hand's movement is consistent with that of a damped linear oscillator driven by white noise. In patients with Parkinson's disease, using the correlation dimension and calculation of Lyapunov exponents, they found evidence that the dynamics are generated by a nonlinear deterministic (and chaotic) process. Recently, they have reconsidered these results and now doubt the determinism behind Parkinsonian (or essential) tremor (Timmer et al. 2000).

Finally, the model presented by Beuter et al. (1993) focused on the dynamics of tremor in relation to delayed visual feedback. They examined experimentally, numerically, and analytically the effect of manipulating delays in two feedback loops. The model based on differential equations with delays was tested over a wide range of parameter values. The authors also explored the effect of an increased gain and noise in one of the loops in order to match more accurately the performance of some patients with Parkinson's disease. They found that the influence of stochastic elements in the equations contributed qualitatively to a more accurate reproduction of the experimental traces of patients.

So far, the attempts made to model tremor in Parkinson's disease have not allowed researchers to isolate the underlying neural dynamics characterizing this disease. However, intense research in this direction is underway at the moment: For example, Edwards et al. (1999) suggested a neural network model to represent the transition from the irregular dynamics of normal physiological tremor to the more regular oscillations of Parkinsonian tremor (further details of this model are in Section 10.5.2). There is a growing agreement among tremor researchers that Parkinsonian resting tremor is *associated* with clusters of neurons, located in subcortical structures such as the thalamus and/or basal ganglia, firing synchronously at a frequency close to tremor frequency (see Tass 2000 and references within). As well, with the recent developments of noninvasive functional imaging techniques (Brooks 2000) and with the use of deep brain stimulation, we can expect significant progress in the understanding of the brain circuitry involved in this disease.

## 10.3    Linear Time Series Analysis Concepts

The previous section summarizes what is currently known about tremor. In this section, we outline some linear time series analysis techniques that have been used to characterize tremor data. First, we discuss the effect of recording tremor in displacement, velocity, or acceleration.

### 10.3.1    Displacement vs. Velocity vs. Acceleration

Tremor is monitored by recording the displacement, velocity, or acceleration of the body part being studied. Most researchers in this field in the past have used accelerometers to record tremor, but it is now possible to make very precise displacement and velocity recordings with laser systems. Beuter et al. (1994) were the first to use displacement lasers to record tremor. Recently, Norman et al. (1999) examined measurements of tremor using a velocity transducing laser and compared these to simultaneous recordings using transducers of displacement, acceleration, and muscle activity (electromyography). Of course, it is possible to differentiate or integrate a signal to convert from one type to another, provided that one is aware of the potential difficulties.

One possible difficulty in differentiating a signal is that the derivatives of the signals give different amplitude results than the originals give (the measurement of amplitude of tremor is discussed in more detail in the next section). For example, if we differentiate two sine waves, the second with half the amplitude but twice the frequency of the first, then the resulting cosine waves will have the same frequencies as before, but their amplitudes will now be equal. If we differentiate again, the second signal will have twice the amplitude of the first. (Recall that if $x(t) = a\sin(\omega t)$, then $dx/dt = a\omega\cos(\omega t)$).

Another difficulty in converting between displacement, acceleration, or velocity in a digital signal is the augmentation of the error made in the approximation of a continuous process by a discrete set of data. We describe the approximation error made when converting one type of data represented in the frequency domain to another (e.g., numerically differentiating displacement data to obtain velocity data). The relationship between a time series and its derivatives can be precisely characterized using a simple function called the transfer function (see Appendix C for more detail). In tremor recording, we may record displacement at 200 Hz, for example, giving us information on frequencies in the tremor up to 100 Hz, but typically, the interesting part of the tremor is below 15 Hz, and here the difference between the continuous signal and its discrete approximation is less than 2%. Figure 10.3 displays the magnitudes of both the discrete and continuous transfer functions for differentiation of a 200 Hz signal. The figure shows that the approximation error made in converting from one signal type to another is negligible in the frequency range of interest for tremor research.

Controlling the noise sufficiently is another difficulty in digitally converting one signal type to another. Whenever a derivative is taken, the relative power (or amplitude) of lower-frequency components in the signal is reduced in comparison to higher frequencies. Integrating will do the reverse. An argument that had been used against deriving accelerations from displacement recordings was that differentiation amplifies whatever high-frequency noise is in the recorded signal. However, it is now possible to record displacements precisely enough to avoid this problem in the range of frequencies of interest in tremor research (in general, up to 15 Hz). On the other hand, acceleration recordings may suppress lower-frequency components to the point of making it difficult to distinguish real oscillations in the 3–4 Hz range, and integrating them twice to obtain displacements will

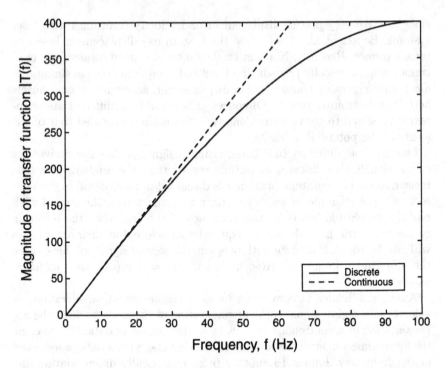

Figure 10.3. The magnitude of the transfer function for differentiation of a 200 Hz displacement signal to obtain velocity data. The dotted line shows the magnitude of the transfer function in the ideal case of a continuous signal, which is $|T(f)| = 2\pi f$. The solid line is the discrete approximation of the magnitude of the transfer function, given by $|T(f)| \approx \sqrt{2}s\sqrt{1 - \cos(2\pi f/s)}$, which is close to exact for frequencies below 15 Hz that are primarily of interest in tremor research.

amplify any imprecision in the recordings at the low-frequency end. Figure 10.4 contains the power spectrum* of a signal and the corresponding power spectrum of the second derivative of the signal, showing the suppression of the lower frequencies (see also Figures 1 and 2 of Beuter and de Geoffroy 1996). This type of imprecision in recordings of tremor data is of great concern, since pathology is expressed in a lowering of frequency.

Whether displacement, velocity, or acceleration is most relevant to tremor research is an unresolved issue, if indeed there is a clear-cut answer. One effect of using velocity or acceleration is essentially to remove the problem of filtering out the drift (trend) that can occur in displacement data. In a sense, if velocity is recorded, for example, a particular filter has already been applied to the data. If detection of significant low-frequency

---

*The power spectrum shows a picture of the frequency content of a signal (see Appendix C for more detail).

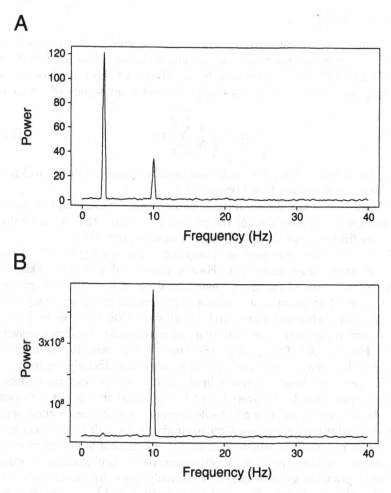

Figure 10.4. (A) Power spectrum of a signal, $x(t)$, composed of two sine waves $x(t) = \sin(2\pi 3t) + 0.5\sin(2\pi 10t)$ plus noise $1.5\eta$, where $\eta$ is Gaussian white noise with mean zero and the same variance as $x$; (B) Power spectrum of the second derivative of $x$, with noise $1.5\nu$, where $\nu$ is Gaussian white noise with mean zero and the same variance as the second derivative of $x$, added after the differentiation. Both spectra were smoothed twice by 11-point Daniell filters (see Appendix C).

components is important, displacement recordings may be more appropriate (see Beuter et al. 1994 and Norman et al. 1999). In the next several sections, we present some measures in the time domain and in the frequency domain that can be used to characterize tremor.

## 10.3.2 Amplitude

The simplest and most obvious measurement of tremor that can be made from its time series is (average) amplitude. If the signal has a mean of zero and no trend (drift), as is the case in recordings of tremor acceleration, for example, then the amplitude is simply the root mean square of the signal:

$$A = \sqrt{\frac{1}{N} \sum_{n=0}^{N-1} x_n^2}.$$  (10.1)

Since the mean is zero, the root mean square is essentially equivalent to the standard deviation (see Appendix C).

Of course, it is not always quite so simple, even with as simple a concept as average amplitude. Two of the practical problems that must be dealt with are filtering and the choice of the measured variable.

Filtering a time series may be thought of as removing (partially or completely) some components of its Fourier transform[†] and then taking the inverse Fourier transform to get a new time series that is a filtered version of the original. For example, a low-pass filter removes frequency components higher than a specified value, and a high-pass filter removes frequencies lower than a specified value. Filtering can also modify amplitude measurement. For example, if a signal is smoothed by a low-pass filter, sharp peaks in oscillations may be rounded off and amplitude artificially reduced.

On the other hand, as often happens in displacement recordings of tremor, there may be an overall drift in the signal that is not considered part of the tremor, and it is preferable to remove low-frequency components of the signal using a high-pass filter to eliminate the drift before calculating tremor amplitude. It is somewhat a matter of judgment to decide which low frequencies to filter. Breathing is often observed in displacement recordings of finger position, for example, and normally has a frequency under 0.35 Hz. Heartbeat can also influence these recordings and has a frequency typically under 1 Hz. Tremor itself is normally considered to have frequencies of at least 3 Hz. Thus, to be cautious, frequencies below 1 or 2 Hz can be filtered to eliminate other, slower, movements, but measurement of tremor amplitude from displacement signals will depend on how this is done. Figure 10.5 displays unfiltered and filtered displacement recordings of normal rest tremor.

Norman et al. (1999) showed that laser recording of tremor velocity is a precise method of measuring finger tremor by verifying correspondence in both amplitude and frequency measures to another proven method.

In the discussion on amplitude above, it has been assumed that drift is not a part of tremor and should therefore be removed. However, it is

---

[†]The Fourier transform is used to represent a time series in the frequency domain. Appendix C explains the Fourier transform in more detail.

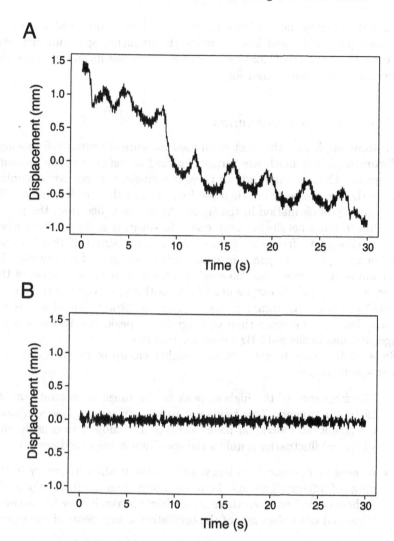

Figure 10.5. (A) Unfiltered and (B) filtered displacement recording of normal rest tremor. The displacement signal was detrended, and then a high-pass filter was applied to remove frequencies below 1.5 Hz.

possible that drift itself may be related to pathology, and we might wish to quantify it also. For example, in a task with visual feedback where a subject is asked to maintain a finger position in relation to a reference point, the overall amount of deviation from the reference point apart from tremor oscillations (i.e., longer scale, slower excursions) might be indicative of pathology. Drift can be quantified as the root mean square amplitude of that part of the Fourier transform of the displacement signal removed in calculating the tremor amplitude, as described above. Of course, there is a somewhat arbitrary choice of how much of the signal is drift and how much is tremor (i.e., where and how to divide the frequency spectrum between the two). See Appendix C for more information on root mean square, power spectra, and the Fourier transform.

### 10.3.3  Frequency Estimation

Aside from amplitude, the most often used measure of tremor is frequency. Unfortunately, it is much less straightforward to define a single measure of frequency than of amplitude. If there is a single, narrow, overwhelming peak in the power spectrum, then the frequency of that peak is essentially the frequency of oscillation in the signal. Appendix C describes the power spectrum in more detail. In many cases, however, it is not so clear what is the predominant frequency of oscillation in the signal, if there is one. In recordings of finger tremor in many normal subjects, for example, the spectrum is quite broad, and no single peak stands out. Estimation of the power spectrum (which can be done by smoothing) is required to identify whether peaks are significant. However, even in subjects with Parkinson's disease there can be more than one significant peak, one in the 4–6 Hz range, and one in the 8–12 Hz range, for example.

Some of the ways to get a single rough measure of frequency from a power spectrum are:

- The frequency of the highest peak in the range of interest (say 3–15 Hz for tremor). This tends not to be very robust, however, since the heights of individual peaks are typically subject to considerable statistical fluctuations, unless the spectrum is smoothed greatly.

- A measure of central tendency, such as the median frequency in the range of interest. This may be more robust than the frequency of the highest peak, but when there are multiple peaks, it may fall between them and not reflect any of the oscillatory components of the signal.

- The "center" of an interval of width, say 1 Hz, which has more power (larger values) than any other such interval in the range of interest. This will isolate the most significant frequency component in the signal, if there is one, and should be more robust than the frequency of the highest peak. It is more or less equivalent to the highest peak in the spectrum smoothed over 1 Hz intervals.

Frequency of oscillations in tremor *does* seem to be related to pathology, but the relationship does not appear to be clear. Many reports suggest that Parkinsonian rest tremor is typically in the range of 4–6 Hz, whereas essential tremor (which is a postural tremor) is in the range of 4–12 Hz (Deuschl et al. 1998). However, there are many exceptions. For example, the frequency of essential tremor can vary, and Parkinsonian postural tremor can be in the same frequency range as essential tremor. Also, normal physiological tremor is typically concentrated in the 8–12 Hz range, but can vary widely (Elble and Koller 1990).

This suggests that it might be more useful to find whether there is a concentration of power in a spectrum in specific frequency ranges, rather than summarizing the frequency information in a single measure. It is possible to quantify the amount of power in a specific range of frequencies in a power spectrum, simply by summing the power at the frequencies concerned. The power in the 4–6 Hz range, as a proportion of the total power in the spectrum, for example, might be expected to be significant in identifying Parkinsonian tremor. The power in the 7–12 Hz band, as a proportion of the total power in the spectrum, might be expected to be lower in many abnormal tremors than in normal physiological tremor. Beuter and Edwards (1999) used these frequency measures, among others, in their examination of frequency domain characteristics to discriminate physiological and Parkinsonian tremor.

### 10.3.4   Closeness to a Sinusoidal Oscillation

Another way to try to identify pathology is to quantify the degree to which the power spectrum is concentrated in narrow ranges of frequencies. Generally speaking, normal tremor is fairly noisy and dispersed in the frequency domain, whereas advanced pathological tremor tends to be concentrated in one or a few narrow peaks. It is possible, therefore, that even when amplitudes and central measures of frequency are similar, some pathological tremors are more concentrated. One way to quantify this, for example, is to take the proportion of the available frequency values required to make up 50% of the total power of the signal, if they are taken in order from largest power to smallest. An approximately sinusoidal signal, or even a sum of several sinusoids, will take only one or a very few frequencies to provide 50% of the power, whereas half the frequencies will be required to make up half the power in a completely flat spectrum. A similar idea is captured by the center of mass frequency concentration of the sorted spectrum (Beuter and Edwards 1999).

### 10.3.5   Amplitude Fluctuations

All of the measurements discussed so far have suppressed any changes over time in tremor oscillations. Measurements based on the power spectrum of an entire time series necessarily average over the time axis. Fluctuations may, however, be important in looking for pathologies reflected in tremor recordings. There is some evidence that symptoms (in Parkinson's disease, for example) may be intermittent, and so they may be present only in part of a time series (Edwards and Beuter 1996). Averaging may dilute them too much. It may therefore be advantageous to measure characteristics of tremor more locally in time. It may also be a good idea to try to quantify how much variation there is in any of these measures (especially amplitude and frequency, but in principle any other measure, too).

The simplest approach is to divide the time series into smaller blocks and calculate each measurement on the blocks separately. One could also pass a moving window across the signal to get a more continuous (in time) measurement. One of the problems with this is that the smaller the window or block is, the poorer the resolution in frequency becomes. Another alternative is to use wavelets, a technique that (like Fourier analysis) decomposes a series into simple components, but the components are not sinusoids, but functions of a particular shape that are localized in time as well as frequency or "scale." If the components are appropriately oscillatory, wavelets allow a frequency analysis that also varies with location in time (Burrus et al. 1998).

There are also standard time series techniques to calculate instantaneous amplitude (or frequency, or phase). For example, to calculate the instantaneous amplitude, one could apply the following procedure to a detrended time series (if necessary, it could be high-pass filtered first to remove drift): Square the values in the time series, making it entirely nonnegative, with mean at half the squared oscillation amplitudes; apply a low-pass filter to eliminate the fast oscillations, leaving a time series with the same mean (the DC component of the Fourier transform); multiply by 2 to get the full amplitudes of the squared oscillations; take the square root of each value. This leaves only an envelope giving an approximate amplitude at each time:

$$x_e = \sqrt{2 \cdot \text{low-pass}\left[\left(\text{high-pass}[x]\right)^2\right]}. \tag{10.2}$$

This is essentially a technique called "complex demodulation" used in signal processing (de Coulon 1986, pp. 389–391).

The standard deviation of this envelope gives a quantification of amount of variation (or drift) in amplitude. Figure 10.6 displays the instantaneous amplitude of static tremor (postural tremor with eyes open) in a subject with Parkinson's disease. (Note: There is a minor problem when the low-pass filtered version of the squares of the signal goes negative, and thus the square root would not be defined. In this case, such values could be set to zero without much adverse effect.) Of course, in order to examine amplitude fluctuations, it is necessary to record tremor for a sufficiently long time. There is not an accepted standard length of time of recording, but we recommend about 60 seconds.

## 10.3.6  Comparison Between Two Time Series

If tremor is primarily a result of a single central neuronal oscillator (in the brain), then one would expect that the two hands might oscillate in synchrony, or at least that the tremor in the two hands would show some

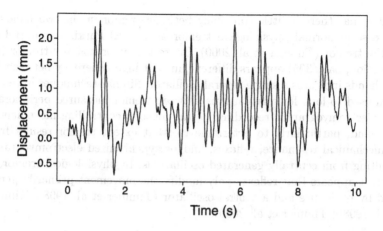

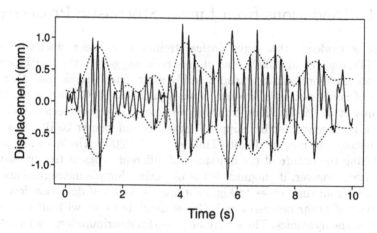

Figure 10.6. Instantaneous amplitude calculated from static tremor (postural tremor with eyes open) of a subject with Parkinson's disease. (A) Raw displacements; (B) high-pass filtered displacement and the "instantaneous amplitude" envelope. High-pass and low-pass filtering were both ramped between 1 and 1.5 Hz.

correlation. If tremor has primarily a peripheral origin, or if there are two independent central oscillators (one for each side), then one might expect little correlation between the movement of the two hands. The simplest measure of relationship between the two hands is coherency, which is essentially a correlation between the Fourier transforms of two time series (see Appendix C for the precise definition). Marsden et al. (1969a) studied the statistical relationship between tremors of the two hands in healthy subjects and found that the overall coherency between hands was 0.033 on average for frequencies between 1 and 20 Hz. Further studies have confirmed that there is little coherency between tremor in the two hands in the case of normal physiological tremor as well as Parkinsonian and essential tremor (Timmer et al. 2000). Interestingly, orthostatic tremor (see definition, page 305) appears different, in that high coherency between the two hands indicates a single central oscillator. Similar techniques have also been used to look for relationships between tremor measured peripherally and electromyographic recordings of the muscles controlling movement in the hand, particularly to determine to what extent tremor results from biomechanical resonance, reflexes, and/or synchronized electromyography resulting from centrally generated oscillations. In physiological tremor, at least, it appears that reflexes only modify the movement primarily generated by resonance and a central oscillator (Timmer et al. 1998a; Timmer et al. 1998b; Timmer et al. 2000).

## 10.4    Deviations from Linear Stochastic Processes

There is evidence that physiological tremor is a linear stochastic process (Timmer et al. 2000); and it has been suggested that pathological tremors are more nonlinear and possibly more deterministic, and that the types of nonlinearities may vary between pathologies (Gantert et al. 1992; Timmer et al. 1993). Aside from the nonlinearity of pathological tremors, which has received further support (Edwards and Beuter 2000), these last conclusions are now in doubt (Timmer et al. 2000). The question of establishing the nature of the dynamics of different types of tremor may be bypassed, however, if diagnosis is the objective. Simple measurements can be made from time series to highlight various types of deviation from the behavior of linear processes with Gaussian white noise, without even considering the dynamics. These may be powerful discriminators, nonetheless. Despite the potential problems with such tests and the ambiguity in interpretation of the results, we can at least treat the statistics used as potential discriminators between types of tremor. Even if we cannot conclude definitely that a particular type of tremor is chaotic or even nonlinear, if we can use a test statistic to separate types of tremor, we have made progress.

### 10.4.1    Deviations from a Gaussian Distribution

If the stochasticity of physiological tremor is Gaussian and the underlying process is linear, then one simple way to look for pathology is to look for distributions of data values that deviate from a Gaussian distribution. A general approach is to look at higher-than-second moments of the distribution. These higher moments can identify deviations of the signal distribution from a Gaussian distribution. Appendix C presents more detail on moments, such as the third moment (skewness) and the fourth moment (peakedness). A regular, symmetric oscillation like a sine wave will have skewness near zero (as does a zero-mean Gaussian distribution) but a much lower peakedness than a Gaussian distribution would have. Figure 10.7 contains histograms of Gaussian white noise, a sine wave, and rest tremor displacement of a control subject and of a subject with Parkinson's disease. For the Gaussian white noise of Figure 10.7A, the skewness is $m_3 = 0.02$ and the peakedness is $m_4 = 3.21$. For the 4.1 Hz sine wave in Figure 10.7B, the skewness is $m_3 = 0.00$ and the peakedness is $m_4 = 0.38$. As can be seen in the figure, the distribution for a control subject resembles that of Gaussian white noise, and the distribution for a subject with Parkinson's disease is somewhat like that of the sine wave.

## 10.4.2 Morphology

Another potentially interesting aspect of tremor is the morphology of the oscillations, regardless of dynamics, in the time domain. There are any number of ways of quantifying aspects of morphology, although few have been explored. Examination of recordings of finger tremor in subjects with Parkinson's disease and with essential tremor suggests some measures to identify specific kinds of deviations from sinusoidal oscillations (Edwards and Beuter 1996; Edwards and Beuter 2000). Figures 10.8 and 10.9 show a morphology of Parkinsonian static tremor in displacement and acceleration data, respectively. It appears that postural Parkinsonian tremor of the index finger in particular sometimes has (displacement) oscillations with a flattened segment on the way up. In the acceleration signal this becomes an extra oscillation that does not go below zero. This makes the acceleration positive more of the time than it is negative, though with less magnitude. Thus, we might measure this positive–negative asymmetry as

$$s = \frac{|\{x_n | x_n > 0\}|}{|\{x_n | x_n < 0\}|} * \frac{\text{rms}\{x_n | x_n < 0\}}{\text{rms}\{x_n | x_n > 0\}}, \tag{10.3}$$

where $*$ indicates multiplication, $|\cdot|$ indicates the number of elements in a set, rms is root mean square, and $x$ here is acceleration. Note that this measures something similar to skewness in the distribution of the acceleration trace, though here the motivation is a particular time asymmetry in the displacement signal.

Another aspect of the particular morphology of Parkinsonian tremor is the extra oscillation in the acceleration trace for each oscillation in the displacement trace. Thus, we could define another measure, "wobble," as the ratio of numbers of extrema (maxima and minima) in acceleration and displacement, or equivalently, the ratio of numbers of zero-crossings in "jerk" (the derivative of acceleration) and velocity:

$$w = \frac{|\{j_n | j_n * j_{n+1} < 0\}|}{|\{v_n | v_n * v_{n+1} < 0\}|}, \tag{10.4}$$

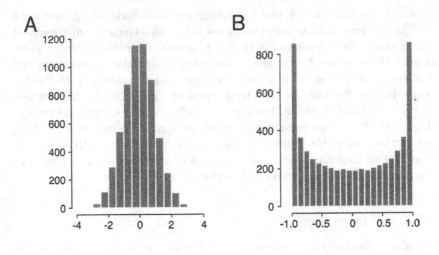

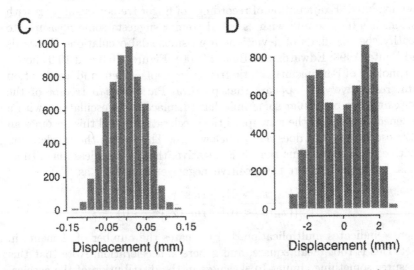

Displacement (mm)                    Displacement (mm)

Figure 10.7. Histograms of 6000 points sampled from (A) Gaussian white noise (mean 0, variance 1); (B) a 4.1 Hz sine wave sampled at 200 Hz; (C) low-amplitude irregular rest tremor of a control subject (displacement, high-pass filtered at 1.0–1.5 Hz); (D) higher amplitude and more regular rest tremor of a subject with Parkinson's disease (displacement, high-pass filtered at 1.0–1.5 Hz).

where $*$ indicates multiplication, $v = dx/dt$ represents velocity, and $j = d^2v/dt^2$ represents jerk. A low-pass filter will have to be applied before this calculation is made, to get rid of high-frequency noise that could produce many extra oscillations in acceleration. "Wobble" is sensitive to the presence of small oscillations of higher frequency that are superimposed on a

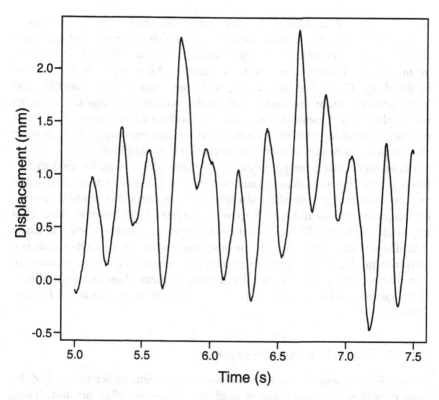

Figure 10.8. Parkinsonian static tremor (postural tremor with eyes open); displacement recording.

higher-amplitude oscillation at a lower frequency (these oscillations must be small enough to modify the shape of the main oscillation but without creating additional extrema). The amplitude of such higher-frequency oscillations is enhanced by differentiating twice, so that additional extrema appear.

## 10.4.3  Deviations from Stochasticity, Linearity, and Stationarity

Determinism (whether linear or nonlinear) can be identified in time series data by a number of techniques, though there is still considerable controversy over their reliability. Prediction error (Kaplan and Glass 1995, pp. 324–330) uses the data to construct a model of the dynamics, and then sees whether the predictions we can make from this deterministic model are accurate. The deterministic versus stochastic plot is an extension of this idea, using local linear models based on different sizes of neighborhoods to make the predictions. The $\delta$-$\epsilon$ method investigates the continuity of the mapping from past to present states. Local slopes of correlation integrals, as well as Lyapunov exponents measuring rates of divergence of nearby trajectories, can be used to distinguish between stochastic and low-dimensional chaotic processes. Poincaré maps or return maps can be examined for signatures of chaotic dynamics. All of these methods were explored by Timmer et al. (2000), and they suggest that essential and Parkinsonian tremor are both second-order nonlinear stochastic processes rather than low-dimensional deterministic (chaotic) processes. The question of distinguishing between nonlinear stochastic processes and *high-order* chaotic ones, however, is currently beyond our capabilities.

Phase-randomized surrogate data (see Appendix C) can be used to test the null hypothesis of linear dynamics driven by Gaussian white noise, using any reasonable discriminating statistic such as nonlinear prediction error or Lyapunov exponents (see Kaplan and Glass 1995, Section 6.8). This approach has been seldom applied to tremor data, and there are still many difficulties with it. For example, inconsistency with the null hypothesis can be due to nonlinear determinism, nonlinear stochasticity, or a non-Gaussian distribution due perhaps to nonlinear measurement. Another possibility is that the underlying stochastic process is simply nonstationary (Timmer 1998).

## 10.4.4  Time-Reversal Invariance

Properties of linear Gaussian processes are invariant under reversal of the time direction, whereas those of nonlinear processes often are not. Thus, a general measure of time asymmetry may be diagnostic of the right sort of nonlinearity. One way to do this is to predict values forwards in time and backwards in time (with the same "lag") from similar values and see how much they differ. This is done for all possible starting values, and the squared differences are summed to give a (lag-dependent) measure of time asymmetry (see Timmer et al. 1993 for details). Another measure of time asymmetry involves cubes of differences between points and lagged points (Edwards and Beuter 2000), given by

$$\max\{|R_j|, \text{time lag } j\}, \quad \text{where } R_j = \frac{\sum_i (x_i - x_{i-j})^3}{\left(\sum_i x_i^2\right)^{3/2}}. \tag{10.5}$$

This measure seems to have some value in discriminating normal physiological tremor from abnormal tremors but may not help in discriminating between essential and Parkinsonian tremor (Timmer et al. 1993; Timmer et al. 2000; Edwards and Beuter 2000).

## 10.4.5  Asymmetric Decay of the Autocorrelation Function

Another characteristic of linear stochastic processes is exponential decay of the extrema of oscillations in the autocorrelation function. Thus, another

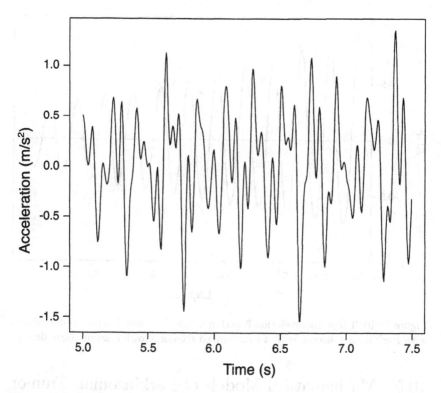

Figure 10.9. Acceleration calculated from second differences of the displacement data of Fig 10.8 and low-pass filtered at 18–20 Hz.

signature of a type of nonlinearity is an asymmetry in the decay of extrema, and in particular, a larger extremum following a smaller one. Timmer et al. (1993) quantified this idea by subtracting the magnitude of the first two extrema of the autocorrelation function (the first minimum and the first maximum). Figure 10.10 shows the autocorrelation function for tremor velocity of a subject with Parkinson's disease with a pronounced tremor. The first minimum is at about −0.73, while the first maximum is bigger at about 0.87. This is asymmetric decay: The minima are decreasing faster than the maxima.

The measure of asymmetric decay of the autocorrelation function has been found useful in discriminating physiological from pathological tremor (Timmer et al. 1993; Edwards and Beuter 2000), though again, its utility in discriminating between essential and Parkinsonian tremor now seems doubtful (Timmer et al. 2000).

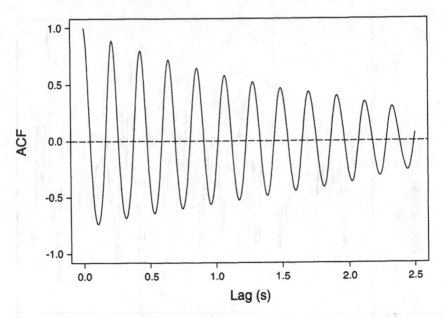

Figure 10.10. The autocorrelation function (ACF) for tremor velocity of a subject with Parkinson's disease with a pronounced tremor showing asymmetric decay.

## 10.5  Mathematical Models of Parkinsonian Tremor and Its Control

### 10.5.1  The Van der Pol Equation

Austin and Tsai (1962) proposed the van der Pol equation as a model of Parkinsonian tremor:

$$\frac{d^2x}{dt^2} + K(x^2 - 1)\frac{dx}{dt} + \beta x = 0, \qquad (10.6)$$

where $x$ is the amplitude of activity in a motoneuron pool; $K = (F/I - 1)$, involving a ratio of facilitation ($F$) to inhibition ($I$) from suprasegmental levels; and $\beta$ is a constant. The van der Pol equation, a classic in nonlinear dynamics, has a globally stable limit cycle for appropriate parameter values.

The van der Pol equation is not capable of a wide variety of behaviors. However, the oscillations of the limit cycle are not sinusoidal, but have flattened segments (for some parameter values), somewhat like those seen in Parkinsonian tremor. In fact, simulated data from this equation with appropriate parameter values have very high values of "wobble," though their symmetric nature prevents them from having high values of "positive/negative asymmetry," as discussed in Section 10.4.2 on morphology.

However, there is nothing particular about the van der Pol equation in this regard. A great many other nonlinear oscillators, with noncircular orbits in phase space, could produce similar behavior. This suggests that Parkinsonian tremor is a nonlinear oscillator, but we cannot argue that this supports the Austin and Tsai model.

## 10.5.2  A Hopfield-Type Neural Network Model

Edwards, Beuter, and Glass (1999) explored the behavior of richly connected inhibitory neural networks under parameter changes that correspond to the weakening of synaptic efficacies between network units. They suggested that this type of model might represent the transition from normal physiological tremor (irregular dynamics) to Parkinsonian tremor (having more regular oscillations) as the synaptic efficacies of dopaminergic neurons in the nigrostriatal pathway are weakened. The $N$-unit neural network that they considered was an additive network like that used by Hopfield (1984) of the form

$$\frac{dy_i}{dt} = -y_i + \sum_{j=1}^{N} w_{ij}g(y_j) - \tau_i, \quad i = 1, \dots, N. \qquad (10.7)$$

Here, $w_{ij}$ represents the "synaptic efficacy" of unit $j$ acting on unit $i$; $\tau_i$ is the threshold level of unit $i$; and $g$ is the response function of a unit to its input. The response function, $g$, could be a sigmoid function, such as

$$g(y_j) = \frac{1 + \tanh \beta y_j}{2}, \qquad (10.8)$$

where $\beta$ controls the slope or gain of the sigmoid, or a Heaviside step function, i.e.,

$$g(y_j) = \begin{cases} 0 & \text{if } y_j < 0, \\ 1 & \text{otherwise.} \end{cases} \qquad (10.9)$$

A "unit" may be a single neuron, but in this context should be interpreted as a larger pool of neurons acting in conjunction whose overall activity can be considered to affect that of other such units. Inherently oscillatory units are clearly not modeled by this formulation. The authors of the study are specifically looking for network effects to produce oscillation.

Their hypothesis was that the onset of a regular oscillation in Parkinson's disease is a change in dynamical regime of the network from a normally aperiodic one to a more regular one as the parameter corresponding to dopamine efficacy is decreased. This implies two things: (1) that tremor in Parkinson's disease and normal physiological tremor are produced by the same motor circuitry but operating in different parameter ranges, and (2) that normal physiological tremor is the output of an aperiodic regime in the (deterministic) network. Their hypothesis differs from one that suggests that normal physiological tremor corresponds to a fixed point perturbed slightly by noise and that regular tremor arises from it by a Hopf bifurcation.

For certain parameter values, the model in equation (10.7) also has a stable fixed point. This could represent akinesia (difficulty in initiation of movement), which is another common symptom of Parkinson's disease. The model suggests that this akinesia could be another mode of operation of the same dynamical system for another value of the varying parameter.

They chose to make all the connections in the network inhibitory and generated random networks in the form of equation (10.7) with the binary response function of equation (10.9). The networks contained $N$ units, and each unit received input from $K$ others: i.e., for each unit, they randomly selected $K$ of the remaining $N - 1$ units to provide inputs to it, and set the corresponding entry in the connection matrix to $-1$ (negative [positive] elements indicate inhibitory [excitatory] connections), and set all other connections to 0. In addition, the connection matrix was adjusted by reducing the entries in the first $d$ columns by a factor $\alpha$, where $\alpha \in [0, 1]$, which corresponds to the weakening of the output of these $d$ units in the network. Since the connection matrices are random, selecting the first $d$ columns is essentially random.

From random initial conditions, they integrated the networks by computing trajectories in terms of points at which units switch state (in this case, change sign) until the trajectory converged to a fixed point or periodic cycle or until a specified maximum number of switchings was reached. Figure 10.11 displays the behavior of one of the units in one of the networks with decreasing values of the parameter $\alpha$: 1.0, 0.5, and 0.2. In general, as $\alpha$ decreases, the behavior becomes more regular.

Equation (10.7) has many parameters (all the entries $w_{ij}$ and $\tau_j$), which allow for a wide variety of behaviors. In particular, a kind of dynamical simplification occurs upon crossing a boundary of the region in parameter space in which all variables can actively change state. The effective dimension of the dynamics is reduced as units become stuck in the "on" or "off" state. Under certain circumstances, this can lead the network to change from irregular to regular behavior. To model a "lesion" in a network, the total strength of output connections of a particular unit (or units) could be decreased sufficiently to allow "stuck" units to become "unstuck" and hence, to allow for aperiodic behavior to reemerge.

If the response function, $g$, is a continuous sigmoid, e.g., in the form (10.8), then the network (10.7) cannot be integrated by computing trajectories and switching times. Instead, the network must be integrated numerically.

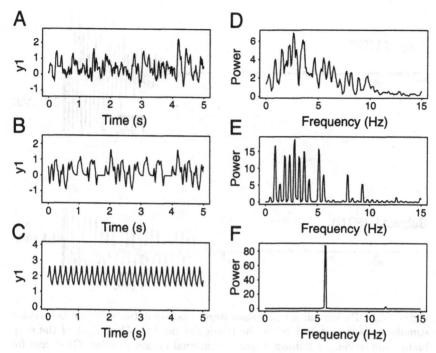

Figure 10.11. Interpolated time series from one unit of each of three random networks ($N = 50$, $K = 10$, $d = 8$) with different values of $\alpha$, after transients have died away: (A)(D) $\alpha = 1.0$, (B)(E) $\alpha = 0.5$, (C)(F) $\alpha = 0.2$. On the left are segments of the time series; on the right are the corresponding power spectra, based on all 6000 points (estimated by smoothing with an 11-point Daniell filter, see Appendix C). The time scale is arbitrary and has been divided by 30 for the plots, so that the frequencies in the spectra are similar to those in the tremor examples. From Edwards, Beuter, and Glass (1999).

## 10.5.3 Dynamical Control of Parkinsonian Tremor by Deep Brain Stimulation

As mentioned in Section 10.2.3, chronic high-frequency electrical deep brain stimulation can suppress tremor in Parkinson's disease. In this surgical technique, an electrode is implanted into subcortical structures in the brain for long-term stimulation. The mechanism by which deep brain stimulation suppresses tremor is unknown, but could involve a gradual change in network properties controlling the generation of tremor. Based on this idea, Titcombe et al. (2001) hypothesized that high-frequency deep brain stimulation induces a qualitative change in the dynamics so that the stable oscillations are destabilized as a parameter affecting the oscillation is modified. One possible qualitative change in the dynamics is a supercritical Hopf bifurcation (Guckenheimer and Holmes 1983; Glass and Mackey 1988): As the control parameter $g$ crosses a critical threshold, $\mu_c$, the dynam-

Subject I (GPi)

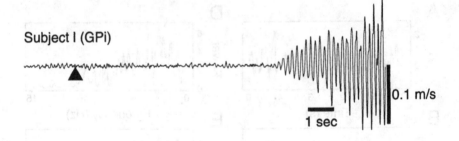

Subject II (STN)

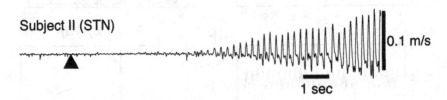

Figure 10.12. Parkinsonian rest tremor finger velocity in two subjects. Deep brain stimulation was switched off at the triangular marker. The target of the deep brain stimulation, for Subject 1 was the internal globus pallidus (GPi), and for Subject 2 was the subthalamic nucleus (STN).

ics change from a fixed point that is nonoscillatory to a stable oscillation (see Section 2.4 on the Hopf bifurcation). Experimental data of rest tremor in subjects with Parkinson's disease receiving high-frequency stimulation showed a gradual increase of tremor amplitude after the stimulation was switched off, as shown in Figure 10.12. This gradual increase of tremor amplitude is typical of transitions occurring as a consequence of a supercritical Hopf bifurcation (Titcombe et al. 2001).

Other hypotheses have been proposed to explain the effects of high-frequency deep brain stimulation on Parkinsonian tremor. Assuming that Parkinsonian tremor is associated with abnormal oscillations in some region of the brain, high-frequency stimulation could block or interfere with the transmission of oscillatory activity to the motor neurons (Montgomery Jr and Baker 2000). With this hypothesis, the abnormal oscillations would still be present in brain structures but would no longer lead to observed tremor. Another hypothesis is that Parkinsonian tremor is associated with an abnormal synchronization of several independent oscillators, but that deep brain stimulation acts to desynchronize these oscillators (Tass 1999; Tass 2000). Yet another hypothesis is that high-frequency stimulation acts via reversible inhibition of the target function, thus mimicking the effects of lesioning the target structure (Benabid et al. 1998). Certain other hypotheses, which are well known in other contexts, have not been considered before in the context of controlling Parkinsonian tremor but still seem equally plausible. For example, high-frequency deep brain stimulation could entrain the abnormal oscillator in a 1:0 rhythm so that the oscillator is effectively arrested as a consequence of repetitive phase resetting; as happens in simple models of cardiac cells or other oscillators (Guevara and Glass 1982). See Chapter 5 for more details on entrainment, especially the section on phase locking of limit cycles by periodic stimulation.

A better understanding of how tremor in Parkinson's disease is controlled, i.e., by high-frequency deep brain stimulation, may provide key information into unlocking the mystery of the generation of Parkinsonian tremor.

## 10.6   Conclusions

The techniques described in Sections 10.3 and 10.4 can be applied to data from rest tremor, proprioceptive tremor (postural tremor with eyes closed), and static tremor (postural tremor with eyes open) and to simulated data from models. They are potentially most useful to aid diagnosis in early stages of pathologies, especially when tremor amplitude itself is not yet large. All of the measures and techniques proposed are potentially discriminators between normal and pathological tremors or between types of pathological tremor. Their utility must be evaluated, however, by testing many subjects with various conditions and answering one of two kinds of questions (for each measure):

- Do the distributions of the measure for each group of subjects show statistically significant differences in means?

- What proportion of subjects can be correctly classified on the basis of the measure?

The first type of question can help to determine which measures are related to particular pathologies or to pathology in general, and should therefore contribute to an understanding of these conditions. However, typically, the distributions for the different groups of subjects overlap considerably, and a statistically significant difference in means may not be particularly helpful in diagnosis, especially since the variability in a normal population in somewhat large. The second type of question helps to identify which measures are most useful in diagnosis, in the sense of making the fewest errors. This type of analysis is actively being pursued (see Beuter and Edwards 1999 and Edwards and Beuter 2000). It is also possible to use neuromotor profiles (eye movements, alternating movements, rapid pointing movements, as well as various tremors) to differentiate neurological conditions (Beuter et al. 2000).

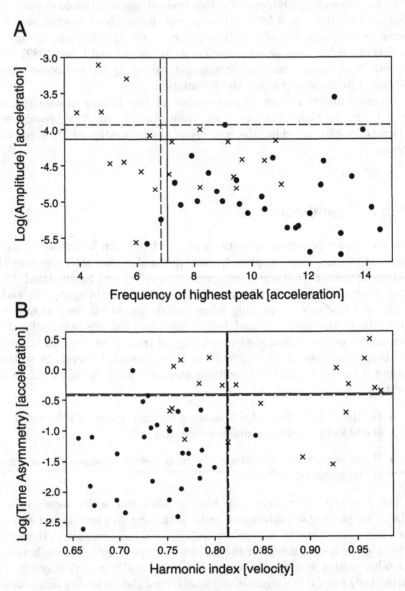

Figure 10.13. Separation of patients with Parkinson's disease (×'s) and control subjects (circles) by pairs of characteristics calculated from postural tremor (no visual feedback). The dotted lines represent the 96.7th percentile for the control group for each characteristic. The solid lines represent the 93.5% level for $z$ scores based on the control group distribution. From Beuter and Edwards (1999).

In practice, it is difficult to assign some subjects to a group with certainty. This is particularly the case with early stages of pathologies, which is where these more sophisticated techniques are potentially of the most value. There are other practical issues that can affect results, such as whether a subject is on medication at the time of testing. Also, the range of variability in the symptoms of these pathologies tends to be very large. In the light of these considerations, a combination of measures may be more useful than any single one. Figure 10.13 shows discrimination between Parkinsonian subjects and control subjects using (a) log amplitude versus frequency of the highest peak in the spectrum, both in acceleration, and (b) harmonic index (in velocity) versus time asymmetry (in acceleration). Details of this discrimination are in Beuter and Edwards (1999). Commonly, abnormalities will show up for a given subject with some of the above techniques that are usually relevant to their condition, but not with others. If a subject shows no abnormalities of the relevant type, we have to consider whether they may be misclassified. Keep in mind, however, that not everyone with Parkinson's disease appears to have a pronounced tremor as a symptom. This, of course, means that they do not have tremor with a higher-than-normal amplitude, but it may or *may not* mean that their tremor is normal in all other respects.

Today, the debate is no longer between central or peripheral mechanisms in tremor generation, since the two are not necessarily mutually exclusive (Britton 1995). Tremor is now increasingly explored by studying in noninvasive ways the relative contributions of the different neural structures under normal and pathological conditions, by examining the encoding of information via the rate or timing of action potentials, and by performing computer simulations. The focus of intense research efforts will be successful if collaborative efforts are undertaken between cerebral imagery specialists, clinical neurologists, neurosurgeons specializing in deep brain stimulation, experimentalists, applied mathematicians, and theoreticians in biology. This chapter has been written in this perspective, and we hope that the approach proposed here will help to generate and explore new testable hypotheses. Much still needs to be done to understand the mechanisms underlying the production of tremor and its numerous transformations in the expression of many neurological disorders.

## 10.7    Computer Exercises: Human Tremor Data Analysis

In the supplied collection of data,[‡] you will find many files containing tremor time series. Each file is a single column containing the $y$-component

---

‡Available on book's website, available via www.springer-ny.com.

(which corresponds to flexion/extension motion of the index finger) of displacement, sampled at 200 Hz. The displacement is measured in millimeters (mm). (**Note:** There is no .dat on the file names.)

There are four "artificial" datasets, three datasets containing tremor time series from normal (control) subjects, three datasets of tremor time series from subjects with Parkinson's disease, and three "mystery" subjects:

| Data type | File | Description |
|---|---|---|
| Artificial data | trema1 | Gaussian white noise |
|  | trema2 | A sine wave at 4.7 Hz |
|  | trema3 | Addition of a sine wave at 4.7 Hz to one with double the amplitude at 8.1 Hz |
|  | trema4 | A sine wave to the 4th power |
| Control data | tremn1 | Control subject 1 |
|  | tremn2 | Control subject 2 |
|  | tremn3 | Control subject 3 |
| Parkinsonian data | tremp1 | Parkinsonian subject 1 |
|  | tremp2 | Parkinsonian subject 2 |
|  | tremp3 | Parkinsonian subject 3 |
| Mystery data | tremm1 | Mystery subject 1 |
|  | tremm2 | Mystery subject 2 |
|  | tremm3 | Mystery subject 3 |

The "artificial" datasets do not contain tremor data at all. They are synthesized data intended to help you to develop intuition about the statistics that will be introduced later in this lab exercise.

The "artificial" datasets are contained in the files **trema1** through **trema4**:

- **trema1** Gaussian white noise;

- **trema2** A sine wave at 4.7 Hz (recall that the sampling frequency of the tremor data is 200 Hz):

  sin(2*pi*4.7*(0:1000)/200)

  gives 5 seconds of this sine wave;

- **trema3** A sine wave at 4.7 Hz added to a sine wave of amplitude 2 at 8.1 Hz;

- **trema4** A sine wave to the fourth power:

  1 - 2*sin(2*pi*2.3*(0:1000)/200).^4

Of the three "mystery" subjects, there is one normal (control) subject, one subject with the trembling form of Parkinson's disease, and one subject with the akineto-rigid form of Parkinson's disease. In Ex. 10.7.2-4, you will use statistics on the time series to classify each of the mystery subjects. The status of the mystery subjects is revealed at the end of the exercises.

## *Software*

There are twelve Matlab[§] programs you will use for the data analysis exercises:

**diff(x)** a built-in function of Matlab that takes finite differences of a time series **x** (i.e., approximates the first derivative);

**dvaspec(displdata)** plots three power spectra in one figure window of (i) **displdata**, the raw displacement data specified, (ii) its numerical first derivative, or velocity, and (iii) its numerical second derivative, or acceleration;

**datafilt(data,filt)** filters **data** in the frequency domain using the filter **filt**;

**makefilt(cut,len,ramp)** creates a low-pass filter in the frequency domain at the cut-off frequency **cut** with a transition of width **ramp** of the same length **len** as the dataset to be filtered;

**tremplot(group,deriv)** plots the differentiated, filtered time series in one **group** ('p'=Parkinsonian, 'n'=Normal (control) group, 'm'=Mystery group, 'a'= the artificial data group), in the specified variable **deriv** (0 = displacement, 1 = velocity, 2 = acceleration, and if no **deriv** is specified, the raw displacement data are plotted, i.e., without any filtering);

**tremstat(group,stat,deriv,first,last)** calculates a specified statistic **stat** on each dataset of one **group** in the variable indicated by **deriv** [optional: **first** and **last** specify the first and last datapoints to use];

**std(x)** a built-in function of Matlab that computes the standard deviation (used as a measure of amplitude of a time series **x**);

**domfreq(x)** finds the dominant frequency in a time series **x**;

**pow4to6(x)** finds the power in the 4–6 Hz frequency range of the data in **x**;

**freqconc(x)** measures the 68th percentile frequency concentration, which is the combined width of the narrowest set of frequency ranges (between 0 and 15 Hz) that contains 68% of the energy of the signal **x**;

**wobble(x)** measures the ratio of numbers of extrema (maxima and minima) in acceleration and displacement, or equivalently, the ratio of numbers of zero-crossings in "jerk" (the derivative of acceleration) and velocity, calculated from the signal **x** (Note: **x** should be a velocity signal);

---

[§]See Introduction to Matlab in Appendix B.

**posneg(x)** is a measure of the positive/negative asymmetry of the time series **x** using the ratio of duration and magnitude of positive and negative parts of the oscillations. See equation (10.3) in Section 10.4.2 on morphology.

### 10.7.1  Exercises: Displacement Versus Velocity Versus Acceleration

The purpose of these exercises is to visualize the raw data, to learn to filter appropriately these data, to compare datasets in the form of displacement, velocity, and acceleration, and to visualize data in the time and frequency domains.

Note: in Ex. 10.7.1-1 and Ex. 10.7.1-2, the data are loaded automatically by the programs **tremplot** and **tremfreqplot**.

Ex. 10.7.1-1. **Plot raw displacement data in the time domain.** An important preliminary step in data analysis is to visualize the data.

First, to plot the raw displacement data for the Parkinsonian group, enter the commands

```
figure(1)
tremplot('p')
```

Next, to plot the raw displacement data for the control (normal) group, enter the commands

```
figure(2)
tremplot('n')
```

You should have two figure windows, each with three tremor time series.

Look for differences in the morphology of the tremor at different time scales, e.g., bursts in 0–25 seconds and regularity in 0–5 seconds in Parkinsonian tremor, versus the irregularity that you see in normal tremor.

Use the Zoom-in Tool on the Matlab Figure window to focus your attention on particular ranges of the data. Alternatively, you can use the built-in **axis** command, which has the form **axis([xmin xmax ymin ymax])**.

Do you see any trends in the data that would indicate that the finger position is drifting?

Do the first and last points of the time series appear to match very well?

**Ex. 10.7.1-2. Plot the raw displacement data in the frequency domain.** It is also helpful to visualize the data in the frequency domain.

To plot the power spectra for the raw displacement data of the Parkinsonian group, enter the commands

```
figure(3)
tremfreqplot('p')
```

Then, to plot the power spectra for the raw displacement data of the control (normal) group, enter the commands

```
figure(4)
tremfreqplot('n')
```

You should have another two figure windows, each with three power spectra. Note that the $x$-axis is now frequency measured in Hz.

Use the Zoom-in Tool on the Matlab Figure window to focus your attention on particular ranges of the spectra.

Look for differences in the spectra: A more concentrated spectrum in Parkinsonian tremor versus a broad spectrum associated with normal tremor.

Do you see a peak (or, more generally, a concentration of energy) in the frequency range associated with tremor (3–15 Hz)? Sections 10.2.2 and 10.2.3 describe frequency ranges of interest in physiological and Parkinsonian tremor. If you are looking at normal tremor, then there is no expectation of a clear peak in the spectrum. You may see a broad hump in the spectrum somewhere between 7 and 12 Hz, but even then, it may be small compared to the power in the low-frequency ranges in the displacement spectrum.

Do you see considerable low-frequency content that is below the frequency range of tremor?

**Ex. 10.7.1-3. Load displacement data; Obtain velocity and acceleration data by numerical differentiation.** Recall that velocity is the derivative of displacement and that acceleration is the derivative of velocity. Using numerical differentiation, one can obtain approximations to velocity and acceleration signals from displacement data. If we let $x$ be the displacement as a function of time $t$, then the velocity $v$ is

$$v = \frac{dx}{dt} = \lim_{\Delta t \to 0} \frac{x(t + \Delta t) - x(t)}{\Delta t}.$$

Thus, we can approximate the velocity using the `Matlab` command `diff`, provided that $\Delta t$ (i.e., the reciprocal of the sampling rate) is sufficiently small.

For example, to obtain the velocity approximation of `tremp1` (displacement data), take its first derivative using the commands

```
load tremp1;
vtremp1 = diff(tremp1)*200;
```

Why multiply by the factor 200? How many datapoints will `vtremp1` contain compared to `tremp1`? What are the units of these velocity data?

Similarly, to obtain the acceleration approximation of `tremp1`, take its second derivative using the command

```
atremp1 = diff(diff(tremp1))*(200^2);
```

This corresponds to measuring acceleration instead of displacement. Why multiply by the factor of $200^2$? What are the units of these acceleration data?

### Ex. 10.7.1-4. Plot raw velocity and acceleration data in the frequency domain.

Plot the power spectra for Parkinsonian subject 1 using the raw displacement data (`tremp1`), raw velocity data (`vtremp1`), and raw acceleration data (`atremp1`). We have provided the function `dvaspec`, which will automatically plot the three spectra in the same figure window. Enter the command

```
dvaspec('tremp1')
```

The function `dvaspec` automatically takes the numerical first and second derivatives of `tremp1` to obtain the approximations for velocity and acceleration, respectively.

Do you see a peak in the tremor range in any or all of these spectra?

Compare the power spectra of the displacement, velocity, and acceleration data. In particular, compare the energy content in a low-frequency range (say, 0–3 Hz) and in a higher-frequency range (say, 15–20 Hz). Do you see a difference between the signal in displacement, velocity, or acceleration? (Remember, these are the same data!)

**Recall:** The relationship between the power spectrum of the acceleration and that of the displacement is described in Section 10.3.1 in the lecture notes.

Repeat Ex. 10.7.1-3 and Ex. 10.7.1-4 for one of the control subjects (e.g., **tremn1**).

**Ex. 10.7.1-5. Design and apply a low-pass filter to the data.** You have just seen that the acceleration has a lot of energy at frequencies above the range of tremor, due to amplification of noise by the double differentiation. This is also true for the velocity signal obtained by differentiating the displacement data. (In all digital measurements, there is quantization noise that is white. We amplify this noise, and any other noise, when taking the second derivative to simulate measuring acceleration. **Note:** If the acceleration had been measured directly by an accelerometer, this amplification would not have occurred. However, low-frequency noise would be amplified if we were to integrate acceleration data to obtain velocity or displacement.)

You will design a low-pass filter to get rid of this amplification of noise using the function **makefilt**. The general form of **makefilt** is

```
makefilt(cut,len,ramp)
```

where **cut** is the cutoff frequency in terms of the sampling frequency, **len** is the length of the dataset to be filtered, and **ramp** is the width of the transition ramp.

Since the sampling frequency is 200 Hz, to eliminate everything above 15 Hz, we can design a filter with **cut** $= 15/200 = 0.075$ and and with **width** $= (20 - 15)/200 = 0.025$ to ramp the filter between 15 and 20 Hz. To create this filter to apply to the velocity signal **vtremp1**, enter the command

```
low1520 = makefilt(0.075,length(vtremp1),0.025);
```

The filter you have created is called **low1520** to indicate a low-pass filter at 15 Hz that ramps to 20 Hz.

Now use the filter **low1520** to clean up the noise amplification in the velocity signal:

```
fvtremp1 = datafilt(vtremp1,low1520);
```

The filtered velocity signal is now in a dataset called **fvtremp1**.

Compare **vtremp1** and **fvtremp1** in the time domain:

```
plot([1:length(vtremp1)]/200,[vtremp1,fvtremp1])
```

The oscillations are much more apparent in **fvtremp1** (use the Zoom-in Tool).

**Ex. 10.7.1-6. Compare amplitude in displacement and acceleration.** The file **accvdisp.dat** contains the amplitude values[¶] of 48 datasets. The file has three columns:

**column 1:** the amplitude of the filtered acceleration measured in m/s², sured in $m/s^2$,

**column 2:** the amplitude of the raw displacement time series,

**column 3:** described in Ex. 10.7.1-8 below.

There is one row in the file **accvdisp.dat** for each of the 48 datasets. Compare the two amplitudes using

```
load accvdisp.dat;
plot(accvdisp(:,1),accvdisp(:,2),'+')
```

Is there a strong correlation between the amplitudes of the filtered acceleration series and those of the raw displacement data? Speculate why there is not.

Note that **vtremp1** and **fvtremp1** are measured in mm/s, so to obtain the amplitude of **vtremp1**, for example, one would use the command **std(vtremp1)/1000** to convert the units to m/s. Similarly for acceleration signals.

**Ex. 10.7.1-7. Displacement data: detrending and highpass filtering.** One of the problems with the displacement data is that the finger position drifts, and so there is considerable low-frequency content that is below the frequency range of tremor. We want to get rid of this by high-pass filtering.

Since the first and last points of the time series do not match very well, filtering in the spectral domain is going to cause problems. The problems arise because complete sine wave oscillations will have the same value at the beginning and end of the dataset and thus will not match the data well at both ends. We can address this issue by *detrending*: subtracting out a trend from the data. After this trend is subtracted, we can filter in the spectral domain.

---

[¶]We will use the standard deviation as a measure of the amplitude; e.g., **std(tremp1)** computes the standard deviation of the data in **tremp1**

- Fit a polynomial of order 4 to the data using the Matlab commands polyfit and polyval:

  ```
  [p,s,mu] = polyfit((1:length(tremp1))',tremp1,4);
  % Note the ' for transpose.
  y = polyval(p,(1:length(tremp1))',[],mu); % Again,
  note the '
  plot([tremp1,y])
  ```

- Subtract the polynomial trend from the data:

  ```
  dtremp1 = tremp1-y;
  ```

- Create a high-pass filter that will eliminate everything below 2 Hz:

  ```
  hp = 1 - makefilt(0.01,length(dtremp1),0.005);
  ```

- Filter the detrended data, and take the standard deviation:

  ```
  fdtremp1 = datafilt(dtremp1,hp);
  std(fdtremp1)
  ```

  Compare this result with the standard deviation (amplitude) of the original data.

- Compare fdtremp1 and the original data tremp1 in both the time and frequency domains to see the effect of the filtering. For example, for the time domain plot,

  ```
  plot([fdtremp1,tremp1]);
  ```

Ex. 10.7.1-8. **Compare amplitude of data in displacement and acceleration, filtered and raw.** The third column of accvdisp contains the standard deviation (amplitude) of the high-pass-filtered, detrended displacement data. Compare this to the standard deviation of the filtered acceleration data and the raw displacement data. See Ex. 10.7.1-6 above.

What conclusions can you draw about the comparability of using acceleration, velocity, or displacement to measure tremor?

## 10.7.2  Exercises: Distinguishing Different Types of Tremor

The purpose of these exercises is to distinguish different types of tremor. You are given three displacement time series from subjects with Parkinson's disease, three from normal (control) subjects, and three from "mystery" subjects. As part of this exercise, it is left to you to classify these mystery subjects: One is a normal (control) subject, one is a subject with the trembling form of Parkinson's disease, and one is a subject with the akineto-rigid form of Parkinson's disease.

## Statistics

Calculate various statistics on these time series or their first or second derivatives to see which ones allow you to distinguish between the control subjects and the subjects with Parkinson's disease. You can develop or use any statistic you like.

Here are some statistics you might like to try (all the `Matlab` programs take the variable holding the time series as an argument, e.g., `std(x)`, where x is the time series):

**Amplitude:** as measured by the standard deviation, `std(x)`;

**Dominant frequency:** `domfreq(x)` the frequency at which the highest peak in a smoothed spectrum occurs;

**Proportion of energy in the 4–6 Hz range:** `pow4to6(x)`;

**Frequency concentration:** `freqconc(x)` measures the combined width of the narrowest set of frequency ranges that contains 68% of the energy of the signal;

**Wobble:** `wobble(x)`, which *must* be applied to velocity data, measures the ratio of numbers of extrema (maxima and minima) in acceleration and displacement. This measure is calculated via the ratio of numbers of zero-crossings in "jerk" (the derivative of acceleration) and velocity;

**Positive/negative asymmetry:** `posneg(x)`, which should be applied to acceleration data, measures the ratio of duration and magnitude of the positive and negative parts of the oscillations. See equation (10.3) in Section 10.4.2 on morphology.

**Note:** Remember that certain statistics may not be appropriate for all types of data signals. For example, the measure `wobble` is applied only to velocity data, and the measure `posneg` is applied only to acceleration data.

## Exercises

Ex. 10.7.2-1. **Compare the statistics to plots of the data.** To convince yourself that the values you calculate make sense, plot the data that you are analyzing. (Note that the statistics are applied to appropriately filtered data.)

The function `tremplot` can also be used to plot filtered data. The first argument is either 'p', 'n', 'm', or 'a', depending on whether you want Parkinsonian data, normal data, mystery data, or the artificial data. The second argument is 0, 1, or 2, depending on whether you want 0 = filtered displacement signal (detrended and high-pass filtered), 1 = filtered velocity signal (low-pass filtered), or 2 = filtered acceleration signal (low-pass filtered).

For example, to plot the filtered velocity data of the Parkinson group, enter the command

```
tremplot('p',1)
```

Or to plot the filtered acceleration data of the normal (control) group, enter the command

```
tremplot('n',2)
```

Ex. 10.7.2-2. **Artificial data.** Try out the statistics listed above on each of the "artificial" datasets trema1 through trema4, and any other statistics that you wish to apply. Organize your results in tables (see example table on page 350). Look at the plots of the data you analyzed as described in Ex. 10.7.2-1. Try to explain to yourself how and why the results are different on the different datasets.

For example, to calculate the dominant frequency (domfreq) on the filtered artificial displacment data, enter the command:

```
tremstat('a','domfreq',0)
```

The results, echoed on the screen, are the values in order for each of the four artificial datasets.

**Note:** The general form of the function tremstat is

```
tremstat(GROUP,STAT,SIGNAL)
```

The first argument, GROUP, is either 'p', 'n', 'm', or 'a', depending on whether you want Parkinsonian data, normal data, mystery data, or the artificial data. The second argument, STAT, is the statistic you want calculated, which must be a function of the form stat(x) and must be between single quotes. The last argument, SIGNAL, is 0, 1, or 2, depending on whether you want $0 =$ displacement signal (detrended and high-pass filtered), $1 =$ velocity signal (low-pass filtered), or $2 =$ acceleration (low-pass filtered).

The function tremstat automatically applies the appropriate filtering to the signal you wish to analyze.

Ex. 10.7.2-3. **Normal and Parkinsonian data.** Repeat Ex. 10.7.2-2 on each of the datasets of the normal (control) group and the Parkinsonian group. Especially, try to figure out what statistics work best to distinguish between Parkinsonian data and data from control subjects, and whether the statistics are better in displacement, velocity, or acceleration.

Ex. 10.7.2-4. **Mystery data.** Now that you are familiar with the statistics and with Parkinsonian and normal tremor, try to classify each of the recordings from the "mystery" dataset. The true nature of the mystery files will be revealed at the end of the exercises.

Example: When examining velocity data (also make tables for displacement and acceleration data), your table might look like the following:

**Velocity data results:**

| File | std | domfreq | pow4to6 | freqconc | wobble |
|---|---|---|---|---|---|
| Artificial trema1 | | | | | |
| trema2 | | | | | |
| trema3 | | | | | |
| trema4 | | | | | |
| Normal tremn1 | | | | | |
| tremn2 | | | | | |
| tremn3 | | | | | |
| Park tremp1 | | | | | |
| tremp2 | | | | | |
| tremp3 | | | | | |
| Mystery tremm1 | | | | | |
| tremm2 | | | | | |
| tremm3 | | | | | |

# 10.8 Computer Exercises: Neural Network Modeling of Human Tremor

*Software*

Here is a list of some software (in Matlab) you will use in the neural network model exercise of this lab:

**plotresponse(beta)** plots the sigmoid response function for a specified gain **beta**;

**sigmnetwork(yinitial,alpha,tau,beta)** computes the solution to the six-unit Hopfield-type neural network with a sigmoid response, using a fourth-order Runge–Kutta integration scheme with 5000 steps of size 0.01, for a given vector of initial values **yinitial**, parameter **alpha**, threshold vector **tau**, and response gain **beta**;

**stepnetwork(yinitial,alpha,tau)** computes the solution to the six-unit Hopfield-type neural network with a step function response, by computing trajectories and switching times of the units, for a given vector of initial values **yinitial**, parameter **alpha**, and threshold vector **tau**.

## *Exercises*

In these exercises, you will examine the behavior of the solution of a six-unit Hopfield-type neural network of the type described in Section 10.5.2. We have written `Matlab` programs `stepnetwork` and `sigmnetwork`, described above, that will compute the solution for you.

**Ex. 10.8-1. The six-unit Hopfield-type neural network.** The six-unit neural network that you will solve for $(y_1, \ldots, y_6)$ is

$$\dot{y}_i = -y_i + \sum_{j=1}^{6} w_{ij} g(y_j) - \tau_i, \quad i = 1, \ldots, 6. \qquad (10.10)$$

The response function, $g(y_j)$, that you use will be either a Heaviside step function, i.e.,

$$g(y_j) = \begin{cases} 0 & \text{if } y_j < 0, \\ 1 & \text{otherwise}, \end{cases} \qquad (10.11)$$

or a sigmoid function of the form

$$g(y_j) = \frac{1 + \tanh \beta y_j}{2}, \qquad (10.12)$$

where $\beta$ controls the slope (gain) of the response. The program `stepnetwork` solves equation (10.10) using the response in equation (10.11), and the program `sigmnetwork` solves equation (10.10) using the response in equation (10.12). The matrix $W$ containing the "synaptic efficacies," $w_{ij}$, of the units of the network in equation (10.10) is set to be

$$W = - \begin{pmatrix} 0 & 1 & 0 & 0 & 0 & 1 \\ 0 & 0 & 0 & 1 & 0 & 1 \\ 0 & 0 & 0 & 1 & \alpha & 0 \\ 1 & 0 & 0 & 0 & 0 & 1 \\ 1 & 1 & 0 & 0 & 0 & 0 \\ 0 & 0 & \alpha & 0 & \alpha & 0 \end{pmatrix}. \qquad (10.13)$$

(The minus sign in front of the matrix indicates that all the connections in this network are inhibitory.) See Figure 10.14. Decreasing the parameter $\alpha$ in equation (10.13) represents a weakening of synaptic efficacy. Note that $\alpha \in [0, 1]$. The vector $\tau = (\tau_1, \ldots, \tau_6)$ in equation (10.10) contains the threshold levels of the units in the network.

**Ex. 10.8-2. Compute the solution to the network with a Heaviside step function response.** In this exercise, you will solve the six-unit

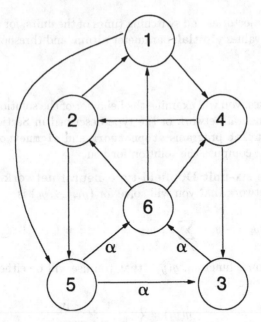

Figure 10.14. The connection structure of the six-unit network. All connections are inhibitory with weight 1 except the three labeled $\alpha$, which are weakened as $\alpha$ decreases. From Edwards, Beuter, and Glass (1999).

system in equation (10.10) with a Heaviside step function response using **stepnetwork**.

To do this, you will need to specify the following parameters:

- the initial value[||] of the vector $y = (y_1, \ldots, y_6)$,

- the parameter $\alpha$ (decreasing its value represents weakening of synaptic efficacy), and

- the thresholds of the units, $\tau = (\tau_1, \ldots, \tau_6)$. For interesting behavior, set each $\tau_i \in (-2, 0)$.

(a) **Specify the parameters: y with random initial values, $\alpha = 1$, and $\tau_i = -1.5$ for $i = 1, \ldots, 6$ (i.e., all the thresholds are the same).** To specify these parameters, enter the commands

```
y0 = randn(6,1);
alpha = 1;
tau = -1.5*ones(6,1);
```

---

[||]For the initial value of the vector $y$, we use the **Matlab** random number (normally distributed) generator **randn**.

(b) **Compute the solution ys, as a function of time, ts, in arbitrary time units.** To compute the solution using the specified parameters, enter the command

```
[ts,ys] = stepnetwork(y0,alpha,tau);
```

You will see the message "Solution computed" when the program has terminated.

(c) **Plot the solution for the first unit, $y_1$.** To plot the solution for the first unit, enter the command

```
plot(ts,ys(1,:))
```

Looking at your plot, consider whether the behavior of the solution is consistent with normal physiological tremor or Parkinsonian tremor.

(d) **Save parameters and solution for later comparison.** Save the initial value y0, the threshold tau, and the solution ys for comparison later with the system with a sigmoidal response. To save these values, enter the commands

```
y0save = y0;
tausave = tau;
ystep = ys;
```

(e) **Examine the effect of $\alpha$ and $\tau$ on the behavior of the solution.** Redo Ex. 10.8-2a–Ex. 10.8-2c with a decreased value of $\alpha$, e.g., $\alpha = 0.7$ (Note: In Ex. 10.8-2a, you need to enter the command only for a new value of **alpha**).

Then, redo Ex. 10.8-2a–Ex. 10.8-2c for different thresholds in the vector $\tau$ and see what happens to the solution, (e.g., `tau=-0.5*[3 1 3 1 1 1];`).

Again, consider the behavior of the solution in terms of normal tremor or Parkinsonian tremor, and compare the result with $\alpha = 1$ to that of $\alpha = 0.7$.

**Ex. 10.8-3. Compute the solution to the network with a sigmoidal response.** Now you will solve the six-unit system in equation (10.10) with a sigmoid function response using **sigmnetwork**. [A small warning: This program is a lot slower than the step response program, so be patient as it computes!]

As well as specifying the parameters as you did above (listed just before Ex. 10.8-2a), you will also need to specify

- the parameter $\beta$, which controls the slope of the response in equation (10.12).

(a) **Examine the effect of $\beta$ on the response function.** Plot the response function, $g(y)$, in equation (10.12) for varying values of $\beta$. For example, to plot the response function for $\beta = 4$, enter the commands

```
beta = 4;
plotresponse(beta);
```

(b) **Specify the parameters: y with random initial values,** $\alpha = 1$, $\tau_i = -1.5$ for $i = 1, \ldots, 6$ (i.e., **all the thresholds are the same), and** $\beta = 10$. To specify these parameters, enter the commands

```
y0 = randn(6,1);
alpha = 1;
tau = -1.5*ones(6,1);
beta = 10;
```

(c) **Compute the solution ys, as a function of time, ts (in arbitrary time units).** To compute the solution to the network with a sigmoidal response using the specified parameters, enter the command

```
[ts,ys] = sigmnetwork(y0,alpha,tau,beta);
```

You will see the message "Computing solution ..." as the program is running, and the message "Solution computed" when it has terminated.

(d) **Plot the solution for the first unit,** $y_1$. To plot the solution for the first unit, enter the command

```
plot(ts,ys(1,:))
```

Again, when looking at your plot, consider whether the behavior of the solution is consistent with normal physiological tremor or Parkinsonian tremor.

(e) **Examine the effect of $\beta$ and $\alpha$ on the solution.** Redo Ex. 10.8-3b–Ex. 10.8-3d for different values of $\alpha$ and $\beta$. For example, try $\beta = 4$ and $\alpha = 0.7$.

(f) **Compare solutions of the network with step function or sigmoidal response.** To compare the solutions, we need first to compute the solution for a sigmoid response using the *same* initial value of $y$, and the same threshold $\tau$, that we used for the step function response (which you saved in y0save and tausave).

To compute the solution for a sigmoid response with the saved parameters, and with $\alpha = 1$ and $\beta = 10$, enter the commands

```
[ts,ysigm] = sigmnetwork(y0save,1,tausave,10);
```
Be patient! The computations in the program **sigmnetwork** take a while.

Then, plot the two solutions (for the first unit) on the same graph. To plot $y_1$ for the step function solution as a solid (blue) line and $y_1$ for the sigmoid solution as a dashed (red) line, enter the commands

```
plot(ts,ystep(1,:),'b',ts,ysigm(1,:),'r--')
legend('Step','Sigmoid') % This legend command
is optional
```

Here, **ts** is a vector containing the time data (in arbitrary time units). Recall that **ystep** is the step function solution that you saved in Ex. 10.8-2.

What conclusions can you draw about the effect of the slope of the response function from the comparison of the two solutions?

# Answers to Exercise: Distinguishing Different Types of Tremor

Mystery subject $1$ = parkinsonian (typical)
Mystery subject $2$ = normal
Mystery subject $3$ = parkinsonian (akineto-rigid form)

# Appendices

Appendices

# AppendixA

# An Introduction to XPP

## Michael C. Mackey

We shall be using the mathematical software package XPP for differential equations in the computer exercises in Sections 3.11, 4.7, 4.8, and for time-delay equations in computer exercises of Section 8.7. In addition, XPP can also handle finite-difference equations, stochastic systems, boundary value problems, etc. One reason why we have selected it for use in this book is that it incorporates a bifurcation analysis package, Auto. The version of XPP incorporating Auto is thus called XPPAUT.

The XPP package was developed by Bard Ermentrout, of the Mathematics Department of the University of Pittsburgh. It is available without charge for both Unix and Windows operating systems from

> http://www.math.pitt.edu/~bard/xpp/download.html

(The XPP manual is also available to download.)

There is a nice tutorial that can be run over the web from the original site at

> http://www.math.pitt.edu/~bard/bardware/tut/start.html

At the end of the XPP manual, there is a brief description of the Auto interface. There is also online documentation at

> http://www.math.pitt.edu/~bard/xpp/help/xpphelp.html

One can also obtain the original Auto program, which was written by Esebius Doedel (a member of the Centre for Nonlinear Dynamics in Physiology and Medicine), free of charge, from

> http://indy.cs.concordia.ca/auto/download/index.html

## A.1 ODE Files

The basic unit for XPP is a single ASCII file that has the equations, parameters, variables, boundary conditions, and functions for your model. The methods of solving the equations, the graphics, and postprocessing are all done within the program using the mouse and various menus and buttons.

You should regard the ODE file as a framework for exploring the system; its main role is to set up the number of variables, parameters, and functions. All XPP files must have the extension .ode, and this holds even for differential delay equations. Here is an example of an ODE file:

```
# Note: in ODE files, comments are preceded by #

# fhn.ode (XPP file):- written by M. Guevara.
# This is an  XPP file to numerically integrate the
# FitzHugh-Nagumo equations.

# EQUATIONS
  dx/dt = c*(x-(x^3)/3+y+stim(t))
  dy/dt = -(x-a+b*y)/c
  aux v=-x
  stim(t) = amplitude*(heav(t-tstart)-heav(t-(tstart+duration)))

# PARAMETERS
  p a=0.7, b=0.8, c=3.0, amplitude=0.0, tstart=20.0, duration=0.0

# INITIAL CONDITIONS
  init x=2.0, y=2.0

# CHANGES FROM XPP'S DEFAULT VALUES
  @ TOTAL=20, DT=0.01, XLO=0.0, XHI=20.0, YLO=-2.5, YHI=2.5

  done
```

The most basic ODE file will include the equations in XPP format and the parameters. XPP will automatically read the number and names of the variables of the model as they are introduced in the equations. The definition of an auxiliary function (aux) allows one to define supplementary variables that are not included in the main equations of the model. The initial conditions and the last section, headed "changes from XPP's default values," are not mandatory but allow one to change values away from the XPP default values for the integration routine and the plotting package.

## A.2   Starting and Quitting XPP

To start XPP, type in xpp filename.ode. Six XPP windows will appear on your screen. Iconify the six windows (these have handles (the title appearing at the bar at the top of each window) XPP >> filename.ode, Equations, Initial Data, Delay ICs, Boundary Conditions, Parameters, and Data Viewer) by clicking on the iconify button (the button with the dot in its middle in the extreme right-hand corner of each window). Deiconify the icon titled xpp by double-clicking on it. The main XPP window, with title XPP >> filename.ode, will now appear.

To quit XPP (any time it is not in the middle of a computation), click on (F)ile to get the FILE menu and then click on (Q)uit to quit the program. You will be asked whether you are sure. Click (Yes) to really quit.

Any time that you want to get out of a particular XPP section all you have to do is hit the <Esc> key. Hitting <Esc> will abort most commands such as saving to a file or integrating equations.

The first thing to do might be to turn off XPP's "beep" sound, which will be annoying after a few iterations. Click on (F)ile and click on Bell off.

## A.3   Time Series

Start numerical integration by clicking on Initialconds (the first item in the menu list at the left of the window), which brings up the Integrate submenu. Click on (G)o to start the process of integration. You will see a curve appear in the plotting window: This is a plot of the variable as a function of time.

Now if you want to look at what the other variables are doing, click on $X_i$ vs t in the main XPP menu. The line Plot vs t: X will appear above the menu. Hit the Backspace key to delete X and then type in the name of the new variable. You will note that the new variable is now plotted as a function of $t$ in the plot part of the window.

Click on Initialconds and then on Last. This restarts an integration run using as initial conditions the values at the end of the last run.

## A.4   Numerics

There are a number of things in the numerics window that you may wish to play with. In the bar above the main screen you will see total:. Click on nUmerics and then Total Backspace to erase the number in the bar and replace this with the length of time that you wish to run the integrations (for example, 100).

Dt sets the integration step size, whose default value is 0.05, which is perfectly adequate for the labs.

Method determines the numerical algorithm used to integrate the differential equation. Click on Method, and a smaller window will appear listing several options. You should see (R)unge-Kutta x, indicating that a Runge-Kutta method is the default. This is usually perfectly adequate, and if it is not selected, do it now. (Surprisingly, the brain-dead Euler method can also work pretty well, but you sometimes have to use small integration step sizes Dt to get accurate results.)

If you click on dElay, in the top bar of the main screen you will see Maximal delay: 0. Use the backspace key to eliminate this and insert something greater that the largest delay that you want to use (in the case of delay differential equations).

## A.5  Graphic Tricks

### A.5.1  Axis

You can expand the range of the Y-axis using Viewaxes (or Window/zoom followed by Fit). You will often need to change the scale of the plot, for example to examine the transient at the start of the trace more closely. In order to do this, click on Viewaxes in the main XPP menu, and then on 2D in the Axes window that appears. This brings up the 2D View window that sets the plot variables, axis ranges, and axis labels. Use the mouse to click on Xmin. Use the backspace key to delete the number appearing here. Type in the new number followed by <Enter>. Then move sequentially between the items in the menu entering the new number for Ymin, Xmax, and for Ymax. (You can also use the <Esc> key to sequentially move among the items in this menu or use the mouse to select individual items). If you wish, you can also enter a label for the X- and Y-axes (Xlabel and Ylabel). Click on Ok to return to the main XPP window.

### A.5.2  Multiplotting

XPP always deletes the data from an integration each time one is run. Thus in order to plot curves with various parameters, you have to Freeze the current curve before changing the parameters and reintegrating.

### A.5.3  Erasing

If the plot window gets too cluttered with traces, they can be erased by clicking on Erase in the main XPP window.

## A.5.4 Printing the Figures

If at any time you wish to get a printed copy of your graphical output, all you have to do is click on **Graphics stuff**, and then on **Postscript**. This will cause the graphics to be exported to a graphics file, and the default name **filename.ode.ps** will appear in the top bar. You can modify this if desired, and then print the file.

# A.6    Examining the Numbers

So far, we have just looked at the actual trajectories. XPP lets you examine the actual numerical values very easily and write them to a text file for further processing. Click on the iconified Data Browser/Viewer to manipulate the numbers you have computed. You will see two columns corresponding to time and to membrane potential. Use the arrow keys to move up and down the data file. The Data Viewer lets you write the numbers as an ASCII file. Click on (Write), type in a file name, and click (OK) to save all the numbers in a file.

# A.7    Changing the Initial Condition

Deiconify the **ICs** icon by clicking on it. You will see that the variables have initial conditions as set up in the ODE file. To change these, hit the Delete key and type in the new value followed by **<Enter>** to change the initial condition on each variable. When you select **Initialconds** and then **Go** from the main XPP window, you will see a second trace appear on the main window.

An easy way to set initial conditions is to click on **Initialconds** and then on **Mouse**. Then, click at the position in the $(x, y)$ plane that you wish to be the initial condition. Again, the trajectories can be erased by clicking on **Erase** in the main XPP window.

You can have XPP explore initial conditions in a systematic way by first erasing the plots and then clicking on **Dir.field/flow** in the main XPP window, and then on **Flow** in the **Two-D Fun** menu that appears. The phrase **Grid:10** will appear at the top of the menu. Press **<Enter>** to accept this default value. XPP will then plot trajectories starting from an evenly spaced $10 \times 10$ grid of initial conditions.

You can also use the window **Range** in **Integrate**, which will allow you to do automatically a series of integrations using a range of initial conditions.

## A.8   Finding the Fixed Points and Their Stability

Click on Sing pts in the main XPP window, and then on Go in the Equilibria window that pops up, and then on NO in response to the question Print Eigenvalues?. The number of positive $(r^+)$ and negative $(r^-)$ real eigenvalues and complex eigenvalues with positive $(c^+)$ and negative $(c^-)$ real parts is given in the Equilibria window.

If you click YES in response to the question Print Eigenvalues?, XPP will calculate the eigenvalues of the equilibrium point. These eigenvalues appear in the xterm window from which you invoked XPP.

## A.9   Drawing Nullclines and Direction Field

Erase the screen and then click on Nullcline in the main XPP window and then on New in the Nullclines menu that pops up. The red curve is the $x$-nullcline (the solution of the equation $dx/dt = 0$), while the green line is the $y$-nullcline. The intersection of the two nullclines gives the steady state or the equilibrium point.

Do this by selecting Dir.field/flow from the main XPP window, clicking on Direct Field in the Two-D Fun menu that pops up, and typing <Enter> in response to the prompt Grid:10 that appears at the top of the window. You will now see arrows drawn in the phase plane at each point on an evenly spaced $10 \times 10$ grid. These arrows are **tangent vectors** to the flow and so give the direction of $dy/dx$. By examining this **direction field** (the collection of tangent vectors), you should be able to understand why trajectories have the shape that they do (try running a few integrations from different initial conditions set with the mouse, so that you have a few trajectories superimposed on the vector field).

## A.10   Changing the Parameters

Deiconify the Par(ameter) window. Enter the new values for each parameter (remember to click on <Enter> after each value and to click on Ok at the end).

To do a range of integrations automatically, click on Phaseplane v.3 and then on (R)ange. You will have the opportunity to enter the name of the parameter in the Range over: box, and following that, select the number of steps, and the starting and ending values of the parameter.

# A.11   Auto

## A.11.1   Bifurcation Diagram

Invoke Auto by clicking on File and then on Auto. This brings up the main Auto window, entitled It's Auto man!. We must now give Auto instruction on how to compute and then plot the bifurcation diagram. Click on Axes and then on hI-lo. You can change the values for Xmin and Ymin, and Xmax and Ymax in the AutoPlot window. These parameters control the length of the $x-$ and $y$-axes of the bifurcation diagram. Click on Ok to exit that window. You will now see that the axes of the bifurcation diagram have been drawn and labeled in the main Auto window. Click on Numerics to bring up the menu that controls the main numerical methods used by Auto. In the AutoNum window that pops up, set Par Min and Par Max (this sets the range of $r$ that will be investigated). The parameters Ds, Dsmin, and Dsmax control the adaptive step size used in the continuation method (essentially, these three parameters control how finely the bifurcation parameter is to be changed), while Ntst is used when periodic orbits are investigated. Click on Ok to return to the main Auto window.

We now have to give Auto a seed from which to start by setting the initial parameter value and initial condition. Auto will use this value of $x$ as the location of the steady state for that value of the bifurcation parameter $r$.

Now go back to the It's Auto man! window and select Run and click on Steady state. This will start the computation of where the steady states lie in the system. A bifurcation curve will now appear on the plot. When you select Run and then Periodic, Auto will generate the periodic branch.

## A.11.2   Scrolling Through the Points on the Bifurcation Diagram

To inspect the points on the bifurcation curve. Click on Grab in the main Auto window. You will see a line of information appearing below the Auto plot, giving us information about the first point (Pt) on the diagram. A crosshairs cursor will also appear at point 1 on the diagram. Press the →  key to scroll through the first few points on the bifurcation diagram. Look at the line printed at the bottom of the window. If a negative sign appears in front of the Pt label, it means that the equilibrium point is stable (a positive sign indicates that it is unstable). You might also notice a small circle appearing in the lower left-hand corner of the screen. This gives the eigenvalue. A transformation has been applied to the eigenvalue, so that an eigenvalue in the left-hand plane (i.e., with negative real part) is now within the unit circle, while one in the right-hand plane (i.e., with positive real part) lies outside of the unit circle.

## A.12    Saving Auto Diagrams

One can save a copy of the bifurcation diagram as a PostScript file to be printed out later. To do this, select File from the main Auto menu (not the main XPP menu!), then Postscript. Then delete auto.ps and enter instead xcrit.auto.ps in response to the request for a filename. Click on Ok to return to the main Auto window. You can also keep a working copy of the bifurcation diagram on file by clicking on File and Save Diagram and giving it a filename. It can be recalled into Auto at a later time (e.g., for further work on extending the diagram or for printing it out) by invoking File and then Load Diagram and entering the name of the file in which it was saved.

# AppendixB

# An Introduction to Matlab

**Michèle S. Titcombe**
**Caroline Haurie**

We will be using Matlab for the computer exercises associated with Chapters 2, 5, 6, 7, 9, and 10.

Matlab is a software tool for numerical computation and displaying graphics. Matlab, short for Matrix Laboratory, runs primarily on vector–matrix algebra.

For a nice overview of Matlab, there is a Matlab Primer (1993) written by Kermit Sigmon (1936–1997) available on the web. The third edition in PDF format can be found (among many other sites) at

> http://ise.stanford.edu/Matlab/matlab-primer.pdf

## B.1 Starting and Quitting Matlab

To start Matlab on Unix-based or Linux operating systems, type matlab at the shell prompt. To start Matlab on the Windows operating system, click on the Matlab icon.

In either system, Matlab commands are typed at the *command prompt*:

> >>

To quit Matlab, you can either type quit at the Matlab command prompt, or select Quit from the File menu at the top left of the Matlab command window.

# B.2    Vectors and Matrices

### B.2.1    Creating Matrices and Vectors

To enter the matrix

$$\begin{pmatrix} 1 & 2 \\ 3 & 4 \end{pmatrix}$$

and store it in a variable **a**, at the Matlab command prompt (>>), type

    a = [1 2; 3 4]

Semicolons are used to separate rows in matrices, while numbers separated by a space or a comma will be entered as elements in a row. The square brackets are used at the beginning and the end of the matrix.

To display the contents of the matrix **a**, just type its name:

    a

To enter a vector (e.g., a one-dimensional matrix), type

    a = [1 2 3 4]

or

    a = [1,2,3,4]

both of which produce a row vector (of dimension $1 \times 4$), whereas the command

    a = [1;2;3;4]

produces a column vector (of dimension $4 \times 1$).

When you are handling vectors, you can use either row vectors (first two commands) or column vectors (third command). Using the single quote symbol

    [1 2]'

will transpose the row vector into a column vector and vice versa.

Assigning a number to a variable (scalar) is done the same way, except that the square brackets are not needed:

    x = 2

There is a specific command to create vectors with regular increments. For example, to create a vector **t** starting at 0 and ending at 10.0 by steps of 0.05, use the form variable = start:by:end, that is, type

    t = 0.0:0.05:10.0;

The vector **t** contains 201 data points (it is a $1 \times 201$ vector). To check this, type

```
whos t
```

# B.3 Suppressing Output to the Screen (the Semicolon!)

The semicolon! Placing ";" at the end of a **Matlab** command (or script file or function file) will execute the command(s) but will suppress the output to the screen. This is *very* helpful if the **Matlab** program you are running has a potentially long output.

For example, the two commands

```
y1=0.1*[1:100];
y2=0.1*[1:100]
```

will both create a $1 \times 100$ vector starting at 0.1 and ending at 10.0 with increments of 0.1. However, the first command will suppress the output to the screen. Try it!

# B.4 Operations on Matrices

Once a matrix or vector (or scalar) has been defined, you can change it by reassigning it a new value or changing some of the elements only. To change the element in the $i$th row and the $j$th column of matrix **a** to a new value, use the form a(i,j)=new value. For example, to change the element in the first row and the third column of **a**, type

```
a(1,3) = 5
```

If you want to assign the same value to **a** as to **b**, type

```
a = b
```

(**Note:** The dimensions of **a** and **b** must be the same, or you will get an error message.)

**Matlab** allows you to perform the usual algebraic operations on matrices and vectors as well as a number of predefined operations. Remember your linear algebra!

To perform matrix multiplication,* division, and exponentiation, the commands are

| matrix multiplication | * |
| matrix division | / |
| matrix exponentiation | ^ |

---

*Type **help times** for more information on matrix multiplication.

To perform **element-by-element**[†] operations on matrices, the commands are

| element-by-element multiplication | .* |
|---|---|
| element-by-element division | ./ |
| element-by-element exponentiation | .^ |

**Note:** the "dot" is very important.

For example, having created a vector t such that t = 0.0:0.05:10.0; you can create a second vector of data, y, to be the sum of the sine of t and the cosine of t. To do so, type

```
y = sin(t)+cos(t);
```

The vector y will have the same dimensions as t. Think: Linear algebra operations! To check this, type

```
whos y
```

# B.5   Programs (M-Files)

Matlab files to execute commands or to execute functions are called "M-files" because the file extension is ".m": for example, testper.m or pupilk.m.

One can create one's own M-Files with a list of commands that can be called in the main command window, as long as they have an extension ".m" There are two types of M-files: **scripts** and **functions**.

## B.5.1   Script Files

A Matlab script file is a plain text file that contains a series of commands to be executed. To run a Matlab script file, e.g., scriptfile.m, at the Matlab command prompt, type

```
scriptfile
```

**Note** that you do not type the ".m" of the filename. Script files are useful when you are performing the same commands over and over, with perhaps only slight modifications.

## B.5.2   Function Files

A Matlab function file is a plain text file that contains operations to be executed, and generally input arguments are passed to the function. In general, there can also be output arguments passed from the function.

---

[†]Type **help mtimes** for more information on element-by-element multiplication.

To run a Matlab script file, e.g., fditer.m, you first need to know the form of the input arguments (if any). A properly written function file will have informative comments that can be displayed with the help command.

For example, the function fditer takes four input values and returns a vector as output:

```
FUNCTION y=fditer(function,xzero,mu,niter)
This program iterates the map specified in 'FUNCTION'
(Note: single quotes must surround the function name.)
There are three additional arguments:
   XZERO is the initial condition;
   MU is the bifurcation parameter mu;
   NITER is the number of iterations.
The output if a vector Y of length NITER containing
the iterated values.
```

To iterate 100 times the quadratic map using the function fditer, with mu=4 and starting at xzero = 0.1, type

```
y=fditer('quadmap',0.1,4,100);
```

It is also possible to assign values to each parameter and then to call the function with the parameter names instead of their value, e.g.,

```
mu=4;
x0=0.1;
y=fditer('quadmap',x0,mu,100);
```

A Matlab function may also have multiple output arguments. This is the case for some programs you will use in the computer exercises of Section 5.9 on resetting curves for the Poincaré oscillator. For example, the function **poincare.m** has four input arguments and two output arguments, as you would see by typing

```
help poincare
```

```
[PHI,BEATS] = POINCARE(PHIZERO,B,TAU,NITER) iterates the
1D Poincare oscillator.
```

So, to run this program with $\phi_0 = 0.3$, $b = 1.13$, $\tau = 0.35$, and $n = 100$ iterations, and to put the output in variables phiout and beatsout, type

```
[phiout,beatsout]=poincare(0.3,1.13,0.35,100);
```

**Note**: You can replace the output variable names with anything you like, e.g., `phiout` instead of `phi`, and `beatsout` instead of `beats`.

Finally, a `Matlab` function may have no output argument but return a graph (`bifurc.m`, for example).

## B.6   The Help Command

The `help` command will give you access to the online Help for `Matlab` functions amd M-files. The command `help` by itself lists all primary help topics. The command `help topic` gives help on the specified topic. The topic can be a function name or a directory name.

The `help` command works for internal `Matlab` functions, and also for any M-files you write yourself. To create online help for your M-files, write text on commented lines at the beginning of the file (after the header line). We created online help for all the `Matlab` programs that were written to accompany this text. For example, to get help on `fditer.m`, at the `Matlab` command prompt, type

```
help fditer
```

## B.7   Loops

The **for** loop in `Matlab` allows you to repeat a command or series of commands for a certain number of iterations. Each loop starts with the command **for** and ends with **end**. (Similarly, there is also a **while** loop.)

For example, if you want to compute the value of the sum

$$x_n = \sum_1^{100} \frac{1}{4^n},$$

type

```
x=0;
for n=0:100
x=x+0.25^n;
end
```

The value of n will start at 0 and be incremented by 1 until it reaches 100, after which the loop stops. At each iteration, the value of $x$ is increased by $0.25^n$. This will give you the sum of the first 101 elements of the series in x.

**Note**: This example shows you a simple loop in `Matlab`. However, computations in `Matlab` are often *much* faster if loops are converted to the vector/matrix equivalent. In this case, the sum $x_n = \sum_1^{100} 1/4^n$ can also be computed using the vector command

```
sum(0.25.^[0:100])
```

Try both ways and verify that the answer is the same.

# B.8  Plotting

To plot data in **Matlab**, the command is **plot**. There are a few ways that you can use this command.

## B.8.1  Examples

Let us illustrate the **plot** command with some simple examples.

### Example B.8.1

Create a "time" vector **t**, starting at 0, ending at 10.0, by steps of 0.05, and a second vector of data, **y**, to be the sine of **t**. That is, type

```
t = 0.0:0.05:10.0;
```

or, equivalently,

```
t = 0.05*[0:200];
```

and

```
y = sin(t);
```

Now we will plot the sinusoid that we have created.

(a) The command

```
plot(y)
```

will plot the coordinate pairs (i,y(i)) for i=1,...,201. The horizontal axis represents the index of the datapoint (i.e., 1 to 201 datapoints) and the vertical axis represents the values of **y**.

(b) The command

```
plot(t,y)
```

will plot the "time" vector **t** (i.e., 0 to 10) along the horizontal axis, and the values of **y** along the vertical axis. It is essential that **t** and **y** have the same dimension.[‡] Otherwise, **Matlab** will give the error

```
??? Error using ==> plot
Vectors must be the same lengths.
```

---

[‡]The command **plot(t,y)** will still work if **t** is a $(1 \times n)$ row vector and **y** is an $(n \times 1)$ column vector.

(c) The command

```
plot(y,'rx')
```

is the same as in part (a) above, except that instead of plotting a
blue line (the default), the data points are plotted as red x's.

(d) The command

```
plot(t,y,'m+')
```

will plot t along the horizontal axis and y along the vertical axis,
with the data points appearing as magenta +'s.

### Example B.8.2

For this example, create another vector, which we will assign to the variable
z. To creates a vector of values $z = \sin(t^2)$, type

```
z = sin(t.*t);
```

**Note**: The . after the first t in the previous command is very important. This "dot" tells **Matlab** to perform element-by-element multiplication
instead of matrix multiplication.

The commands

```
plot(t,y,'r-')
hold on
plot(t,z,'g--')
```

will plot the first curve y(t) as a solid red line, then hold it there, and then
plot the second curve z(t) as a dashed green line on the same set of axes.

Another way to do this is

```
plot(t,y,'r-',t,z,'g--')
```

### Example B.8.3

To make multiple graphs in the same window, each with its own set of axes,
use the **subplot** command. The general format of the subplot command is
subplot(m,n,i), to create an m-row, n-column grid of graphs, and i is the
index that runs from 1 to m*n.

The commands

```
subplot(2,1,1)
plot(t,y,'b.-')
subplot(2,1,2)
plot(t,z,'m--')
```

will plot two graphs one above the other (i.e., a two-row, one-column grid
of graphs).

## B.8.2   Clearing Figures and Opening New Figures

To clear a figure, type

```
clf
```

To open a new figure window, the command has the form `figure(n)`, where n the number of the figure (1, 2, 3, etc.). For example, to open Figure number 1, type

```
figure(1)
```

## B.8.3   Symbols and Colors for Lines and Points

Some data plot symbols (markers) are

| Marker symbol |
|---|
| x |
| + |
| * |
| . |
| square |
| ^ |
| v |
| diamond |

The colors for lines/points on plots are

| Color | code |
|---|---|
| black | k |
| blue | b |
| red | r |
| green | g |
| magenta | m |
| cyan | c |
| yellow | y |
| white | w |

For example, `plot(y,'ksquare')` plots a black square for each data point in the vector y.

# B.9   Loading Data

The default for a data file created in `Matlab` is a binary file with a ".mat" extension (also called a "MAT-file"). If a data file has an extension other than `.mat`, `Matlab` treats it as plain text.

### B.9.1  Examples

**Example B.9.1**

To load a `Matlab` data file called **mydata.mat**, type

```
load mydata
```

This will create a variable called "mydata" in `Matlab`'s working memory.

**Example B.9.2**

To load a `Matlab` data file called `mydata.mat` and to assign its contents to the variable x, type

```
x = load('mydata');
```

Note that the ";" suppresses output to the screen.

**Example B.9.3**

To load a non-`Matlab` (plain text) data file, for example a data file called `yourdata.dat`, type

```
load yourdata.dat
```

or equivalently,

```
load('yourdata.dat')
```

This will create a variable called "yourdata" in `Matlab`'s working memory.

**Example B.9.4**

To load a non-`Matlab` data file, `yourdata.dat` for example, and to assign its contents to the variable y, type

```
y = load('yourdata.dat');
```

For more information on the `load` command, type `help load` at the `Matlab` command prompt.

# B.10  Saving Your Work

To save your work, the command is **save**. See **help save** for more information.

# AppendixC

## Time Series Analysis

**Roderick Edwards**
**Michèle S. Titcombe**

We can think of a physiological recording, such as finger position, in the context of tremor (or blood pressure, membrane potential, or ...), as a continuous variable that changes in time, $x(t)$. Of course, we usually record it only at a sequence of discrete times, usually equally spaced, so we have $x_n = x(t_n)$ for $n = 1, \ldots, N$. This is called a **time series**, and the **sampling rate** is the frequency of recorded points, e.g., 12 per day or 200 Hz (hertz, or data points per second). While we may be interested in simply characterizing the information in the time series itself, it is often worthwhile to consider it as a single realization of an underlying process that involves a random component (a small amount of "noise" in the system), called a **stochastic process**. In that case, different recordings would look slightly different in detail, but should share statistical characteristics that are determined by the underlying process, and we are most interested in estimating these characteristics rather than the details depending on the random variation from recording to recording.

In this appendix we summarize briefly some of the main concepts of time series analysis. There are many texts that give more detail, such as Kaplan and Glass (1995) and Bloomfield (1976) at a fairly elementary level; Brockwell and Davis (1991) and Diggle (1990) at a somewhat more advanced level.

## C.1 The Distribution of Data Points

The simplest kind of information about a time series depends only on the distribution of its data points, irrespective of their arrangement in time. One can simply plot a histogram of the data points or quantify aspects of the distribution, such as its mean and variance as well as higher **moments**, which characterize the distribution's shape. The **mean** is, of course,

$$\bar{x} = \sum_{n=1}^{N} x_n ,$$

and the **variance**, $\sigma^2$, can be expressed as

$$\frac{1}{N} \sum_{n=1}^{N} (x_n - \bar{x})^2 .$$

If we are interested in a hypothetical underlying stochastic process, then it turns out that the best estimate of the true variance is slightly different,

$$\sigma^2 = \text{var}(x) = \frac{1}{N-1} \sum_{n=1}^{N} (x_n - \bar{x})^2 ,$$

but the discrepancy is small if the number of data points, $N$, is large. The **standard deviation** is just the square root of the variance. We are sometimes interested in the "average" amplitude of the data points in the time series, one measure of which is the **root mean square (rms)**,

$$\text{rms}(x) = \sqrt{\frac{1}{N} \sum_{n=1}^{N} x_n} ,$$

which is equivalent to the standard deviation when the mean of the series is 0. Higher moments are useful for characterizing the shape of the distribution of points in a time series. While the mean (first moment) characterizes the center of the distribution, and the variance (second moment) characterizes the amount of spread in the distribution, the **skewness** (third moment)

$$m_3(x) = \frac{1}{N} \sum_{n=1}^{N} (x_n - \bar{x})^3$$

characterizes the symmetry of the distribution, and the **peakedness** (fourth moment)

$$m_4(x) = \frac{1}{N} \sum_{n=1}^{N} (x_n - \bar{x})^4$$

characterizes the degree to which it is sharply peaked in the center. Higher moments are more sensitive to outliers in the data, so that a small number of erroneous values can significantly affect them. Another way to characterize the shape (peakedness) of a distribution is the mean absolute value

$$m_a(x) = \frac{1}{N} \sum_{n=1}^{N} |x_n - \bar{x}| ,$$

which is less sensitive to outliers (Timmer et al. 1993; Edwards and Beuter 2000).

## C.2  Linear Processes

If a physiological signal has some pattern or meaningful temporal arrangement of data points, then we expect that data points will be correlated in some way with points earlier or later in the time series. We can calculate the **autocorrelation** of points that are $k$ data points apart as

$$R(k) = \frac{\sum_{n=1}^{N-k} x_{n+k} x_n}{\sum_{n=1}^{N-k} x_n x_n},$$

where $R$ is called the **autocorrelation function** for the time series. The **cross-correlation function** of two time series, $x_n$ and $y_n$, has a similar structure. One model of autocorrelated processes is the linear **autoregressive (AR)** model

$$x_{n+1} = \sum_{k=0}^{p-1} a_k x_{n-k} + \nu_n,$$

wherein each data point is presumed to be determined by a linear combination of the $p$ previous points, except for a random offset, $\nu_n$, at each point, typically taken to be independent and normally distributed (Gaussian white noise). The choice of coefficients in such a model determines the autocorrelation function, and from noisy data it is possible to estimate the coefficients: The estimates will be the values that fit the data optimally (with least error). Of course, it is quite possible for a nonlinear process to have the same autocorrelation structure as a particular linear autoregressive model. However, AR processes can produce time series with fairly complex structure, such as oscillations of different frequencies superimposed on each other with amplitudes growing or decaying at different rates. Oscillatory behavior in a time series is most clearly revealed by a related approach called Fourier analysis.

## C.3  Fourier Analysis

One of the most important aspects of physiological signals is often their oscillatory components. While they can be complicated, there may be one or several frequencies inherent in their dynamic structure. Fourier analysis of recorded signals is a way of extracting information about their oscillatory structure. It is a remarkable fact that any time series (or any continuous function) can be represented as a sum of sinusoidal oscillations of different frequencies. This may seem hard to imagine; some of Joseph Fourier's influential contemporaries certainly thought so. More precisely, for a continuous signal $x(t)$, the contribution of oscillations at a frequency $f$ (in hertz, or cycles per second) is given in its most compact form by

$$X(f) = \int_{-\infty}^{\infty} x(t)e^{-2\pi fti} \, dt, \tag{C.1}$$

where $i$ is the imaginary unit of complex numbers. The original signal is built up from the frequencies according to

$$x(t) = \int_{-\infty}^{\infty} X(f)e^{2\pi fti} \, df. \tag{C.2}$$

These equations define the **Fourier transform**, equation (C.1), and its inverse, equation (C.2). Written in this form, it may not be clear where the sinusoidal oscillations are, but they are present in the complex exponentials as a result of Euler's formula

$$e^{\pm 2\pi fti} = \cos(2\pi ft) \pm i\sin(2\pi ft).$$

This may be clearer in the discrete version of the Fourier transform (the **discrete Fourier transform**, or **DFT**), applied to a sampled (discrete-time) recording $x_n$:

$$X_k = \sum_{n=0}^{N-1} x_n e^{-2\pi nki/N}, \quad k = 1, \ldots, N. \tag{C.3}$$

Again the original signal can be rebuilt from sinusoids of each frequency from 1 to $k$ via

$$x_n = \sum_{k=0}^{N-1} X_k e^{2\pi nki/N}, \quad n = 1, \ldots, N,$$

where the sinusoids are again "hiding" in the complex exponentials. The Fourier transform is a complex-valued function. Recall that complex numbers can be represented either in terms of real and imaginary components, as in Euler's formula above, or in terms of phase (angle, argument) and amplitude (magnitude, modulus): $X = a + ib = |X|e^{i\theta}$. The magnitude of each of the Fourier coefficients $X_k$ of a discrete signal, or rather the square of the magnitude, $|X_k|^2$, defines what is called the **power spectrum**

$$|X_k|^2 = \left(\sum_{n=1}^{N} x_n \cos(2\pi nk/N)\right)^2 + \left(\sum_{n=1}^{N} x_n \sin(2\pi nk/N)\right)^2,$$

as expressed in terms of the sinusoids. The power spectrum represents the amplitudes of the sinusoidal oscillations of each frequency that make up the original signal. The first Fourier coefficient, $X_0$, is special, since it represents the component with frequency zero, which is a constant and corresponds to the mean of the time series. The **phase spectrum** is given by the phase part of the Fourier coefficients.

To get an idea of how this works, consider first a simple sinusoidal signal, $x(t) = A\cos(12\pi t)$, so that when sampled at discrete times $t_n = n/s$, where $s$ is the sampling rate, we have $x_n = A\cos(12\pi t_n)$. If the frequency of this oscillation (6 Hz if $t$ is in seconds) is one of those resolved by the discrete Fourier transform (so that an exact number of oscillations is completed by the end of the time series), then the power spectrum will be zero at each $k$ except for the $X_k$ with the exact frequency of the sine wave, namely, $k = 6N/s$. If $x_n$ is made up of a sum of two sinusoids, then the power spectrum will contain two such peaks. A typical complicated signal will have contributions of varying amplitudes from *all* available frequencies.

The **cross-spectrum** of two time series is defined as the Fourier transform of the cross-correlation function. If $\hat{X}_k = E(X_k)$ and $\hat{Y}_k = E(Y_k)$ are estimates of the Fourier coefficients, equation (C.3), of two time series, and $\hat{Z}_k \approx E(X_k \bar{Y}_k)$ is an estimate of the cross-spectrum at the $k$th frequency (where the bar over $Y_k$ indicates the complex conjugate), then the **coherency** is defined as

$$C_k = \frac{|\hat{Z}_k|}{|\hat{X}_k||\hat{Y}_k|} \approx \left|\mathrm{CORR}(\hat{X}_k, \hat{Y}_k)\right|. \qquad (C.4)$$

A combination of oscillations at frequencies up to half the sampling rate, (i.e., the **Nyquist frequency**) can produce any discrete time series. Because of the discrete sampling, oscillations with frequencies higher than the Nyquist frequency cannot be distinguished from oscillations below this frequency. A sinusoid at a frequency 2 Hz below the sampling frequency (assuming that the sampling frequency is greater than 4 Hz), for example, will look identical to a sinusoid at 2 Hz (a phenonenon called **aliasing**). Clearly, in order to see oscillations at a given frequency in our recording, we must sample at a frequency at least twice as large.

If we suspect that a recorded time series is generated by a system that could be described by a linear autoregressive process, then we need to estimate the true correlations underneath the noise, or in terms of Fourier analysis, to estimate the true Fourier coefficients, $X_k$. It can be shown that one way to estimate the amplitudes of the Fourier coefficients, i.e., to estimate the power spectrum, is to smooth the raw Fourier amplitudes (this unmodified power spectrum calculated directly from the time series is called the **periodogram**). One such smoothing procedure is the Daniell filter, which takes a weighted average of the values in the power spectrum over a range of frequencies, with half the weight assigned to the first and last frequencies in the range. For example, with an 11-point Daniell filter, the estimated $X_k$ will be

$$\hat{X}_k = \frac{1}{10}\left(\frac{1}{2}X_{k-5} + \sum_{j=-4}^{4} X_{k+j} + \frac{1}{2}X_{k+5}\right).$$

The power spectrum, which contains the amplitude information but not the phase information in the Fourier transform of the time series, can be shown to be the Fourier transform of the autocorrelation function. Thus, these two objects contain the same information about the time series.

Filtering a time series is useful if we have reason to believe that a certain range of frequencies contains the meaningful information, as in the case of tremor. We can think of filtering as removing unwanted frequency components from a signal. The basic idea is simple: Calculate the (discrete) Fourier transform, set coefficients, $X_k$, for unwanted frequencies to zero, and invert the transform to recreate a (now filtered) time series. Using a sharp divide between coefficients that are present and those that are completely removed can have undesirable consequences on the recreated time series, so in practice, ramping the coefficients to zero over some range of frequencies is sometimes preferred. A **high-pass filter** is one in which frequencies higher than a specified value are retained and lower frequencies are removed. A **low-pass filter** is one in which lower frequencies are retained and frequencies above some specified value are removed.

An extension of the Fourier power spectrum, developed to treat unevenly sampled data, is called the Lomb periodogram. Specifically, let $x_j$ be the number of a particular type of cell as measured at times $t_j$, where $j = 1, \ldots, N$ and $N$ is the number of data points. The mean $\bar{x}$ and variance $\sigma^2$ of the data values are given on page 378. Then the **Lomb normalized periodogram** $P(T)$ at a period $T$ is defined by

$$P(T) \equiv \frac{1}{\sigma^2}\left\{\frac{\left[\sum_{j=1}^{N}(x_j - \bar{x})\cos\frac{2\pi(t_j-\tau)}{T}\right]^2}{\sum_{j=1}^{N}\cos^2\frac{2\pi(t_j-\tau)}{T}}\right.$$

$$\left. + \frac{\left[\sum_{j=1}^{N}(x_j - \bar{x})\sin\frac{2\pi(t_j-\tau)}{T}\right]^2}{\sum_{j=1}^{N}\sin^2\frac{2\pi(t_j-\tau)}{T}}\right\}, \qquad (C.5)$$

where the constant $\tau$ is defined implicitly by

$$\tan\left(\frac{4\pi\tau}{T}\right) = \frac{\sum_{j=1}^{N}\sin\left(4\pi t_j/T\right)}{\sum_{j=1}^{N}\cos\left(4\pi t_j/T\right)}. \qquad (C.6)$$

The value of $P(T)$ indicates the likelihood of a periodicity with period $T$ in the data set. Given that the null hypothesis is that the values $x_j$ are independent Gaussian random noise, and that $P(T)$ has an exponential probability distribution with unit mean, the significance level ($p$ value) of any peak is given by

$$p \equiv 1 - \left(1 - e^{-P(T)}\right)^{M},$$  (C.7)

where $M \approx N$ (Press et al. 1992). When the significance level $p$ is small (significant), equation (C.7) can be approximated to give

$$p \approx Me^{-P(T)}.$$  (C.8)

The estimation of the significance level of $P(T)$ is straightforward as long as some criteria are satisfied for the choice of the range and the number of periods that are scanned (Press et al. 1992; Scargle 1982). Here, a data set is considered periodic if the significance level $p$ of the principal peak in the periodogram satisfies $p \leq 0.05$ (5%).

# Bibliography

Abkowitz, J. L., R. D. Holly, and W. P. Hammond (1988). Cyclic hematopoiesis in dogs: Studies of erythroid burst-forming cells confirm an early stem cell defect. *Exp. Hematol. 16*, 941–945.

Abraham, R. H. and C. D. Shaw (1982). *Dynamics: The Geometry of Behavior.* Santa Cruz: Aerial Press.

Abramson, S., R. G. Miller, and R. A. Phillips (1977). The identification in adult bone marrow of pluripotent and restricted stem cells of the myeloid and lymphoid systems. *J. Exp. Med. 145*, 1567–1575.

Ackerman, E., L. C. Gatewood, J. W. Rosevar, and G. D. Molnar (1969). Blood glucose regulation and diabetes. In F. Heinmets (Ed.), *Concepts and Models of Biomathematics.* New York: Marcel Dekker.

Adamson, J. W., D. C. Dale, and R. J. Elin (1974). Hematopoiesis in the grey collie dog: Studies of the regulation of erythropoiesis. *J. Clin. Invest. 54*, 965–973.

Aihara, K. and G. Matsumoto (1983). Two stable steady states in the Hodgkin–Huxley axons. *Biophys. J. 41*, 87–89.

Alberts, W. W. (1972). A simple view of Parkinsonian tremor. Electrical stimulation of cortex adjacent to the Rolandic fissure in awake man. *Brain Res. 44*, 357–369.

Alexander, J. C., E. J. Doedel, and H. G. Othmer (1990). On the resonance structure in a forced excitable system. *SIAM J. Appl. Math. 50*, 1373–1418.

Allessie, M. A., F. U. M. Bonke, and F. J. G. Shopman (1973). Circus movement in rabbit atrial muscle as a mechanism of tachycardia. *Circ. Res. 33*, 54–62.

Allessie, M. A., F. U. M. Bonke, and F. J. G. Shopman (1976). Circus movement in rabbit atrial muscle as a mechanism of tachycardia II. The role of nonuniform recovery of excitability in the occurrence of unidirectional block, as studied with multiple microelectrodes. *Circ. Res. 39*, 168–177.

Allessie, M. A., F. U. M. Bonke, and F. J. G. Shopman (1977). Circus movement in rabbit atrial muscle as a mechanism of tachycardia III. The 'leading circle' concept: A new model of circus movement in cardiac tissue without the involvement of an anatomical obstacle. *Circ. Res. 41*, 9–18.

Alligood, K., T. Sauer, and J. A. Yorke (1997). *Chaos: An Introduction to Dynamical Systems*. New York: Springer.

Alonso, A. and R. Klink (1993). Differential responsiveness of stellate and pyramidal-like cells of medial entorhinal cortex layer II. *J. Neurophysiol. 70*, 128–143.

American Diabetes Association (1995). Office guide to diagnosis and classification of Diabetes Mellitus and other categories of glucose intolerance. *Diabetes Care 18(Supplement 1)*, 4–7.

an der Heiden, U., A. Longtin, M. C. Mackey, J. G. Milton, and R. Scholl (1990). Oscillatory modes in a nonlinear second-order differential equation with delay. *J. Dynam. Diff. Eqns. 2*, 423–449.

an der Heiden, U. and M. C. Mackey (1982). The dynamics of production and destruction: Analytic insight into complex behavior. *J. Math. Biol. 16*, 75–101.

an der Heiden, U. and K. Reichard (1990). Multitude of oscillatory behavior in a nonlinear second-order differential-difference equation. *Z. Angew. Math. Mech. 70*, T621–T624.

Arnold, V. I. (1983). *Geometrical Methods in the Theory of Ordinary Differential Equations*. New York: Springer-Verlag.

Arnold, V. I. (1986). *Catastrophe Theory*. Berlin: Springer-Verlag.

Astumian, R. D. and F. Moss (1998). The constructive role of noise in fluctuation driven transport and stochastic resonance. *Chaos 8*, 533–538.

Auger, P. M. and A. Bardou (1988). Computer simulation of ventricular fibrillation. In *Mathematical Computer Modelling*, pp. 813–822. Pergamon.

Austin, G. and C. Tsai (1962). A physiological basis and development of a model for Parkinsonian tremor. *Confinia Neurologica 22*, 248–258. 1st International Symposium of Stereoencephalotomy 1961.

Avalos, B. R., V. C. Broudy, S. K. Ceselski, B. J. Druker, J. D. Griffen, and W. P. Hammond (1994). Abnormal response to granulocyte colony-stimulating factor (G-CSF) in canine cyclic hematopoiesis is not caused by altered G-CSF receptor expression. *Blood 84*, 789–794.

Baer, M. and M. Eiswirth (1993). Turbulence due to spiral breakup in a continuous excitable medium. *Phys. Rev. E 48*, 1635–1637.

Bain, P. (1993). A combined clinical and neurophysiological approach to the study of patients with tremor. *J. Neurol. 56*(8), 839–844.

Baker, P. F., A. L. Hodgkin, and T. I. Shaw (1961). Replacement of the protoplasm of a giant nerve fibre with artificial solutions. *Nature 190*(4779), 885–887.

Bayer, W. and U. an der Heiden (1998). Oscillation types and bifurcations of a nonlinear second-order differential–difference equation. *J. Dynam. Diff. Eqns. 10*, 303–326.

Becker, A., E. McCulloch, and J. Till (1963). Cytological demonstration of the clonal nature of spleen colonies derived from transplanted mouse marrow cells. *Nature 197*, 452–454.

Beeler, G. W. and H. Reuter (1977). Reconstruction of the action potential of ventricular myocardial fibers. *J. Physiol. (Lond.) 28*, 177–210.

Benabid, A. L., A. Benazzouz, D. Hoffmann, P. Limousin, P. Krack, and P. Pollack (1998). Long-term electrical inhibition of deep brain targets in movement disorders. *Movement Disord. 13*(Supplement 3), 119–125.

Beresford, C. H. (1988). Time: A biological dimension. *J. Roy. Coll. Phys. Lond. 22*, 94–96.

Bernstein, N. (1967). *The Co-ordination and Regulation of Movements.* Oxford: Pergamon. Chapter 4.

Bernstein, R. C. and L. H. Frame (1990). Reentry around a fixed barrier: resetting with an advancement in an in vitro model. *Circ. 81*, 267–280.

Berson, D. M., F. A. Dunn, and M. Takao (2002). Phototransduction by retinal ganglion cells that set the circadian clock. *Science 295*, 1070–1073.

Beuter, A., J. Bélair, and C. Labrie (1993). Feedback and delays in neurological diseases: a modeling study using dynamical systems. *Bull. Math. Biol. 55*(3), 525–541.

Beuter, A. and A. de Geoffroy (1995). A dynamical approach to normal and Parkinsonian tremor. In E. Mosekilde and O. G. Mouritsen (Eds.), *Modelling the Dynamics of Biological Systems.* Berlin: Springer-Verlag.

Beuter, A. and A. de Geoffroy (1996). Can tremor be used to measure the effect of chronic mercury exposure in human subjects? *Neurotoxicology 17*(1), 213–228.

Beuter, A., A. de Geoffroy, and P. Cordo (1994). The measurement of tremor using simple laser systems. *J. Neurosci. Meth. 53*, 47–54.

Beuter, A. and R. Edwards (1999). Using frequency domain characteristics to discriminate physiologic and Parkinsonian tremors. *J. Clin. Neurophysiol. 16*(5), 484–494.

Beuter, A., R. Edwards, and D. Lamoureux (2000). Neuromotor profiles: what are they and what can we learn from them? *Brain Cognition 43*, 39–44.

Beuter, A., M. S. Titcombe, F. Richer, C. Gross, and D. Guehl (2001). Effect of deep brain stimulation on amplitude and frequency characteristics of rest tremor in Parkinson's disease. *Thalamus and Related Systems 1*(3), 203–211.

Beuter, A. and K. Vasilakos (1995). Tremor: Is Parkinson's disease a dynamical disease? *Chaos 5*, 35–42.

Biktashev, V. N. (1997). Control of re-entrant vortices by electrical stimulation. In A. V. Panfilov and A. V. Holden (Eds.), *Computational Biology of the Heart*, pp. 155–170. Wiley.

Binet, L. (1920, Jan. 31). The laws of tremor. *French Supplement to the Lancet*, 265–266.

Birgens, H. S. and H. Karl (1993). Reversible adult-onset cyclic haematopoiesis with a cycle length of 100 days. *Brit. J. Hematol. 83*, 181–186.

Bloomfield, P. (1976). *Fourier Analysis of Time Series: An Introduction*. NewYork/London: Wiley.

Blythe, S. P., R. M. Nisbet, and W. S. C. Gurney (1984). The dynamics of population models with distributed maturation periods. *Theor. Popul. Biol. 25*, 289–311.

Bonhoeffer, K. F. (1948). Activation of passive iron as a model for the excitation in nerve. *J. Gen. Physiol. 32*, 69–91.

Borgdorff, P. (1975). Respiratory fluctuations in pupil size. *Amer. J. Physiol. 228*, 1094–1102.

Boyett, M. R. and B. R. Jewell (1978). A study of factors responsible for rate-dependent shortening of the action potential in mammalian ventricular muscle. *J. Physiol. (Lond.) 285*, 359–380.

Brandt, L., O. Forssman, R. Mitelman, H. Odeberg, T. Oloffson, I. Olsson, and B. Svensson (1975). Cell production and cell function in human cyclic neutropenia. *Scand. J. Haematol. 15*, 228–240.

Braun, H. A., K. Schäfer, and H. Wissing (1990). *Thermoreception and Temperature Regulation*, Chapter Theories and models of temperature transduction, pp. 23. Berlin: Springer Verlag.

Braun, H. A., H. Wissing, K. Schäfer, and M. C. Hirsch (1994). Oscillation and noise determine signal transduction in shark multimodal sensory cells. *Nature 367*, 270–273.

Breda, E., M. K. Cavaghan, G. Toffolo, K. S. Polonsky, and C. Cobelli. (2001). Oral glucose tolerance test minimal model indexes of $\beta$-cell function and insulin sensitivity. *Diabetes 50*, 150–158.

Bressloff, P. C. and C. V. Wood (1997). Spontaneous oscillation in a nonlinear delayed-feedback shunting model of the pupil light reflex. *Phys. Rev. E 58*, 3597–3606.

Bressloff, P. C., C. V. Wood, and P. A. Howarth (1996). Nonlinear shunting model of the pupil light reflex. *Proc. Roy. Soc. B 263*, 953–960.

Britton, N. F. (1986). *Reaction–Diffusion Equations and Their Applications to Biology*. London: Academic Press.

Britton, T. C. (1995). Central and peripheral mechanisms in tremorgenesis. In L. J. Findley and W. D. Koller (Eds.), *Handbook of Tremor Disorders*. Marcel Dekker, Inc.

Brockwell, P. and R. Davis (1991). *Time Series: Theory and Methods*. New York/Berlin: Springer-Verlag.

Bronchud, M. H., J. H. Scarffe, N. Thatcher, D. Crowther, L. M. Souza, N. K. Alton, N. G. Testa, and T. M. Dexter (1987). Phase I/II study of recombinant human granulocyte colony-stimulating factor in patients receiving intensive chemotherapy for small cell lung cancer. *Brit. J. Cancer 56*, 809 –813.

Brooks, D. J. (2000). Imaging basal ganglia function. *J. Anat. 196*, 543–554.

Bulsara, A., E. Jacobs, T. Zhou, F. Moss, and L. Kiss (1991). Stochastic resonance in a single neuron model: Theory and analog simulation. *J. Theor. Biol. 154*, 531–555.

Bulsara, A. R., T. C. Elston, C. R. Doering, S. B. Lowen, and K. Lindenberg (1996). Cooperative behavior in periodically driven noisy integrate-and-fire models of neuronal ensembles. *Phys. Rev. E 53*, 3958–3969.

390    Bibliography

Burrus, C. S., R. A. Gopinath, and H. Guo (1998). *Introduction to Wavelets and Wavelet Transforms: A Primer*. New Jersey: Prentice Hall.

Cabo, C., A. M. Pertsov, W. T. Baxter, J. M. Davidenko, R. A. Gray, and J. Jalife (1994). Wave-front curvature as a cause of slow conduction and block in isolated cardiac muscle. *Circ. Res. 75*, 1014–1028.

Cabo, C., A. M. Pertsov, J. M. Davidenko, , and J. Jalife (1998). Electrical turbulence as a result of the critical curvature for propagation in cardiac tissue. *Chaos. 8*, 116–126.

Calcagnini, G., S. Lino, F. Censi, G. Calcagnini, and S. Cerutti (1997). Cardiovascular autonomic rhythms in spontaneous pupil fluctuations. *Computers in Cardiology 24*, 133–136.

Campbell, S. A., J. Bélair, T. Ohira, and J. G. Milton (1995a). Complex dynamics and multi-stability in a damped harmonic oscillator with delayed negative feedback. *Chaos 5*, 640–645.

Campbell, S. A., J. Bélair, T. Ohira, and J. G. Milton (1995b). Limit cycles, tori and complex dynamics in a second-order differential equation with delayed negative feedback. *J. Dynamics Diff. Eqns. 7*, 213–225.

Capildeo, R. (1984). Parkinson's disease complex – restyling an old overcoat! In L. J. Findley and R. Capildeo (Eds.), *Movement Disorders: Tremor*, Chapter 20, pp. 286. MacMillan.

Casti, J. L. (1989). *Alternate Realities, Mathematical Models of Nature and Man*. N.Y: Wiley.

Cebon, J. and J. Layton (1984). Measurement and clinical significance of circulating hematopoietic growth factor levels. *Curr. Opin. Hematol. 1*, 228–234.

Chacron, M. J., A. Longtin, M. St-Hilaire, and L. Maler (2000). Suprathreshold stochastic firing dynamics with memory in p-type electroreceptors. *Phys. Rev. Lett. 85*, 1576–1579.

Chaos Focus Issue (2001). Molecular, metabolic, and genetic control. *Chaos 11*(1), 81–292.

Chapeau-Blondeau, F., X. Godivier, and N. Chambet (1996). Stochastic resonance in a neuron model that transmits spike trains. *Phys. Rev. E 53*, 1273–1275.

Charcot, J. M. (1877). Lecture V on Paralysis Agitans. In G. Sigerson (Ed.), *Lectures on the Diseases of the Nervous System*. London: The New Sydenham Society.

Chay, T. R. and Y. S. Lee (1985). Phase resetting and bifurcation in the ventricular myocardium. *Biophys. J. 47*, 641–651.

Chialvo, D. R. and V. Apkarian (1993). Modulated noisy biological dynamics: Three examples. *J. Stat. Phys. 70*, 375–391.

Chialvo, D. R., A. Longtin, and J. Müller-Gerking (1997). Stochastic resonance in models of neuronal ensembles. *Phys. Rev. E 55*, 1798–1808.

Chialvo, D. R., D. C. Michaels, and J. Jalife (1990). Supernormal excitability as a mechanism of chaotic activation in cardiac Purkinje fibers. *Circ. Res. 66*, 525–545.

Chikkappa, G., G. Borner, H. Burlington, A. D. Chanana, E. P. Cronkite, S. Ohl, M. Pavelec, and J. S. Robertson (1976). Periodic oscillation of blood leukocytes, platelets, and reticulocytes in a patient with chronic myelocytic leukemia. *Blood 47*, 1023 –1030.

Chikkappa, G., A. Chanana, P. Chandra, E. P. Cronkite, and K. Thompson (1980). Cyclic oscillation of blood neutrophils in a patient with multiple myeloma. *Blood 55*, 61–66.

Clay, J. R. (1977). Monte Carlo simulation of membrane noise: an analysis of fluctuation in graded excitation of nerve membrane. *J. Theor. Biol. 64*, 671–680.

Clay, J. R. (1998). Excitability of the squid giant axon revisited. *J. Neurophysiol. 80*, 903–913.

Cole, K. and H. Curtis (1939). Electrical impedance of the squid giant axon during activity. *J. Gen. Physiol. 22*, 649–670.

Cole, K. S. (1968). *Membranes, Ions and Impulses; A Chapter of Classical Biophysics*. University of California Press, Berkeley.

Collins, J. J., C. Chow, and T. T. Imhoff (1995a). Stochastic resonance without tuning. *Nature 376*, 236–238.

Collins, J. J., C. C. Chow, and T. Imhoff (1995b). Aperiodic stochastic resonance. *Phys. Rev. E 52*, R3321–R3324.

Collins, J. J., T. T. Imhoff, and P. Grigg (1996). Noise-enhanced tactile sensation. *Nature 383*, 770.

Comtois, P. and A. Vinet (1999). Curvature effects on activation speed and repolarisation in an ionic model of cardiac myocytes. *Phys. Rev. E 60*, 4619–4627.

Cooke, K. and Z. Grossman (1982). Discrete delay, distributed delay and stability switches. *J. Math. Anal. Appl. 86*, 592–627.

Cordo, P. J., J. T. Inglis, J. J. Collins, S. M. P. Verschueren, D. M. Merfeld, S. Buckley, and F. Moss (1996). Noise-enhanced information transmission in human muscle spindles via stochastic resonance. *Nature 383*, 769–770.

Courtemanche, M. (1995). Wave propagation and curvature effects in a model of excitable medium. *Chaos Soliton. Fract. 5*, 527–542.

Courtemanche, M., L. Glass, and J. P. Keener (1993). Instabilities of a propagating pulse in a ring of excitable media. *Phys. Rev. Lett. 70*, 2182–2185.

Courtemanche, M., L. Glass, and J. P. Keener (1996). A delay equation representation of pulse circulation on a ring of excitable media. *SIAM J. Appl. Math. 56*, 119–142.

Courtemanche, M. and A. T. Winfree (1991). Re-entrant rotating waves in a Beeler–Reuter based model of two-dimensional cardiac activity. *Int. J. Bifurcat. Chaos 1*, 431–444.

Crutchfield, J., D. Farmer, N. Packard, R. Shaw, G. Jones, and R. J. Donnelly (1980). Power spectral analysis of a dynamical system. *Phys. Lett. A 76*, 1–4.

Dale, D. C., D. Alling, and S. M. Wolff (1973). Application of time series analysis to serial blood neutrophil counts in normal individuals and patients receiving cyclophosphamide. *Brit. J. Haematol. 24*, 57–64.

Dale, D. C., D. W. Alling, and S. M. Wolff (1972). Cyclic hematopoiesis: the mechanism of cyclic neutropenia in grey collie dogs. *J. Clin. Invest. 51*, 2197–2204.

Dale, D. C., M. Bonilla, M. Davis, A. Nakanishi, W. Hammond, J. Kurtzberg, W. Wang, A. Jakubowski, E. Winton, P. Lalezari, W. Robinson, J. Glaspy, S. Emerson, J. Gabrilove, M. Vincent, and L. Boxer (1993). A randomized controlled phase III trial of recombinant human granulocyte colony stimulating factor (filgrastim) for treatment of severe chronic neutropenia. *Blood 81*, 2496–2502.

Dale, D. C., C. Brown, P. Carbone, and S. M. Wolff (1971). Cyclic urinary leukopoietic activity in gray collie dogs. *Science 173*, 152–153.

Dale, D. C. and R. G. Graw (1974). Transplantation of allogenic bone marrow in canine cyclic neutropenia. *Science 183*, 83–84.

Dale, D. C. and W. P. Hammond (1988). Cyclic neutropenia: A clinical review. *Blood Rev. 2*, 178–185.

Dale, D. C., S. B. Ward, H. R. Kimball, and S. M. Wolff (1972). Studies of neutrophil production and turnover in grey collie dogs with cyclic neutropenia. *J. Clin. Invest. 51*, 2190–2196.

Dancey, J. T., K. A. Deubelbeiss, L. A. Harker, and C. A. Finch (1976). Neutrophil kinetics in man. *J. Clin. Invest. 58*, 705–715.

Date, I. and T. Ohmoto (1999). Neural transplantation in Parkinson's disease. *Cell. Mol. Neurobiol. 19*(1), 67–78.

Daum, K. M. and G. A. Fry (1981). The component of physiological pupillary unrest correlated with respiration. *Amer. J. Optom. Physiol. Optics 58*, 831–840.

Daum, K. M. and G. A. Fry (1982). Pupillary micro movements apparently related to pulse frequency. *Vision Res. 22*, 173–177.

Davidenko, J. M., A. V. Pertsov, R. Salomonsz, W. Baxter, and J. Jalife (1992). Stationary and drifting spiral waves of excitation in isolated cardiac muscle. *Nature 355*, 349–350.

De Bakker, J. M. T., F. J. L. Van Capelle, M. J. Janse, A. A. M. Wilde, R. Coronel, A. E. Becker, K. P. Dingemans, N. M. Van Hemel, and R. N. W. Hauer (1988). Reentry as a cause of ventricular tachycardia in patients with chronic ischemic heart disease: Electrophysiological and anatomic correlation. *Circulation 77*, 589–606.

de Coulon, F. (1986). *Signal Theory and Processing*. Dedham, MA: Artech House.

de la Boë (Sylvius), F. (1675). *A New Idea of the Practice of Physic*. London: Brabazon Aylmer. transl. Gower, R. (contemporary translation of the first book of *Praxeos medicae idaea*).

De Mello, W. (1989). Cyclic nucleotides, Ca and cell junctions. In N. Sperelakis and W. Cole (Eds.), *Cell Interactions and Gap Junctions*. CRC Press.

DeFelice, L. J. (1981). *Introduction to Membrane Noise*. New York: Plenum.

Deiber, M.-P., P. Pollak, R. Passingham, P. Landais, C. Gervason, L. Cinotti, K. Friston, R. Frackowiak, F. Mauguiere, and A. L. Benabid (1993). Thalamic stimulation and suppression of Parkinsonian tremor. Evidence of a cerebellar deactivation using positron emission tomography. *Brain 116*, 267–279.

Deubelbeiss, K. A., J. T. Dancey, L. A. Harker, and C. A. Finch (1975). Neutrophil kinetics in the dog. *J. Clin. Invest. 55*, 833–839.

Deuschl, G., P. Bain, and M. Brin (1998). Consensus statement of the Movement Disorder Society on tremor. *Movement Disord. 13*(Suppl. 3), 2–23.

Devaney, R. (1989). *An Introduction to Chaotic Dynamical Systems, 2nd edition*. Perseus Publishing Co.

Diggle, P. (1990). *Time Series: A Biostatistical Introduction*. Oxford: Oxford University Press.

Douglass, J. K., L. Wilkens, E. Pantazelou, and F. Moss (1993). Noise-enhancement of information transfer in crayfish mechanoreceptors by stochastic resonance. *Nature 365*, 337–340.

Dunn, C. D. R. (1983). Cyclic hematopoiesis: The biomathematics. *Exp. Hematol. 11*, 779–791.

Dunn, C. D. R., J. D. Jolly, J. B. Jones, and R. D. Lange (1978). Erythroid colony formation *in vitro* from the marrow of dogs with cyclic hematopoiesis: Interrelationship of progenitor cells. *Exp. Hematol. 6*, 701–708.

Dunn, C. D. R., J. B. Jones, and R. D. Lange (1977). Progenitor cells in canine cyclic hematopoiesis. *Blood 50*, 1111–1120.

Duval, C., M. Panisset, G. Bertrand, and A. F. Sadikot (2000). Evidence that ventrolateral thalamotomy may eliminate the supraspinal component of both pathological and physiological tremors. *Exp. Brain Res 132*, 216–222.

Edelstein-Keshet, L. (1988). *Mathematical Models in Biology*. New York: Random House.

Edwards, R. and A. Beuter (1996). How to handle intermittent phenomena in time series analysis: a case study of neuromotor deficit. *Brain Cognition 32*, 262–266.

Edwards, R. and A. Beuter (2000). Using time domain characteristics to discriminate normal and Parkinsonian tremor. *J. Clin. Neurophysiol. 17*(1), 87–100.

Edwards, R., A. Beuter, and L. Glass (1999). Parkinsonian tremor and simplification in network dynamics. *Bull. Math. Biol. 61*(1), 157–177.

Efimov, I. R., V. I. Krinsky, and J. Jalife (1995). Dynamics of rotating vortices in the Beeler–Reuter model of cardiac tissue. *Chaos Soliton. Fract. 5*, 513–526.

Elble, R. J. (1986). Physiologic and essential tremor. *Neurology 36*, 225–231.

Elble, R. J. (1996). Central mechanisms of tremor. *J. Clin. Neurophysiol. 13*(2), 133–144.

Elble, R. J. and W. C. Koller (1990). *Tremor*. Baltimore: Johns Hopkins University Press.

Elble, R. J. and J. E. Randall (1976). Motor-unit activity responsible for the 8–12 hz component of human physiological finger tremor. *J. Neurophysiol. 39*, 370–383.

Elek, J. M., R. Dengler, A. Konstanzer, S. Hesse, and W. Wolf (1991). Mechanical implications of paired motor unit discharges in pathological and voluntary tremor. *Electroen. Clin. Neuro. 81*, 779–783.

Elharar, V. and B. Surawicz (1983). Cycle length effect on the restitution of action potential duration in dog cardiac fibers. *Am. J. Physiol. 244*, H782–H792.

Erying, H., R. Lumry, and J. Woodbury (1949). Some applications of modern rate reaction theory to physiological systems. *Record Chem. Prog. 10*, 100–111.

Eshner, A. A. (1897). A graphic study of tremor. *J. Exp. Med. 20*, 301–312.

Eurich, C. W. and J. G. Milton (1996). Noise-induced transitions in human postural sway. *Phys. Rev. E 54*, 6681–6684.

Fargue, D. (1973). Reductibilité des systèmes héréditaires à des systèmes dynamiques. *Comptes Rendus de l'Académie des Sciences B277*, 471–473.

Fargue, D. (1974). Reductibilité des systèmes héréditaires. *Inter. J. Nonlin. Mech. 9*, 331–338.

Farmer, J. D. (1982). Chaotic attractors of an infinite-dimensional dynamical system. *Physica 4D*, 366–393.

Fast, V., I. Efimov, and V. Krinsky (1991). Stability of vortex rotation in excitable cellular media. *Physica D 49*, 75–81.

Fatt, P. and B. Katz (1952). Spontaneous subthreshold activity at motor nerve endings. *J. Physiol. (Lond.) 117*, 109–128.

Feigenbaum, M. (1980). Universal behavior in nonlinear systems. *Los Alamos Sci. 1*, 4–27.

Findley, L. J. and L. Cleeves (1989). *Classification of tremor. Disorders of Movement*. Academic Press.

Fisher, G. (1993). An introduction to chaos theory and some haematological applications. *Comp. Haematol. Int. 3*, 43–51.

FitzHugh, R. (1960). Thresholds and plateaus in the Hodgkin–Huxley nerve equations. *J. Gen. Physiol. 43*, 867–897.

FitzHugh, R. (1961). Impulses and physiological states in theoretical models of nerve membrane. *Biophys. J. 1*, 445–466.

FitzHugh, R. (1969). Mathematical models of excitation and propagation in nerve. In H. P. Schwan (Ed.), *Biological Engineering*. New York: McGraw-Hill.

Fortin, P. and M. C. Mackey (1999). Periodic chronic myelogenous leukemia: Spectral analysis of blood cell counts and etiological implications. *Brit. J. Haematol. 104*, 336–345.

Foss, J., A. Longtin, B. Mensour, and J. G. Milton (1996). Multistability and delayed recurrent loops. *Phys. Rev. Lett. 76*, 708–711.

Foss, J. and J. Milton (2000). Multistability in recurrent neural loops arising from delay. *J. Neurophysiol. 84*, 975–985.

Foss, J., F. Moss, and J. G. Milton (1997). Noise, multistability and delayed recurrent loops. *Phys. Rev. E 55*, 4536–4543.

Fox, J. J., J. L. McHarg, and R. F. Gilmour (2002). Ionic mechanisms of electrical alternans. *Am. J. Physiol. 282*, H516–H530.

Fox, R. F., I. R. Gatland, R. Roy, and G. Vemuri (1988). Fast, accurate algorithm for numerical simulation of exponentially correlated colored noise. *Phys. Rev. A 38*, 5938–5942.

Frame, L. H. F., R. L. Page, and B. F. Hoffman (1986). Atrial reentry around an anatomical barrier with a partially refractory excitable gap. *Circ. Res. 58*, 495–511.

Frame, L. H. F. and M. B. Simson (1988). Oscillations of conduction, action potential duration, and refractoriness: A mechanism for spontaneous termination of reentrant tachycardias. *Circulation 77*(6), 1277–1287.

Franz, M. R. (1991). Method and theory of monophasic action potential recording. *Prog. Cardiovasc. Dis. 33*(6), 347–368.

Frazier, D. W., P. D. Wolf, J. M. Wharton, A. S. Tang, W. M. Smith, and R. E. Ideker (1989). Stimulus-induced critical point: mechanism of electrical initiation of reentry in normal canine myocardium. *J. Clin. Invest. 83*, 1039–1052.

French, A. S., A. V. Holden, and R. B. Stein (1972). The estimation of the frequency response function of a mechanoreceptor. *Kybernetik 11*, 15–23.

Friedlander, W. J. (1956). Characteristics of postural tremor in normal and various abnormal states. *Neurology 6*, 714–724.

Friedman, N., A. Vinet, and F. Roberge (1996). A study of a new model of the cardiac ventricular cell incorporating myoplasmic calcium regulation. In *22nd CMBES Conference.*

Fukumoto, I. (1986). Computer simulation of Parkinsonian tremor. *J. Biomed. Eng. 8*, 49–55.

Gamlin, P. D. R., H. Zhang, and J. J. Clarke (1995). Luminance neurons in the pretectal olivary nucleus mediate the pupillary light reflex in the rhesus monkey. *Exp. Brain Res. 106*, 177–180.

Gammaitoni, L. (1995). Stochastic resonance and the dithering effect in threshold physical systems. *Phys. Rev. Lett. 52*, 4691–4698.

Gammaitoni, L., P. Hänggi, P. Jung, and F. Marchesoni (1998). Stochastic resonance. *Rev. Mod. Phys. 70*, 223–288.

Gang, H., T. Ditzinger, C. Ning, and H. Haken (1994). Stochastic resonance without external periodic forcing. *Phys. Rev. Lett. 71*, 807.

Gantert, C., J. Honerkamp, and J. Timmer (1992). Analyzing the dynamics of hand tremor time series. *Biol. Cybern. 66*, 479–484.

Gardiner, C. (1985). *Handbook of Stochastic Methods for Physics, Chemistry and the Natural Sciences*. Berlin: Springer-Verlag.

Garfinkel, A. et al. (1997). Quasiperiodicity and chaos in cardiac fibrillation. *J. Clin. Invest. 99*, 305–314.

Garrison, F. H. (1969). *History of Neurology*. Springfield, IL: Charles C. Thomas. Rev. and enl. with a bibliography of classical, original, and standard works in neurology, by Lawrence C. McHenry, Jr.

Gatica, J. A. and P. Waltman (1982). A threshold model of antigen antibody dynamics with fading memory. In V. Lakshmikantham (Ed.), *Nonlinear Phenomena in Mathematical Sciences*. New York: Academic Press.

Gatica, J. A. and P. Waltman (1988). A system of functional differential equations modeling threshold phenomena. *Appl. Anal. 28*, 39–50.

Gedeon, T. and L. Glass (1998). Continuity of resetting curves for FitzHugh–Nagumo equations on the circle. In *Differential Equations with Applications to Biology*, pp. 225–236. Fields Institute Communications.

Gerhardt, M., H. Shuster, and J. J. Tyson (1990). A cellular automaton model of excitable media. *Physica D 46*, 392–415.

Gerstein, G. and B. Mandelbrot (1964). Random walk models for the spike activity of a single neuron. *Biophys. J. 4*, 41–68.

Geselowitz, D. and W. Miller (1983). A bidomain model for anisotropic cardiac muscle. *Ann. Biomed. Eng. 11*, 191–206.

Giaquinta, A., S. Boccaletti, and F. Arecchi (1996). Superexcitability induced spiral breakup in excitable systems. *Int. J. Bifurcat. Chaos 6*, 1753–1759.

Gibson, C. M., C. W. Gurney, E. O. Gaston, and E. L. Simmons (1984). Cyclic erythropoiesis in the $S1/S1^d$ mouse. *Exp. Hematol. 12*, 343–348.

Gibson, C. M., C. W. Gurney, E. L. Simmons, and E. O. Gaston (1985). Further studies on cyclic erythropoiesis in mice. *Exp. Hematol. 13*, 855–860.

Gidáli, J., E. István, and I. Fehér (1985). Long-term perturbation of hemopoiesis after moderate damage to stem cells. *Exp. Hematol. 13*, 647–651.

Gilmour, R. F. (1990). Phase resetting of circus movement reentry in cardiac tissue. In D. P. Zipes and J. Jalife (Eds.), *Cardiac Electrophysiology: From Cell to Bedside*, pp. 396–402. Philadelphia: W. B. Saunders.

Glantz, S. A. (1979). *Mathematics for Biomedical Applications*. Berkeley and Los Angeles: University of California Press.

Glass, L. (1969). Moiré effect from random dots. *Nature 223*, 578–580.

Glass, L. (1997). The topology of phase resetting and the entrainment of limit cycles. In H. G. Othmer, F. R. Adler, M. A. Lewis, and C. Dallon (Eds.), *Case Studies in Mathematical Modeling – Ecology, Physiology, and Cell Biology*, pp. 255–276. New York: Prentice-Hall.

Glass, L., C. Graves, G. A. Petrillo, and M. C. Mackey (1980). Unstable dynamics of a periodically driven oscillator in the presence of noise. *J. Theor. Biol. 86*, 455–475.

Glass, L., M. R. Guevara, J. Bélair, and A. Shrier (1984). Global bifurcations of a periodically forced biological oscillator. *Phys. Rev. A 29*, 1348–1357.

Glass, L., M. R. Guevara, and A. Shrier (1987). Universal bifurcations and the classification of cardiac arrhythmias. *Ann. N.Y. Acad. Sci. 504*, 168–178.

Glass, L., M. R. Guevara, A. Shrier, and R. Perez (1983). Bifurcation and chaos in a periodically stimulated cardiac oscillator. *Physica D 7*, 89–101.

Glass, L. and M. E. Josephson (1995). Resetting and annihilation of reentrant abnormally rapid heartbeat. *Phys. Rev. Lett. 75*, 2059–2063.

Glass, L. and M. C. Mackey (1979). Pathological conditions resulting from instabilities in control systems. *Ann. N.Y. Acad. Sci. 316*, 214–235.

Glass, L. and M. C. Mackey (1988). *From Clocks to Chaos: The Rhythms of Life*. Princeton: Princeton University Press.

Glass, L., Y. Nagai, K. Hall, M. Talajic, and S. Nattel (2002). Predicting the entrainment of reentrant cardiac waves using phase resetting curves. *Phys. Rev. E 65*, 021908.

Glass, L. and R. Perez (1973). Perception of random dot interference patterns. *Nature 246*, 360–362.

Glass, L. and J. Sun (1994). Periodic forcing of a limit cycle oscillator: Fixed points, Arnold tongues, and the global organization of bifurcations. *Phys. Rev. E 50*, 5077–5084.

Glass, L. and A. T. Winfree (1984). Discontinuities in phase-resetting experiments. *Am. J. Physiol. 246*, R251–R258.

Goldbeter, A. and J.-L. Martiel (1985). Birhythmicity in a model for the cyclic AMP signalling system of the slime mold *dictyostelium discoideum*. *FEBS Lett. 191*, 149–153.

Goldman, D. E. (1943). Potential, impedance and rectification in membranes. *J. Gen. Physiol. 27*, 37–60.

Goodman, D. and J. A. S. Kelso (1983). Exploring the functional significance of physiological tremor: a biospectroscopic approach. *Exp. Brain Res. 49*, 419–431.

Goodwin, B. C. (1963). *Temporal Organization in Cells: A Dynamic Theory of Cellular Control Processes*. New York: Academic.

Gourley, S. A. and S. Ruan (2000). Dynamics of the diffusive Nicholson's blowflies equation with distributed delay. *Proc. Royal Soc. Edinburgh 130A*, 1275–1291.

Greenberg, J. M., N. D. Hassard, and S. P. Hastings (1978). Pattern formation and periodic structures in systems modeled by reaction diffusion equation. *B. Am. Math. Soc. 84*, 1296–1326.

Greenberg, J. M. and S. P. Hastings (1978). Spatial patterns for discrete models of diffusion in excitable media. *SIAM J. Appl. Math. 34*, 515–523.

Gresty, M. and D. Buckwell (1990). Spectral analysis for tremor: understanding the results. *J. Neurol. Neurosur. Ps. 53*, 976–981.

Grignani, F. (1985). Chronic myelogenous leukemia. *Crit. Rev. Oncol. Hematol. 4*, 31–66.

Guckenheimer, J. (1975). Isochrons and phaseless sets. *J. Math. Biol 1*, 259–273.

Guckenheimer, J. and P. Holmes (1983). *Nonlinear Oscillations, Dynamical Systems and Bifurcations of Vector Fields*. New York: Springer-Verlag.

Guerry, D., J. W. Adamson, D. C. Dale, and S. M. Wolff (1974). Human cyclic neutropenia: Urinary colony-stimulating factor and erythropoietin levels. *Blood 44*, 257–262.

Guerry, D., D. C. Dale, M. Omine, S. Perry, and S. M. Wolff (1973). Periodic hematopoiesis in human cyclic neutropenia. *J. Clin. Invest. 52*, 3220–3230.

Guevara, M. R. (1987). Afterpotentials and pacemaker oscillations in an ionic model of cardiac Purkinje fibres. In L. Rensing, U. an der Heiden, and M. C. Mackey (Eds.), *Temporal Disorder in Human Oscillatory Systems*, pp. 126–133. Berlin: Springer-Verlag.

Guevara, M. R., F. Alonso, D. Jeandupeux, and A. van Ginneken (1989). Alternans in periodically stimulated isolated ventricular myocytes: Experiment and model. In A. Goldbeter (Ed.), *Cell to Cell Signalling: From Experiments to Theoretical Models*, pp. 551–563. London: Academic Press.

Guevara, M. R. and L. Glass (1982). Phase-locking, period-doubling bifurcations and chaos in a mathematical model of a periodically driven biological oscillator: A theory for the entrainment of biological oscillators and the generation of cardiac dysrhythmias. *J. Math. Biol. 14*, 1–23.

Guevara, M. R., L. Glass, and A. Shrier (1981). Phase locking, period doubling bifurcations and irregular dynamics in periodically stimulated cardiac cells. *Science 214*, 1350–1353.

Guevara, M. R., D. Jeandupeux, F. Alonso, and N. Morissette (1990). Wenckebach rhythms in isolated ventricular heart cells. In S. Pnevmatikos, T. Bountis, and S. Pnevnatikos (Eds.), *Singular Behaviours and Nonlinear Dynamics*, pp. 629–642. Singapore: World Scientific.

Guevara, M. R. and H. Jongsma (1992). Three ways of abolishing automaticity in sinoatrial node: ionic modeling and nonlinear dynamics. *Am. J. Physiol. 262*, H1268–H1286.

Guevara, M. R., A. Shrier, and L. Glass (1986). Phase resetting of spontaneously beating embryonic ventricular heart cell aggregates. *Am. J. Physiol. 251*, H1298–H1305.

Guevara, M. R., A. Shrier, and L. Glass (1988). Phase locked rhythms in periodically stimulated heart cell aggregates. *Am. J. Physiol. 254*, H1–H10.

Guevara, M. R., A. Shrier, and L. Glass (1990). Chaotic and complex cardiac rhythms. In D. Zipes and J. Jalife (Eds.), *Cardiac Electrophysiology: From Cell to Bedside*, pp. 192–201. Philadelphia: Saunders.

Guevara, M. R., G. Ward, A. Shrier, and L. Glass (1984). Electric alternans and period doubling bifurcations. *IEEE Comput. in Cardiol.*, 167–170.

Guillemin, V. and A. Pollack (1975). *Differential Topology*. Englewood Cliffs, NJ: Prentice Hall.

Guillouzic, S., I. L'Heureux, and A. Longtin (1999). Small delay approximation of stochastic delay differential equations. *Phys. Rev. E 59*, 3970–3982.

Gurfinkel, V. S. and S. M. Osovets (1973). Mechanism of generation of oscillations in the tremor form of Parkinsonism. *Biofizika 18*(4), 781–790.

Gurney, C. W., E. L. Simmons, and E. O. Gaston (1981). Cyclic erythropoiesis in $W/W^v$ mice following a single small dose of $^{89}Sr$. *Exp. Hematol. 9*, 118–122.

Guttman, R., S. Lewis, and J. Rinzel (1980). Control of repetitive firing in squid axon membrane as a model for a neuron oscillator. *J. Physiol. (Lond.) 305*, 377–395.

Hakim, V. and A. Karma (1999). Theory of spiral wave dynamics in weakly excitable media: Asymptotic reduction to kinematic model and applications. *Phys. Rev. E 60*, 5073–5105.

Hale, J. K. and S. M. V. Lunel (1993). *Introduction to Functional Differential Equations,*. Springer-Verlag, New York.

Hall, G. M., S. Bahar, and D. J. Gauthier (1999). Prevalence of rate-dependent behaviors in cardiac muscle. *Phys. Rev. Lett. 82*, 2995–2998.

Hallett, M. (1991). Classification and treatment of tremor. *JAMA 266*, 115–117.

Hammond, W. P., T. C. Boone, R. E. Donahue, L. M. Souza, and D. C. Dale (1990). Comparison of treatment of canine cyclic hematopoiesis with recombinant human granulocyte-macrophage colony-stimulating factor (GM-CSF), G-CSF, Interleukin-3, and Canine G-CSF. *Blood 76*, 523–532.

Hammond, W. P., G. S. Chatta, R. G. Andrews, and D. C. Dale (1992). Abnormal responsiveness of granulocyte committed progenitor cells in cyclic neutropenia. *Blood 79*, 2536–2539.

Hammond, W. P. and D. C. Dale (1982). Cyclic hematopoiesis: Effects of lithium on colony forming cells and colony stimulating activity in grey collie dogs. *Blood 59*, 179–184.

Hammond, W. P., T. H. Price, L. M. Souza, and D. C. Dale (1989). Treatment of cyclic neutropenia with granulocyte colony stimulating factor. *New Eng. J. Med. 320*, 1306–1311.

Hardy, G. (1940). *A Mathematician's Apology*. Cambridge: Cambridge University Press.

Hartline, H. K. (1974). *Studies on Excitation and Inhibition in the Retina*. New York: Rockefeller University Press.

Hattar, S., H.-W. Liao, M. Takao, D. M. Berson, and K.-W. Yau (2002). Melanopsin-containing retinal ganglion cells: Architecture, projections, and intrinsic photosensitivity. *Science 295*, 1065–1070.

Haurie, C., M. C. Mackey, and D. C. Dale (1998). Cyclical neutropenia and other periodic hematological diseases: A review of mechanisms and mathematical models. *Blood 92*, 2629–2640.

Haurie, C., M. C. Mackey, and D. C. Dale (1999). Occurrence of periodic oscillations in the differential blood counts of congenital, idiopathic, and cyclical neutropenic patients before and during treatment with G-CSF. *Exper. Hematol. 27*, 401–409.

Haurie, C., R. Person, M. C. Mackey, and D. C. Dale (1999). Hematopoietic dynamics in grey collies. *Exper. Hematol. 27*, 1139–1148.

Hearn, T., C. Haurie, and M. C. Mackey (1998). Cyclical neutropenia and the peripherial control of white blood cell production. *J. Theor. Biol. 192*, 167–181.

Henriquez, C. (1993). Simulating the electrical behavior of cardiac tissue using the bidomain model. *Crit. Rev. Biomed. Eng. 21*, 1–77.

Hill, A. V. (1965). *Trails and Trials in Physiology*. Baltimore: Williams and Wilkins.

Hille, B. (2001). *Ionic Channels of Excitable Membranes* (3rd ed.). Sinauer, Sunderland.

Hochmair-Desoyer, I., E. Hochmair, H. Motz, and F. Rattay (1984). A model for the electrostimulation of the nervus acusticus. *Neuroscience 13*, 553–562.

Hodgkin, A. L. (1958). The Croonian Lecture: Ionic movements and electrical activity in giant nerve fibres. *Proc. Roy. Soc. London, Series B 148*(930), 1–37.

Hodgkin, A. L. (1964). *The Conduction of the Nervous Impulse.* Liverpool: Liverpool University Press.

Hodgkin, A. L. and A. F. Huxley (1952). A quantitative description of membrane current and its application to conduction and excitation in nerve. *J. Physiol. (Lond.) 117*, 500–544.

Hodgkin, A. L. and R. D. Keynes (1956). Experiments on the injection of substances into squid giant axons by means of a microsyringe. *J. Physiol. (Lond.) 131*(592).

Hoffman, H. J., D. Guerry, and D. C. Dale (1974). Analysis of cyclic neutropenia using digital band-pass filtering techniques. *J. Interdiscipl. Cycle 5*, 1–18.

Honeycutt, R. L. (1992). Stochastic Runge–Kutta algorithms. I. White noise. *Phys. Rev. A 45*, 600.

Hopfield, J. J. (1984). Neurons with graded responses have collective computational properties like those of two-state neurons. *Proc. Natl. Acad. Sci. USA 81*, 3088–3092.

Hoppensteadt, F. C. and J. P. Keener (1982). Phase locking of biological clocks. *J. Math. Biol. 15*, 339–349.

Horsthemke, W. and R. Lefever (1984). Noise-induced transitions. Theory and applications in physics, chemistry and biology. In *Springer Series in Synergetics, edited by H. Haken, Vol. 15.* Berlin: Springer-Verlag.

Howarth, P. A. (1989). *Flicker and the Pupil.* University of California at Berkeley: Ph.D. thesis.

Howarth, P. A., I. L. Bailey, S. M. Berman, G. Heron, and D. S. Greenhouse (1991). Location of the nonlinear processes within the pupillary pathway. *Appl. Optics 30*, 2100–2105.

Hunter, J. D., J. G. Milton, H. Lüdtke, B. Wilhelm, and H. Wilhelm (2000). Spontaneous fluctuations in pupil size are not triggered by lens accommodation. *Vision Res. 40*, 567–573.

Huxley, A. F. (1957). Muscle structure and theories of contraction. *Prog. Biophys. Biop. Ch. 7*, 255–318.

Inoue, T. (1980). The response of rabbit ciliary nerve to luminance intensity. *Brain Res. 201*, 206–209.

Ito, H. and L. Glass (1991). Spiral breakup in a new model of discrete excitable media. *Phys. Rev. Lett. 66*, 671–674.

Ito, H. and L. Glass (1992). Theory of reentrant excitation on a ring of cardiac tissue. *Physica D 56*, 84–106.

Ivey, C., V. Apkarian, and D. R. Chialvo (1998). Noise-induced tuning curve changes in mechanoreceptors. *J. Neurophysiol.* 79, 1879–1890.

Jacobsen, N. and H. E. Broxmeyer (1979). Oscillations of granulocytic and megakaryocytic progenitor cell populations in cyclic neutropenia in man. *Scand. J. Haematol.* 23, 33–36.

Jalife, J. and C. Antzelevitch (1979). Phase resetting and annihilation of pacemaker activity in cardiac tissue. *Science* 206, 695–697.

Jamaleddine, R. and A. Vinet (1999). Role of gap junction resistance in rate-induced delay in conduction in a cable model of the atrioventricular node. *J. Biol. Syst.* 4, 475–490.

Jamaleddine, R., A. Vinet, and F. Roberge (1994). Simulation of frequency entrainment in a pair of cardiac cell. In *Proceedings of the XVII IEEE/EMBS Conference*, pp. 1186–87.

Jones, J. B. and J. D. Jolly (1982). Canine cyclic haematopoiesis: bone marrow adherent cell influence of CFU-C formation. *Brit. J. Haematol.* 50, 607–617.

Jones, J. B., R. D. Lange, T. J. Yang, H. Vodopick, and E. S. Jones (1975). Canine cyclic neutropenia: Erythropoietin and platelet cycles after bone marrow transplantation. *Blood* 45, 213–219.

Jones, J. B., T. J. Yang, J. B. Dale, and R. D. Lange (1975). Canine cyclic haematopoiesis: Marrow transplantation between littermates. *Brit. J. Haematol.* 30, 215 –223.

Josephson, M. E., D. Callans, J. M. Almendral, B. G. Hook, and R. B. Kleiman (1993). Resetting and entrainment of ventricular tachycardia associated with infarction: Clinical and experimental studies. In M. E. Josephson and H. J. J. Wellens (Eds.), *Tachycardia: Mechanisms and Management*, pp. 505–536. Mount Kisco: Futura.

Joyner, R. W. (1982). Effects on the discrete pattern of electrical coupling on propagation through an electric syncytium. *Circ. Res.* 50, 192–200.

Joyner, R. W. (1986). Modulation of repolarization by electrotonic interactions. *Jpn. Heart J.* 27, 167–183.

Kandel, E. R., J. H. Schwartz, and T. M. Jessell (1991). *Principles of Neural Science* (3rd ed.). New York: Elsevier.

Kao, C. and B. Hoffman (1958). Graded and decremental response in heart muscle fibers. *Am. J. Physiol.* 194, 187–196.

Kaplan, D. and L. Glass (1995). *Understanding Nonlinear Dynamics*, Volume 91 of *Texts in Applied Mathematics*. New York: Springer-Verlag.

Kaplan, D. T., J. R. Clay, T. Manning, L. Glass, M. R. Guevara, and A. Shrier (1996). Subthreshold dynamics in periodically stimulated squid giant axons. *Phys. Rev. Lett.* **76**, 4074–4076.

Kaplan, D. T., J. M. Smith, B. E. H. Saxberg, and R. Cohen (1988). Nonlinear dynamics in cardiac conduction. *Math. Biosci.* **90**, 19–48.

Karma, A. (1993). Spiral breakup in a model equations of action potential progagation in cardiac tissue. *Phys. Rev. Lett.* **71**, 1103–1106.

Karma, A. (1994). Electrical alternans and spiral wave breakup in cardiac tissue. *Chaos* **4**, 461–472.

Karma, A., H. Levine, and X. Zou (1994). Theory of pulse instabilities in electrophysiological models of excitable tissues. *Physica D* **73**, 113–127.

Katzung, B. G., L. M. Hondeghem, W. C. M. Craig, and T. Matsubura (1985). Mechanisms for selective actions and interactions of antiarrhythmic drugs. In D. Zipes and J. Jalife (Eds.), *Cardiac Electrophysiology and Arrhythmias*, pp. 199–206. N.Y.: Grune & Stratton.

Kawato, M. (1981). Transient and steady state phase response curves of limit cycle oscillators. *J. Math. Biol.* **12**, 13–30.

Kawato, M. and R. Suzuki (1978). Biological oscillators can be stopped. Topological study of a phase response curve. *Biol. Cybern.* **30**, 241–284.

Kazarinoff, N. D. and P. van den Driessche (1979). Control of oscillations in hematopoiesis. *Science* **203**, 1348–1350.

Kearns, C. M., W. C. Wang, N. Stute, J. N. Ihle, and W. E. Evans (1993). Disposition of recombinant human granulocyte colony stimulating factor in children with severe chronic neutropenia. *J. Pediatr.* **123**, 471–479.

Keener, J. P. (1987). Propagation and its failure in coupled systems of discrete excitable cells. *SIAM J. Appl. Math.* **4**, 556–572.

Keener, J. P. (1988a). A mathematical model for the vulnerable phase in myocardium. *Math. Biosci.* **90**, 3–18.

Keener, J. P. (1988b). On the formation of circulating patterns of excitation in anisotropic media. *J. Math. Biol.* **26**, 41–56.

Keener, J. P. (1991). Am eikonal-curvature equation for action potential propagation in myocardium. *J. Math. Biol.* **29**, 629–651.

Keener, J. P. (1996). Direct activation and defibrillation of cardiac tissue. *J. Theor. Biol. 178*, 313–324.

Keener, J. P. and L. Glass (1984). Global bifurcations of a periodically forced oscillator. *J. Math. Biol. 21*, 175–190.

Keener, J. P., F. C. Hoppensteadt, and J. Rinzel (1981). Integrate-and-fire models of nerve membrane response to oscillatory input. *SIAM J. Appl. Math. 41*, 503–517.

Keener, J. P. and F. Phelps (1989). Lectures on maths in life sciences. In *Consequence of the Cellular Anisotropic Structure of the Myocardium*, Volume 21. AMS.

Kendall, D. G. (1948). On the role of variable generation time in the development of a stochastic birth process. *Biometrika 35*, 316–330.

Kennedy, B. J. (1970). Cyclic leukocyte oscillations in chronic myelogenous leukemia. *Blood 35*, 751–760.

King-Smith, E. A. and A. Morley (1970). Computer simulation of granulopoiesis: Normal and impaired granulopoiesis. *Blood 36*, 254–262.

Kloeden, P. E., E. Platen, and H. Schurz (1991). The numerical solution of nonlinear stochastic dynamical systems: a brief introduction. *Int. J. Bifurcat. Chaos 1*, 277–286.

Knight, B. (1972). Dynamics of encoding in a population of neurons. *J. Gen. Physiol. 59*, 734–766.

Koller, M. L., M. L. Riccio, and R. F. Gilmour (1998). Dynamic restitution of action potential duration during electrical alternans and ventricular fibrillation. *Am. J. Physiol. 275*, H1635–42.

Kopell, N. (1999). We got rhythm: Dynamical systems of the nervous system. *Notices Amer. Math. Soc. 47*, 6–16.

Kosobud, R. and W. O'Neill (1972). Stochastic implications of orbital implications of orbital asymptotic stability of a nonlinear trade cycle. *Econometrica 40*, 69–86.

Köster, B., M. Lauk, J. Timmer, T. Winter, B. Guschlbauer, F. X. Glocker, A. Danek, G. Deuschl, and C. H. Lücking (1998). Central mechanisms in human enhanced physiological tremor. *Neurosci. Lett. 241*, 135–138.

Koury, M. J. (1992). Programmed cell death (apoptosis) in hematopoiesis. *Exp. Hematol. 20*, 391–394.

Kowtha, V. C., A. Kunysz, J. R. Clay, L. Glass, and A. Shrier (1994). Ionic mechanisms and nonlinear dynamics of embryonic chick heart cell aggregates. *Prog. Biophys. Mol. Bio. 61*, 255–281.

Krance, R. A., W. E. Spruce, S. J. Forman, R. B. Rosen, T. Hecht, W. P. Hammond, and G. Blume (1982). Human cyclic neutropenia transferred by allogeneic bone marrow grafting. *Blood 60*, 1263–1266.

Krenz, W. and L. Stark (1984). Neuronal population model for the pupil-size effect. *Math. Biosci. 68*, 247–265.

Krenz, W. and L. Stark (1985). Systems model for the pupil size effect. II. Feedback model. *Biol. Cybern. 51*, 391–397.

Krinsky, V. I. and K. I. Agladze (1983). Interaction of rotating waves in an active chemical medium. *Physica D*, 50–56.

Kucera, J. P. and Y. Rudy (2001). Mechanistic insights into very slow conduction in branching cardiac tissue: a model study. *Circ Res. 89*, 799–806.

Kunysz, A., L. Glass, and A. Shrier (1997). Bursting behavior during fixed delay stimulation of spontaneously beating chick heart cell aggregates. *Am. J. Physiol. 273*, C331–C346.

Lamarre, Y., A. J. Joffroy, M. Dumont, and C. DeMontigny (1975). Central mechanisms of tremor in some feline and primate models. *Can. J. Neurol. Sci. 2*, 227–233.

Layton, J. E., H. Hockman, W. P. Sheridan, and G. Morstyn (1989). Evidence for a novel *in vivo* control mechanism of granulopoiesis: Mature cell-related control of a regulatory growth factor. *Blood 74*, 1303–1307.

Lee, S. and S. Kim (1999). Parameter dependence of stochastic resonance in the stochastic Hodgkin–Huxley neuron. *Phys. Rev. E 60*, 826–830.

Lemischka, I. R., D. H. Raulet, and R. C. Mulligan (1986). Developmental potential and dynamic behavior of hemopoietic stem cells. *Cell 45*, 917–927.

Lengyel, I. and I. R. Epstein (1991). Diffusion-induced instability in chemically reacting systems: Steady-state multiplicity, oscillation, and chaos. *Chaos*, 69–76.

Leon, L. J., F. A. Roberge, and A. Vinet (1994). Simulation of two-dimensional anisotropic cardiac reentry: effects of the wavelength on the reentry characteristics. *Ann. Biomed. Eng. 22*, 592–609.

Levin, J. E. and J. P. Miller (1996). Broadband neural encoding in the cricket cercal sensory system enhanced by stochastic resonance. *Nature 380*, 165–168.

Levinson, N. (1944). Transformation theory of non-linear differential equations of the second order. *Ann. Math. 45*, 723–737.

Lewis, M. L. (1974). Cyclic thrombocytopenia: A thrombopoietin deficiency. *J. Clin. Path.* *27*, 242–246.

Li, T.-Y. and J. Yorke (1975). Period three implies chaos. *Am. Math. Monthly 82*, 985–992.

Lindner, B. and L. Schimansky-Geier (2000). Coherence and stochastic resonance in a two-state system. *Phys. Rev. E 61*, 6103–6110.

Lippold, O. C. J. (1970). Oscillation in the stretch reflex arc and the origin of the rhythmical, 8-12 c/s component of physiological tremor. *J. Physiol. (Lond.) 206*, 359–382.

Lit, A. (1960). The magnitude of the Pulfrich stereo-phenomenon as a function of target velocity. *J. Exp. Psychol. 59*, 165–175.

Loewenfeld, I. E. (1958). Mechanism of reflex dilation of the pupil: Historical review. *Doc. Ophthal. 12*, 185–448.

Loewenfeld, I. E. (1993). *The Pupil: Anatomy, Physiology and Clinical Applications*. Ames, Iowa: Iowa State University Press.

Loewenfeld, I. E. and D. A. Newsome (1971). Iris mechanics. I. Influence of pupil size on dynamics of pupillary movements. *Amer. J. Ophthalmol. 71*, 347–362.

Lomb, N. R. (1976). Least-squares frequency analysis of unequally spaced data. *Astrophys. Space Sci. 39*, 447–462.

Longtin, A. (1991a). Noise-induced transitions at a Hopf bifurcation in a first-order delay-differential equation. *Phys. Rev. A 44*, 4801–4813.

Longtin, A. (1991b). Nonlinear dynamics of neural delayed feedback. In L. Nadel and D. Stein (Eds.), *1990 Lectures in Complex Systems, Santa Fe Institute Studies in the Sciences of Complexity, Lect. Vol. III*, pp. 391–405. Redwood City, Ca: Addison-Wesley.

Longtin, A. (1993). Stochastic resonance in neuron models. *J. Stat. Phys. 70*, 309–327.

Longtin, A. (1995). Mechanisms of stochastic phase locking. *Chaos 5*, 209–215.

Longtin, A. (1997). Autonomous stochastic resonance in bursting neurons. *Phys. Rev. E 55*, 868–876.

Longtin, A. (1998). Firing dynamics of electroreceptors. In S. Usui and T. Omori (Eds.), *Proc. Int. Conf. Neural Info. Processing ICONIP98, KitaKyushu, Japan*, pp. 27–30. IOS Press, Inc.

Longtin, A. (2000). Effect of noise on the tuning properties of excitable systems. *Chaos Soliton. Fract. 11*, 1835–1848.

Longtin, A., A. R. Bulsara, and F. Moss (1991). Time interval sequences in bistable systems and the noise-induced transmission of information by sensory neurons. *Phys. Rev. Lett. 67*, 656–659.

Longtin, A. and D. R. Chialvo (1998). Stochastic and deterministic resonances in excitable systems. *Phys. Rev. Lett. 81*, 4012–4015.

Longtin, A. and K. Hinzer (1996). Encoding with bursting, subthreshold oscillations and noise in mammalian cold receptors. *Neural Comput. 8*, 215–255.

Longtin, A. and J. G. Milton (1988). Complex oscillations in the human pupil light reflex with "mixed" and delayed feedback. *Math. Biosci. 90*, 183–199.

Longtin, A. and J. G. Milton (1989a). Insights into the transfer function, gain, and oscillation onset for the pupil light reflex using nonlinear delay-differential equations. *Biol. Cybern. 61*, 51–58.

Longtin, A. and J. G. Milton (1989b). Modelling autonomous oscillations in the human pupil light reflex using non-linear delay-differential equations. *Bull. Math. Biol. 51*, 605–624.

Longtin, A., J. G. Milton, J. E. Bos, and M. C. Mackey (1990). Noise and critical behavior of the pupil light reflex at oscillation onset. *Phys. Rev. A. 41*, 6992.

Lord, B. I., M. H. Bronchud, S. Owens, J. Chang, A. Howell, L. Souza, and T. M. Dexter (1989). The kinetics of human granulopoiesis following treatment with granulocyte colony stimulating factor in vivo. *Proc. Natl. Acad. Sci. USA 86*, 9499–9503.

Lord, B. I., G. Molineux, Z. Pojda, L. M. Souza, J. J. Mermod, and T. M. Dexter (1991). Myeloid cell kinetics in mice treated with recombinant interleukin-3, granulocyte colony-stimulating factor (CSF), or granulocyte-macrophage CSF in vivo. *Blood 77*, 2154–2159.

Lothrop, C. D., D. J. Warren, L. M. Souza, J. B. Jones, and M. A. Moore (1988). Correction of canine cyclic hematopoiesis with recombinant human granulocyte colony-stimulating factor. *Blood 72*, 1324–1328.

Lucas, R. J., R. H. Douglas, and R. G. Foster (2001). Characterization of an ocular photopigment capable of driving pupillary constriction in mice. *Nat. Neurosci. 4*, 621–626.

Luo, C. and Y. Rudy (1994). A dynamic model of cardiac ventricular action potential. *Circ. Res. 74*, 1071–1096.

MacDonald, N. (1978). Cyclical neutropenia: Models with two cell types and two time lags. In A. Valleron and P. Macdonald

(Eds.), *Biomathematics and Cell Kinetics*, pp. 287–295. Amsterdam: Elsevier/North-Holland.

MacDonald, N. (1989). *Biological Delay systems: Linear Stability Theory.* New York: Cambridge University Press.

Mackey, M. C. (1978). A unified hypothesis for the origin of aplastic anemia and periodic haematopoiesis. *Blood 51*, 941–956.

Mackey, M. C. (1979). Dynamic haematological disorders of stem cell origin. In J. G. Vassileva-Popova and E. V. Jensen (Eds.), *Biophysical and Biochemical Information Transfer in Recognition*, pp. 373–409. New York: Plenum Publishing Corp.

Mackey, M. C. (1996). Mathematical models of hematopoietic cell replication control. In H. Othmer, F. Adler, M. Lewis, and J. Dallon (Eds.), *Case Studies in Mathematical Modeling*, pp. 149–178. New York: Prentice Hall.

Mackey, M. C. and L. Glass (1977). Oscillation and chaos in physiological control systems. *Science 197*, 287–289.

Mannella, R. and V. Palleschi (1989). Fast and precise algorithm for computer simulation of stochastic differential equations. *Phys. Rev. A 40*, 3381–3386.

Markus, M. and B. Hess (1990). Isotropic cellular automaton for modelling excitable media. *Nature 347*, 56–57.

Marmont, G. (1949). Studies on the axon membrane: I. A new method. *J. Cell. Comp. Physiol. 34*, 351–382.

Marsden, C. D., J. C. Meadows, G. W. Lange, and R. S. Watson (1967). Effect of deafferentation on human physiological tremor. *Lancet 2*, 700–702.

Marsden, C. D., J. C. Meadows, G. W. Lange, and R. S. Watson (1969a). The relation between physiological tremor of the two hands in healthy subjects. *Electroen. Clin. Neuro. 27*, 179–185.

Marsden, C. D., J. C. Meadows, G. W. Lange, and R. S. Watson (1969b). The role of the ballistocardiac impulse in the genesis of physiological tremor. *Brain 92*, 647–662.

Masani, P. (Ed.) (1976). *Norbert Wiener: Collected Works.* Cambridge: MIT Press.

McAuley, J. H. and C. D. Marsden (2000). Physiological and pathological tremors and rhythmic central motor control. *Brain 123*, 1545–1576.

McCulloch, W. S. and W. Pitts (1943). A logical calculus of the ideas immanent in nervous activity. *Bull. Math. Biophys. 5*, 115–133.

Mempel, K., T. Pietsch, T. Menzel, C. Zeidler, and K. Welte (1991). Increased serum levels of granulocyte colony stimulating factor in patients with severe congenital neutropenia. *Blood 77*, 1919–1922.

Meron, E. R. (1991). The role of curvature and wavefront interactions in spiral-waves dynamics. *Physica D 49*, 98–106.

Metcalf, D., G. R. Johnson, and A. W. Burgess (1980). Direct stimulation by purified GM-CSF of the proliferation of multipotential and erythroid precursor cells. *Blood 55*, 138–147.

Migliaccio, A. R., G. Migliaccio, D. C. Dale, and W. P. Hammond (1990). Hematopoietic progenitors in cyclic neutropenia: Effect of granulocyte colony stimulating factor in vivo. *Blood 75*, 1951–1959.

Mikhailov, A. S., V. A. Davydov, and V. S. Zykov (1994). Complex dynamics of spiral waves and motion of curves. *Physica D 70*, 1–39.

Milton, J. (1996). Providence, Rhode Island: American Mathematical Society.

Milton, J., W. Bayer, and U. an der Heiden (1998). Modeling the pupil light reflex using differential delay equations. *Z. Angew. Math. Mech.*, S625–S628.

Milton, J. G. and J. Foss (1997). Oscillations and multistability in delayed feedback control. In H. G. Othmer, F. R. Adler, M. A. Lewis, and J. C. Dallon (Eds.), *The Art of Mathematical Modeling: Case Studies in Ecology, Physiology, and Cell Biology*, New York, pp. 179–198. Prentice Hall.

Milton, J. G. and A. Longtin (1990). Evaluation of pupil constriction and dilation from cycling measurements. *Vision Res. 30*, 515–525.

Milton, J. G., A. Longtin, A. Beuter, M. C. Mackey, and L. Glass (1989). Complex dynamics and bifurcations in neurology. *J. Theor. Biol. 138*, 129–147.

Milton, J. G., A. Longtin, T. H. Kirkham, and G. S. Francis (1988). Irregular pupil cycling as a characteristic abnormality in patients with demyelinative optic neuropathy. *Am. J. Ophthalmol. 105*, 402–407.

Milton, J. G. and M. C. Mackey (1989). Periodic haematological diseases: Mystical entities or dynamical disorders? *J. Roy. Coll. Phys. (Lond.) 23*, 236–241.

Milton, J. G., T. Ohira, J. Steck, J. Crate, and A. Longtin (1993). Oscillations and latency in the clamped pupil light reflex. *Proc. SPIE 2036*, 198–203.

Mines, G. R. (1913). On dynamic equilibrium in the heart. *J. Physiol. (Lond.) 46*, 349–383.

Moe, G. K., W. C. Rheinbolt, and J. A. Abildskov (1964). A computer model of atrial fibrillation. *Am. Heart J. 67*, 200–220.

Montgomery Jr, E. B. and K. B. Baker (2000). Mechanisms of deep brain stimulation and future technical developments. *Neurol. Res. 22*, 259–266.

Moore, M., G. Spitzer, D. Metcalf, and Penington (1974). Monocyte production of colony stimulating factor in familial cyclic neutropenia. *Brit. J. Haematol. 27*, 47–55.

Morley, A. (1969). Blood-cell cycles in polycythaemia vera. *Aust. Ann. Med. 18*, 124–137.

Morley, A. (1979). Cyclic hemopoiesis and feedback control. *Blood Cells 5*, 283–296.

Morley, A., E. A. King-Smith, and F. Stohlman (1969). The oscillatory nature of hemopoiesis. In F. Stohlman (Ed.), *Hemopoietic Cellular Proliferation*, pp. 3–14. New York: Grune & Stratton.

Morley, A. and F. Stohlman (1970). Cyclophosphamide induced cyclical neutropenia. *New Engl. J. Med. 282*, 643 –646.

Morley, A. A., A. Baikie, and D. Galton (1967). Cyclic leukocytosis as evidence for retention of normal homeostatic control in chronic granulocytic leukaemia. *Lancet 2*, 1320–1323.

Morley, A. A., E. A. King-Smith, and F. Stohlman (1970). The oscillatory nature of hemopoiesis. In F. Stohlman (Ed.), *Hemopoietic Cellular Proliferation*, pp. 3–14. New York: Grune and Stratton.

Morse, R. P. and E. F. Evans (1996). Enhancement of vowel coding for cochlear implants by addition of noise. *Nat. Med. 2*, 928–932.

Naccarelli, G. V., D. L. Wolbrette, J. T. Dell'Orfano, H. M. Patel, and J. C. Luck (1998). A decade of clinical trial developments in postmyocardial infarction, congestive heart failure, and sustained ventricular tachyarrhythmia patients: from CAST to AVID and beyond. *J. Cardiovasc. Electrophysiol. 9*, 864–891.

Narabayashi, H. (1982). Surgical approach to tremor. In C. Marsden and S. Fahn (Eds.), *Movement Disorders*, pp. 292– 299. London: Butterworth.

Nature Neuroscience (2000). Computational Approaches to Brain Function. *Nat. Neurosci. 3*(supplement), November.

Necas, E. (1992). Triggering of stem cell (CFU-S) proliferation after transplantation into irradiated hosts. *Exp. Hematol. 20*, 1146–1148.

Necas, E., V. Znojil, and J. Vacha (1988). Stem cell number versus the fraction synthesizing DN. *Exp. Hematol. 16*, 231–234.

Neher, E. and B. Sakmann (1976). Single-channel currents recorded from membrane of denervated frog muscle fibres. *Nature 260*, 799–802.

Neiman, A., A. Silchenko, V. Anishchenko, and L. Schimansky-Geier (1998). Stochastic resonance: noise-enhanced phase coherence. *Phys. Rev. E 58*, 7118–7125.

Nernst, W. (1888). Zur Kinetik der in Lösung befindlichen Körper: Theorie der Diffusion. *Zeit. Physik. Chem. 2*, 613–637.

Nernst, W. (1889). Die elektromotorische Wirksamkeit der Ionen. *Zeit. Physik. Chem. 4*, 129–181.

Nicholls, J. G., A. R. Martin, B. G. Wallace, and P. A. Fuchs (Eds.) (2001). *From Neuron to Brain* (3rd ed.). Sinauer, Sunderland.

Nicholson, A. J. (1954). An outline of the dynamics of animal populations. *Aust. J. Zool. 3*, 9–65.

Noble, D. (1995). Ionic mechanisms in cardiac electrical activity. In D. P. Zipes and J. Jalife (Eds.), *Cardiac Electrophysiology: From Cell to Bedside*, pp. 305–313. Philadelphia: W.B. Saunders.

Noble, D. and A. E. Hall (1963). The condition of initiating "all-or-nothing" repolarization in cardiac muscle. *Biophys. J. 3*, 261–274.

Noma, A. and H. Irisawa (1975). Effects of $Na^+$ and $K^+$ on the resting membrane potential of the rabbit sinoatrial node cell. *Jpn. J. Physiol. 25*, 287–302.

Nomura, T. and L. Glass (1996). Entrainment and termination of reentrant wave propagation in a periodically stimulated ring of excitable media. *Phys. Rev. E 53*, 6353–6360.

Nordin, C. and Z. Ming (1995). Computer model of current-induced early afterdepolarization in guinea pig ventricular myocytes. *Am. J. Physiol. 28*, 2440–2459.

Norman, K. E., R. Edwards, and A. Beuter (1999). The measurement of tremor using a velocity transducer: comparison to simultaneous recordings using transducers of displacement, acceleration and muscle activity. *J. Neurosci. Meth. 92*(1-2), 41–54.

Ogawa, M. (1993). Differentiation and proliferation of hematopoietic stem cells. *Blood 81*, 2844–2853.

Okajima, M., T. Fujimo, T. Kobayshi, and K. Yamada (1968). Computer simulation of the propagation process in excitation of the ventricules. *Circ. Res. 23*, 203–211.

Olby, R. (1974). *The Path to the Double Helix.* Seattle: University of Washington Press.

Olsen, L. F. and H. Degn (1985). Chaos in biological systems. *Q. Rev. Biophys. 18*, 165–225.

O'Neill, W. D., A. M. Oroujeh, A. P. Keegan, and S. L. Merritt (1996). Neurological pupillary noise in narcolepsy. *J. Sleep Res. 5*, 265–271.

O'Neill, W. D., A. M. Oroujeh, and S. L. Merritt (1998). Pupil noise is a discriminator between narcoleptics and controls. *IEEE Trans. Biomed. Eng. 45*, 314–322.

Orr, J. S., J. Kirk, K. Gray, and J. Anderson (1968). A study of the interdependence of red cell and bone marrow stem cell populations. *Brit. J. Haematol. 15*, 23–34.

Packard, N. H., J. P. Crutchfield, J. D. Farmer, and R. S. Shaw (1980). Geometry from a time series. *Phys. Rev. Lett. 45*, 712–716.

Panfilov, A. V. and P. Hogeweg (1993). Spiral breakup in a modified FitzHugh–Nagumo model. *Phys. Rev. A 176*, 295–299.

Panfilov, A. V. and A. V. Holden (1991). Spatio-temporal chaos in a model of cardiac electrical activity. *Int. J. Bifurcat. Chaos 1*, 219–225.

Pantazelou, E., F. Moss, and D. R. Chialvo (1993). Noise sampled signal transmission in an array of schmitt triggers. In P. H. Handel and A. L. Chung (Eds.), *Proc. XIIth Intern. Conf. on Noise in Physical Systems and 1/f Fluctuations*, pp. 549–552. New York: American Institute of Physics.

Park, J. R. (1996). Cytokine regulation of apoptosis in hematopoietic precursor cells. *Curr. Opin. Hematol. 3*, 191–196.

Parkinson, J. (1817). *An Essay on the Shaking Palsy.* London: Sherwood, Neely and Jones.

Patt, H. M., J. E. Lund, and M. A. Maloney (1973). Cyclic hematopoiesis in grey collie dogs: A stem-cell problem. *Blood 42*, 873–884.

Pavlidis, T. (1973). *Biological Oscillators: Their Mathematical Analysis.* New York: Academic Press.

Paydarfar, D. and D. M. Buerkel (1995). Dysrhythmias of the respiratory oscillator. *Chaos 5*, 18–29.

Perkel, D. H., J. H. Schulman, T. H. Bullock, G. P. Moore, and J. P. Segundo (1964). Pacemaker neurons: Effects of regularly spaced synaptic input. *Science 145*, 61–63.

Perry, S., J. H. Moxley, G. H. Weiss, and M. Zelen (1966). Studies of leukocyte kinetics by liquid scintillation counting in normal individuals and in patients with chronic myelogenous leukemia. *J. Clin. Invest. 45*, 1388–1399.

Pertsov, A. M., J. M. Davidenko, R. Salomonsz, W. T. Baxter, and J. Jalife (1993). Spiral waves of excitation underlie reentrant activity in isolated cardiac muscle. *Circ. Res. 72*, 631–650.

Pikovsky, A. and J. Kurths (1997). Coherence resonance in a noise-driven excitable system. *Phys. Rev. Lett. 78*, 775–778.

Planck, M. (1890a). Über die Erregung von Elektricität und Wärme in Elektrolyten. *Ann. Phys. u. Chem. Neue Folge 39*, 161–186.

Planck, M. (1890b). Über die Potentialdifferenz zwischen zwei verdünnten Lösungen binärer Elektrolyte. *Ann. Phys. u. Chem. Neue Folge 40*, 561–576.

Plant, R. E. (1981). Bifurcation and resonance in a model for bursting nerve cells. *J. Math. Biol. 11*, 15–32.

Powell, E. O. (1955). Some features of the generation times of individual bacteria. *Biometrika 42*, 16–44.

Powell, E. O. (1958). An outline of the pattern of bacterial generation times. *J. Gen. Microbiol. 18*, 382–417.

Press, W. H., S. A. Teukolsky, W. T. Vetterling, and B. P. Flannery (1992). *Numerical Recipes in C* (2nd ed.). Cambridge University Press.

Price, T. H., G. S. Chatta, and D. C. Dale (1996). Effect of recombinant granulocyte colony stimulating factor on neutrophil kinetics in normal young and elderly humans. *Blood 88*, 335–340.

Prigogine, I. and R. Lefever (1968). Symmetry breaking instabilities in dissipative systems. *J. Chem. Phys. 48*, 1695–1700.

Pumir, A. and V. I. Krinsky (1996). How does an electric field defibrillate cadiac muscle? *Physica D. 91*, 205–219.

Qu, Z., A. Garfinkel, P. S. Chen, and J. N. Weiss (2000). Mechanisms of discordant alternans and induction of reentry in simulated cardiac tissue. *Circulation 102*, 1664–1670.

Qu, Z., F. Xie, A. Garfinkel, and J. N. Weiss (2000). Origins of spiral wave meander and breakup in a two-dimensional cardiac tissue model. *Ann. Biomed. Eng. 28*, 755–771.

Quan, W. and Y. Rudy (1990a). Unidirectional block and reentry: A model study. *Circ. Res. 66*, 367–382.

Quan, W. and Y. Rudy (1990b). Unidirectional block and reentry of cardiac excitation: A model study. *Circ. Res. 66*, 367–382.

Quan, W. and Y. Rudy (1991). Termination of reentrant propagation by a single stimulus: A model study. *PACE 14*, 1700–1706.

Quesenberry, P. J. (1983). Cyclic hematopoiesis: Disorders of primitive hematopoietic stem cells. *Immunol. & Hematol. Res. Mono. 1*, 2–15.

Ranlov, P. and A. Videbaek (1963). Cyclic haemolytic anaemia synchronous with Pel-Ebstein fever in a case of Hodgkin's disease. *Acta Med. Scand. 100*, 429–435.

Rapoport, A., C. Abboud, and J. DiPersio (1992). Granulocyte-macrophage colony stimulating factor (GM-CSF) and granulocyte colony stimulating factor (G-CSF): Receptor biology, signal transduction, and neutrophil activation. *Blood Rev. 6*, 43–57.

Reeve, J. (1973). An analogue model of granulopoiesis for the analysis of isotopic and other data obtained in the non-steady state. *Brit. J. Haematol. 25*, 15–32.

Reimann, H. A. (1963). *Periodic Diseases*. Philadelphia: F.A. Davis Company.

Reulen, J. P. H., J. T. Marcus, M. J. van Gilst, D. Koops, J. E. Bos, G. Tiesinga, F. R. de Vries, and K. Boshuizen (1988). Stimulation and recording of the dynamic pupillary reflex: The IRIS technique. Part 2. *Med. Biol. Eng. Comput. 26*, 27–32.

Rinzel, J. and G. B. Ermentrout (1989). Analysis of neural excitability and oscillations. In C. Koch and I. Segev (Eds.), *Neuronal Modeling: From Synapse to Networks*. Cambridge, MA: MIT Press.

Robinson, R. B., P. A. Boyden, B. F. Hoffman, and R. W. Hewlet (1987). Electrical restitution process in dispersed canine Purkinje and ventricular cells. *Am. J. Physiol. 253*, H1018–1025.

Rondot, P., H. Korn, and J. Sherrer (1968). Suppression of an entire limb tremor by anesthetizing a selective muscular group. *Arch. Neurol.-Chicago 19*, 421–429.

Roper, P., P. Bressloff, and A. Longtin (2000). A phase model of temperature-dependent mammalian cold receptors. *Neural Comput. 12*, 1087–1113.

Rose, J., J. Brugge, D. Anderson, and J. Hind (1967). Phase locked response to low frequency tones in single auditory nerve fibers of the squirrel monkey. *J. Neurophysiol. 30*, 769–777.

Rosenblueth, A. (1958). Mechanism of the Wenckebach-Luciani cycles. *Am. J. Physiol. 194*, 491–494.

Rudolph, M. and A. Destexhe (2001). Correlation detection and resonance in neural systems with distributed noise sources. *Phys. Rev. Lett. 86*, 3662–3665.

Rudy, Y. (1995). Reentry: Insights from theoretical simulations in a fixed pathway. *J. Cardiovasc. Electrophysiol. 6*, 294–312.

Rudy, Y. (2000). Multiple interactions determine cellular electrical processes in the multicellular tissue. *Cardiovasc. Res. 51*, 1–3.

Rushton, W. A. H. (1937). Initiation of a propagated disturbance. *P. Roy. Soc. Lond. B 124*, 210–243.

Ruskin, J. N., J. P. DiMarco, and H. G. Garan (1980). Out-of-hospital cardiac arrest. *New Engl. J. Med. 303*, 607–613.

Sachs, L. (1993). The molecular control of hemopoiesis and leukemia. *CR Acad. Sci. III - Vie 316*, 882–891.

Sachs, L. and J. Lotem (1994). The network of hematopoietic cytokines. *P. Soc. Exp. Biol. Med. 206*, 170–175.

Saito, T., M. Otoguto, and T. Matsubara (1978). Electrophysological studies of the mechanism of electrically induced sustained rhythmic activity in the rabbit right atrium. *Circ. Res. 42*, 199–206.

Sánchez, J., J. Dani, D. Siemen, and B. Hille (1986). Slow permeation of organic cations in acetylcholine receptor channels. *J. Gen. Physiol. 87*, 985–1001.

Santillan, M., J. Bélair, J. M. Mahaffy, and M. C. Mackey (2000). Regulation of platelet production: The normal response to perturbation and cyclical platelet disease. *J. Theor. Biol. 206*, 585–603.

Saxberg, B. E. H. and R. J. Cohen (1991). Cellular automata models of cardiac conduction. In L. Glass, P. Hunter, and A. McCulloch (Eds.), *Theory of Heart*. New York, NY: Springer-Verlag.

Scargle, J. D. (1982). Studies in astronomical time series analysis. II. Statistical aspects of spectral analysis of unevenly spaced data. *Astrophys. J. 263*, 835–853.

Scheich, H., T. Bullock, and R. Hamstra Jr (1973). Coding properties of two classes of afferent nerve fibers: high-frequency electroreceptors in the electric fish, *eigenmannia. J. Neurophysiol. 36*, 39–60.

Schmitz, S. (1988). *Ein mathematisches Modell der zyklischen Haemopoese*. Ph. D. thesis, Universität Köln.

Schmitz, S., H. Franke, J. Brusis, and H. E. Wichmann (1993). Quantification of the cell kinetic effects of G-CSF using a model of human granulopoiesis. *Exp. Hematol. 21*, 755–760.

Schmitz, S., H. Franke, M. Loeffler, H. E. Wichmann, and V. Diehl (1994). Reduced variance of bone-marrow transit time of granulopoiesis: A possible pathomechanism of human cyclic neutropenia. *Cell Proliferat. 27*, 655–667.

Schmitz, S., H. Franke, H. E. Wichmann, and V. Diehl (1995). The effect of continuous G-CSF application in human cyclic neutropenia: A model analysis. *Brit. J. Haematol. 90*, 41–47.

Schmitz, S., M. Loeffler, J. B. Jones, R. D. Lange, and H. E. Wichmann (1990). Synchrony of bone marrow proliferation and maturation as the origin of cyclic haemopoiesis. *Cell Tissue Kinet. 23*, 425–441.

Schreiber, I., M. Dolnik, P. Choc, and M. Marek (1988). Resonance behaviour in two-parameter families of periodically forced oscillators. *Phys. Lett. A 128*, 66–70.

Schrödinger, E. (1944). *What Is Life?* Cambridge: Cambridge University Press.

Segundo, J. P., J. F. Vibert, K. Pakdaman, M. Stiber, and O. D. Martinez (1994). Noise and the neurosciences: A long history, a recent revival and some theory. In K. H. Pribram (Ed.), *Origins: Brain and Self-Organization*, pp. 299–331. Hillsdale, NJ: Lawrence Erlbaum Associates.

Semmlow, J. L. and D. C. Chen (1977). A simulation model of the human pupil light reflex. *Math. Biosci. 33*, 5–24.

Seydel, R. (1988). *From Equilibrium to Chaos: Practical Bifurcation and Stability Analysis.* New York: Elsevier.

Seydel, R. (1994). *From Equilibrium to Chaos: Practical Bifurcation and Stability Analysis.* New York: Springer-Verlag.

Shimokawa, T., K. Pakdaman, and S. Sato (1999). Time-scale matching in the response of a leaky integrate-and-fire neuron model to periodic stimulus with additive noise. *Phys. Rev. E 59*, 3427–3443.

Shvitra, D., R. Laugalys, and Y. S. Kolesov (1983). Mathematical modeling of the production of white blood cells. In G. Marchuk and L. N. Belykh (Eds.), *Mathematical Modeling in Immunology and Medicine*, Amsterdam, pp. 211–223. North-Holland.

Sigmon, K. (1993). *The Matlab Primer* (3rd ed.). Department of Mathematics, University of Florida.

Silva, M., D. Grillot, A. Benito, C. Richard, G. Nunez, and J. Fernandez-Luna (1996). Erythropoietin can promote erythroid progenitor survival by repressing apoptosis through Bcl-1 and Bcl-2. *Blood 88*, 1576–1582.

Singe, D. H., C. M. Baumgarten, and R. E. Ten Eick (1987). Cellular electrophysiology of ventricular and other dysrhythmia: Studies on disease and ischemic heart. *Prog. Cardiovasc. Dis. 24*, 97–156.

Smith, J. D., G. A. Masek, L. Y. Ichinose, T. Watanabe, and L. Stark (1970). Single neuron activity in the pupillary system. *Brain Res. 24*, 219–234.

Smith, J. M. and R. J. Cohen (1984). Simple finite-element model accounts for a wide range of cardiac diysrhythmias. *Proc. Natl. Acad. Sci. USA 81*, 233–237.

Sokol, S. (1976). The Pulfrich stereo-illusion as an index of optic nerve dysfunction. *Surv. Ophthalmol. 20*, 432–434.

Spach, M. S. (1990). The discontinuous nature of propagation in cardiac muscle: Consideration of a quantitative model incorporating the membrane ionic properties and the structural complexities. *Ann. Biomed. Eng. 11*, 209–261.

Spach, M. S. and P. C. Dolber (1985). The relation between discontinuous propagation in anisotropic tissue and the vulnerable period in reentry. In D. P. Zipes and J. Jalife (Eds.), *Cardiac Electrophysiology and Arrhythmias*, pp. 241–252. New York, NY: Grune & Stratton.

Spach, M. S. and J. M. Kootsey (1983). The nature of electrical propagation in cardiac muscle. *Am. J. Physiol. 244*, H3–H22.

Spach, M. S., W. T. Miller, D. B. Geselowitz, R. C. Barr, J. M. Kootsey, and E. Johnson (1981). The discontinuous nature of propagation in normal canine cardiac muscle. *Circ. Res. 48*, 39–54.

Spekreijse, H. (1969). Rectification in the goldfish retina: Analysis by sinusoidal and auxiliary stimulation. *Vision Res. 9*, 1461–1472.

Stacey, W. C. and D. M. Durand (2001). Synaptic noise improves detection of subthreshold signals in hippocampal CA1 neurons. *J. Neurophysiol. 86*, 1104–1112.

Stanten, S. F. and L. Stark (1966). A statistical analysis of pupil noise. *IEEE Trans. Biomed. Eng. 13*, 140–152.

Stark, L. (1959). Stability, oscillations, and noise in the human pupil servomechanism. *IRE 47*, 1925–1939.

Stark, L. (1962). Enironmental clamping of biological systems: Pupil servomechanism. *J. Opt. Soc. Amer. 52*, 925–930.

Stark, L. (1968). *Neurological Control Systems: Studies in Bioengineering*. New York: Plenum.

Stark, L. (1984). The pupil as a paradigm for neurological control systems. *IEEE Trans. Biomed. Eng. 31*, 919–924.

Stark, L., F. W. Campbell, and J. Atwood (1958). Pupil unrest: An example of noise in a biological servomechanism. *Nature 182*, 857–858.

Stark, L. and P. M. Sherman (1957). A servoanalytical study of the consensual pupil reflex to light. *J. Neurophysiol. 20*, 17–26.

Starmer, C. F., V. N. Biktahev, D. N. Romashko, M. R. Stepanov, O. N. Makarova, and V. Krinsky (1993). Vulnerability in an excitable medium: Analytical and numerical study of initiating unidirectional propagation. *Biophys. J. 65*, 1775–1787.

Stein, R. B. and M. N. Oguztoreli (1976). Tremor and other oscillations in neuromuscular systems. *Biol. Cybern. 22*, 147–157.

Stern, H. J. (1944). A simple method for the early diagnosis of abnormality of the pupillary reaction. *Brit. J. Ophthalmol. 28*, 275–276.

Stiles, R. N. (1975). Acceleration time series resulting from repetitive extension-flexion of the hand. *J. Appl. Physiol. 38*, 101–107.

Stiles, R. N. and R. S. Pozos (1976). A mechanical-reflex oscillator hypothesis for Parkinsonian hand tremor. *J. Appl. Physiol. 40*, 990–998.

Strogatz, S. H. (1994). *Nonlinear Dynamics and Chaos*. Reading, MA: Addison Wesley.

Stuart, D. G., E. Eldred, and W. O. Wild (1966). Comparisons between the physiological, shivering and Parkinsonian tremors. *J. Appl. Physiol. 21*, 1918–1924.

Sun, F. C., W. C. Krenz, and L. Stark (1982). A systems model for the pupil size effect. I. Transient data. *Biol. Cybern. 48*, 101–108.

Sun, F. C. and L. Stark (1983). Pupillary escape intensified by large pupillary size. *Vision Res. 23*, 611–615.

Sun, F. C., P. Tauchi, and L. Stark (1982). Dynamic pupillary response controlled by the pupil size effect. *Exp. Neurol. 82*, 313–324.

Swinburne, J. and M. C. Mackey (2000). Cyclical thrombocytopenia: Characterization by spectral analysis and a review. *J. Theor. Med. 2*, 81–91.

Takatani, H., H. Soda, M. Fukuda, M. Watanabe, A. Kinoshita, T. Nakamura, and M. Oka (1996). Levels of recombinant human granulocyte colony stimulating factor in serum are inversely correlated with circulating neutrophil counts. *Antimicrob. Agents Ch. 40*, 988–991.

Tan, R. and R. Joyner (1990). Electrotonic influences on action potential from isolated ventricular cells. *Circ. Res. 67*, 1071–1081.

Tasaki, I. (1959). Demonstration of two stable states of the nerve membrane in potassium-rich media. *J. Physiol. (Lond.) 148*, 306–331.

Tass, P. A. (1999). *Phase Resetting in Medicine and Biology: Stochastic Modelling and Data Analysis*. Berlin, New York: Springer.

Tass, P. A. (2000). Stochastic phase resetting: A theory for deep brain stimulation. *Prog. Theor. Phys. Supp. 139*, 301–313.

Terdiman, J., J. D. Smith, and L. Stark (1969). Pupil response to light and electrical stimulation: Static and dynamic characteristics. *Brain Res. 16*, 288–292.

Thompson, J. M. T. and H. B. Stewart (1986). *Nonlinear Dynamics and Chaos*. Chichester: Wiley.

Timmer, J. (1998). Power of surrogate data testing with respect to nonstationarity. *Phys. Rev. E 58*, 5153–5156.

Timmer, J., C. Gantert, G. Deuschl, and J. Honerkamp (1993). Characteristics of hand tremor time series. *Biol. Cybern. 70*, 75–80.

Timmer, J., S. Häussler, M. Lauk, and C. H. Lücking (2000). Pathological tremors: Deterministic chaos or nonlinear stochastic oscillators? *Chaos 10*, 278–288.

Timmer, J., M. Lauk, B. Köster, B. Hellwig, S. Häussler, B. Guschlbauer, V. Radt, M. Eichler, G. Deuschl, and C. H. Lücking (2000). Cross-spectral analysis of tremor time series. *Int. J. Bifurcat. Chaos 10*(11), 2595–2610.

Timmer, J., M. Lauk, W. Pfleger, and G. Deuschl (1998a). Cross-spectral analysis of physiological tremor and muscle activity. I. Theory and application to unsynchronized electromyogram. *Biol. Cybern. 78*, 349–357.

Timmer, J., M. Lauk, W. Pfleger, and G. Deuschl (1998b). Cross-spectral analysis of physiological tremor and muscle activity. II. Application to synchronized electromyogram. *Biol. Cybern. 78*, 359–368.

Titcombe, M. S., L. Glass, D. Guehl, and A. Beuter (2001). Dynamics of Parkinsonian tremor during deep brain stimulation. *Chaos 11*, 766–773.

Travis, L. E. (1929). The relation of voluntary movement to tremors. *J. Exp. Psychol. 12*, 515–524.

Treutlein, H. and K. Schulten (1985). Noise-induced limit cycles of the Bonhoeffer–Van der Pol model of neural pulses. *Ber. Bunsenges. Phys. Chem. 89*, 710.

Tyson, J. and J. P. Keener (1987). Spiral waves in a model of myocardium. *Physica D 29*, 215–222.

UPDRS (1987). *Handbook of Parkinson's Disease*, Chapter United Parkinson's disease rating scale, pp. 482–488. New York: Marcel Dekker.

Usui, S. and Y. Hirata (1995). Estimation of autonomic nervous activity using the inverse dynamic model of the pupil muscle plant. *Ann. Biomed. Eng. 23*, 375–387.

Usui, S. and L. Stark (1978). Sensory and motor mechanisms interact to control amplitude of pupil noise. *Vision Res. 18*, 505–507.

van Someren, E., R. Lazeron, B. Vonk, M. Mirmiran, and D. Swaab (1996). Gravitational artefact in frequency spectra of movement acceleration: Implications for actigraphy in young and elderly subjects. *J. Neurosci. Meth. 65*, 55–62.

van Someren, E., B. Vonk, W. Thijssen, J. Speelman, P. Schuurman, M. Mirmiran, and D. Swaab (1998). A new actigraph for long-term registration of the duration and intensity of tremor and movement. *IEEE Trans. Biomed. Eng. 45*(3), 386–395.

Varghese, A. and R. L. Winlow (1994). Dynamics of abnormal pacemaking activity in cardiac Purkinje fibers. *J. Theor. Biol. 18*, 407–420.

Varjú, D. (1969). Human pupil dynamics. In W. Reichardt (Ed.), *Processing of optical data by organisms and machines*, New York, pp. 442–464. Academic Press.

Vilis, T. and J. Hore (1980). Central neural mechanisms contributing to cerebellar tremor produced by limb perturbations. *J. Neurophysiol. 43*, 279–291.

Vinet, A. (1995). Non-linear dynamics of propagation in models of cardiac tissue in cardiac electrophysiology. In D. P. Zipes and J. Jalife (Eds.), *Cardiac Electrophysiology: From Cell to Bedside* (Second ed.)., pp. 371–378. Philadelphia: W. B. Saunders.

Vinet, A. (1999). Memory and bistability in a one-dimensional loop model of cardiac cells. *J. Biol. Syst. 7*, 451–473.

Vinet, A. (2000). Quasiperiodic circus movement in a loop of cardiac tissue: Multistability and low dimensional equivalence. *Ann. Biomed. Eng. 28*, 704–720.

Vinet, A., D. R. Chialvo, D. C. Michaels, and J. Jalife (1990). Nonlinear dynamics of rate-dependent activation in models of single cardiac cells. *Circ. Res. 67*, 1510–1524.

Vinet, A. and F. A. Roberge (1990). A model study of stability and oscillations of the myocardial cell membrane. *J. Theor. Biol. 147*, 377–412.

Vinet, A. and F. A. Roberge (1994a). Analysis of an iterative equation model of the cardiac cell membrane. *J. Theor. Biol. 147*, 201–214.

Vinet, A. and F. A. Roberge (1994b). The dynamics of sustained reentry in a ring model of cardiac tissue. *Ann. Biomed. Eng. 22*, 568–591.

Vinet, A. and F. A. Roberge (1994c). Excitability and repolarization in an ionic model of the cardiac cell membrane. *J. theor. Biol. 170*, 183–199.

von Helmholtz, H. (1867). *Handbuch der Physiologischen Optik.* Hamburg und Leipzig: Verlag von Leopold Voss.

von Helmholtz, H. (1874). *Mechanism of the Ossicles and the Membrana Tympani.* London: The New Sydenham Society.

von Helmholtz, H. (1898). *Vorlesungen über Theoretische Physik.* Leipzig: Johan Barth Verlag.

von Schulthess, G. K. and N. A. Mazer (1982). Cyclic neutropenia (CN): A clue to the control of granulopoiesis. *Blood 59*, 27–37.

Wachs, H. and B. Boshes (1966). Tremor studies in normals and Parkinsonism. *Arch. Neurol.-Chicago 4*, 66–82.

Walshe, F. M. R. (1924). Observations on the nature of the muscular rigidity of paralysis agitans, and on its relationship to tremor. *Brain 47*, 159–177.

Watanabe, M., N. Otani, and R. Gilmour (1995). Biphasic restitution of action potential duration and complex dynamics in ventricular myocardium. *Circ. Res. 76*, 915–921.

Watanabe, M. A., F. H. Fenton, S. J. Evans, H. M. Hastings, and A. Karma (2001). Mechanisms for discordant alternans. *J. Cardiovasc. Electrophysiol. 12*, 196–206.

Watari, K., S. Asano, N. Shirafuji, H. Kodo, K. Ozawa, F. Takaku, and S. Kamachi (1989). Serum granulocyte colony stimulating factor levels in healthy volunteers and patients with various disorders as estimated by enzyme immunoassay. *Blood 73*, 117–122.

Webster, D. O. (1968). Critical analysis of the disability of Parkinson's disease. *Mod. Trends 5*, 257–282.

Weiden, P. L., B. Robinett, T. C. Graham, J. Adamson, and R. Storb (1974). Canine cyclic neutropenia. *J. Clin. Invest. 53*, 950–953.

Weimar, J. R., J. J. Tyson, and L. T. Watson (1992). Diffusion and wave propagation in cellular automaton models of excitable media. *Physica D 55*, 309–329.

Welpe, E. (1967). Untersuchungen über den galvanischen Pupillenreflex beim Menschen. *von Graefes Arch. Ophthalmol. 174*, 49–62.

Wichmann, H. E., M. Loeffler, and S. Schmitz (1988). A concept of hemopoietic regulation and its biomathematical realization. *Blood Cells 14*, 411–429.

Wiener, N. (1948). *Cybernetics: Control and Communication in the Animal and the Machine*. Cambridge: MIT Press.

Wiener, N. (1950). *The Human Use of Human Beings*. Cambridge: MIT Press.

Wiener, N. and A. Rosenblueth (1946). The mathematical formulation of the problem of conduction of impulses in a network of connected excitable elements specifically in cardiac muscle. *Arch. Latinoam. de Card. y Hemat. 26*, 205–236.

Wiesenfeld, K. (1985). Noisy precursors of nonlinear instabilities. *J. Stat. Phys. 38*, 1071.

Wiesenfeld, K., D. Pierson, E. Pantazelou, C. Dames, and F. Moss (1994). Stochastic resonance on a circle. *Phys. Rev. Lett. 72*, 2125–2128.

Wiggins, S. (1988). *Global Bifurcations and Chaos*. New York: Springer-Verlag.

Wiggins, S. (1990). *Introduction to Applied Nonlinear Dynamical Systems and Chaos*. New York: Springer-Verlag.

Wikswo, J. P. (1995). Tissue anisotropy, the cardiac bidomain, and the virtual cathode effect. In D. P. Zipes and J. Jalife (Eds.), *Cardiac Electrophysiology: From Cell to Bedside* (Second ed.)., pp. 348–362. Philadelphia: W. B. Saunders.

Wilhelm, B., H. Wilhelm, H. Lüdtke, P. Streicher, and M. Adler (1998). Pupillographic assessment of sleepiness in sleep-deprived healthy subjects. *Sleep 21*, 258–268.

Wilhelm, H., H. Lüdtke, and B. Wilhelm (1998). Pupillographic sleepiness test applied in hypersomniacs and normals. *Graefe's Arch. Clin. Exp. Ophthalmol. 236*, 725–729.

Williams, G. and C. Smith (1993). Molecular regulation of apoptosis: Genetic controls on cell death. *Cell 74*, 777–779.

Williams, G. T., C. A. Smith, E. Spooncer, T. M. Dexter, and D. R. Taylor (1990). Haemopoietic colony stimulating factors promote cell survival by suppressing apoptosis. *Nature 353*, 76–78.

Winfree, A. T. (1987). *When Time Breaks Down: The Three Dimensional Dynamics of Electrochemical Waves and Cardiac Arrhythmias.* Princeton, N.J.: Princeton Univ. Press.

Winfree, A. T. (1989). Electrical instability in cardiac muscle: Phase singularities and rotors. *J. Theor. Biol. 138*, 353–405.

Winfree, A. T. (1991). Varieties of spiral wave behavior: An experimentalist's approach to the theory of excitable media. *Chaos 1*, 303–334.

Winfree, A. T. (2000). *The Geometry of Biological Time* (Second ed.). New York: Springer-Verlag.

Wit, A. L. and P. F. Cranefield (1977). Triggered activity in the canine coronary sinus. *Circ. Res. 41*, 435–445.

Wit, A. L. and P. F. Cranefield (1978). Reentrant excitation as a cause of cardiac arrhythmias. *Am. J. Physiol. 235*, H1–H17.

Wolfram, S. (1986). *Theory and Application of Cellular Automata.* Singapore: World Scientific.

Woodbury, J. (1962). Cellular electrophysiology of the heart. In W. Hamilton (Ed.), *Handbook of Physiology: Section 2, Circulation I*, Washington, D.C., pp. 237–286. American Physiological Society.

Woodcock, A. and M. Davis (1978). *Catastrophe Theory.* New York: Dutton.

Wright, D. G., R. F. Kenney, D. H. Oette, V. F. LaRussa, L. A. Boxer, and H. L. Malech (1994). Contrasting effects of recombinant human granulocyte-macrophage colony-stimulating factor (CSF) and granulocyte CSF treatment on the cycling of blood elements in childhood-onset cyclic neutropenia. *Blood 84*, 1257–1267.

Xie, F., Z. Qu, and A. Garfinkel (1998). Dynamics of reentry around a circular obstacle in cardiac tissue. *Phys. Rev. E 58*, 6355–6358.

Yehia, A. R., D. Jeandupeux, F. Alonso, and M. R. Guevara (1999). Hysteresis and bistability in the direct transition from 1:1 to 2:1 rhythm in periodically driven single ventricular cells. *Chaos 9*, 916–931.

Yoshida, H., K. Yana, F. Okuyama, and T. Tokoro (1994). Time-varying properties of respiratory fluctuations in pupil diameter of human eyes. *Meth. Inform. Med. 33*, 46–48.

Yoss, R. E., N. J. Moyer, and R. W. Hollenhurst (1993). Pupil size and spontaneous pupillary waves associated with alertness, drowsiness, and sleep. *Neurology 20*, 545–554.

Zeng, W., M. Courtemanche, L. Sehn, A. Shrier, and L. Glass (1990). Theoretical computation of phase locking in embryonic atrial heart cell aggregates. *J. Theor. Biol. 145*, 225–244.

Zeng, W., L. Glass, and A. Shrier (1992). The topology of phase response curves induced by single and paired stimuli. *J. Biol. Rhythm 7*, 89–104.

Zykov, V. S. (1984). *Simulation of Wave Processes in Excitable Media.* Manchester, United Kingdom: Manchester University Press. Translated 1987.

Zykov, V. S. (1988). Computer simulation of ventricular fibrillation. In *Mathematical Computer Modelling*, Volume II, pp. 813–822. Pergamon Press.

# Index

# Interdisciplinary Applied Mathematics